Methods in Molecular Biology

Series Editor
John M. Walker
School of Life and Medical Sciences
University of Hertfordshire
Hatfield, Hertfordshire, UK

For further volumes:
http://www.springer.com/series/7651

For over 35 years, biological scientists have come to rely on the research protocols and methodologies in the critically acclaimed *Methods in Molecular Biology* series. The series was the first to introduce the step-by-step protocols approach that has become the standard in all biomedical protocol publishing. Each protocol is provided in readily-reproducible step-by step fashion, opening with an introductory overview, a list of the materials and reagents needed to complete the experiment, and followed by a detailed procedure that is supported with a helpful notes section offering tips and tricks of the trade as well as troubleshooting advice. These hallmark features were introduced by series editor Dr. John Walker and constitute the key ingredient in each and every volume of the *Methods in Molecular Biology* series. Tested and trusted, comprehensive and reliable, all protocols from the series are indexed in PubMed.

Characterization of Nanoparticles Intended for Drug Delivery

Third Edition

Edited by

Jeffrey D. Clogston, Rachael M. Crist, Marina A. Dobrovolskaia and Stephan T. Stern

Nanotechnology Characterization Laboratory, Cancer Research Technology Program, Frederick National Laboratory for Cancer Research, Frederick, MD, USA

Humana Press

Editors
Jeffrey D. Clogston
Nanotechnology Characterization Laboratory
Cancer Research Technology Program
Frederick National Laboratory for Cancer Research
Frederick, MD, USA

Rachael M. Crist
Nanotechnology Characterization Laboratory
Cancer Research Technology Program
Frederick National Laboratory for Cancer Research
Frederick, MD, USA

Marina A. Dobrovolskaia
Nanotechnology Characterization Laboratory
Cancer Research Technology Program
Frederick National Laboratory for Cancer Research
Frederick, MD, USA

Stephan T. Stern
Nanotechnology Characterization Laboratory
Cancer Research Technology Program
Frederick National Laboratory for Cancer Research
Frederick, MD, USA

ISSN 1064-3745 ISSN 1940-6029 (electronic)
Methods in Molecular Biology
ISBN 978-1-0716-3785-2 ISBN 978-1-0716-3786-9 (eBook)
https://doi.org/10.1007/978-1-0716-3786-9

This Humana imprint is published by the registered company Springer Science+Business Media, LLC, part of Springer Nature.
The registered company address is: 1 New York Plaza, New York, NY 10004, U.S.A.

Paper in this product is recyclable.

Preface

The third edition of this book comes 13 years after the original publication in 2011. The Nanotechnology Characterization Laboratory (NCL), operating since 2004, has characterized hundreds of nanoparticles, witnessed the trends in formulation development over the last 20 years, and contributed to many novel concepts advancing to human clinical trials and the commercial market. Keeping pace with the ever-changing landscape of nanomedicine formulations, the NCL continually optimizes and develops new protocols to address the complex questions developers and regulatory figures alike are asking. Twenty-four protocols were introduced in the first edition, followed up 7 years later with a combination of 19 updated and new protocols. For this newest edition, 27 protocols are described, spanning topics such as endotoxin contamination, complex physicochemical properties, in vitro immunotoxicity traits, and in vitro drug release properties.

Eight chapters are dedicated to discussing a variety of physicochemical characterization techniques. These span from relatively newer techniques such as asymmetric-flow field-flow fractionation, single particle inductively coupled plasma mass spectrometry, and resistive pulse sensing, to the tried-and-true techniques such as elemental analysis and gas and liquid chromatography. There are 18 chapters discussing the immunotoxicity of nanomaterials. Topics include discussion of the microbial contaminants endotoxin and beta-glucans, anti-PEG antibodies, autoimmunity, immunosuppressive properties, adjuvant properties, delayed-type hypersensitivity reactions, and several mechanistic immunotoxicology protocols. Finally, two pharmacology protocols are presented. This first of these discusses a novel technique to assess drug release, which can also be used for lot release criteria, and the second describes a tissue distribution assay using PEG immunohistochemistry.

The development of these protocols has been a long-term effort by the NCL team members, many supportive collaborators, and of course, the dedicated research groups that continue to submit their unique concepts and challenging characterization questions to the NCL. We would like to thank all of those involved in these efforts. Current and former NCL staff that have worked tirelessly to develop, test, and optimize these protocols include: Edward Cedrone, Dr. Matthew Hansen, Dr. Enping Hong, Dr. Yingwen Hu, Dr. Anna Ilinskaya, Dr. Krishna Kattel, Barry Neun, Dr. Hannah Newton, Timothy Potter, Dr. Ankit Shah, Sarah Skoczen, Alison Vermilya, Teagan Ware, and Jie Xu. This work would also not be possible without the support and expertise of those within the Frederick National Laboratory, including Donna Butcher, Dr. Simone Difilippantonio, Dr. Elijah Edmondson, Christina Robinson, and Chelsea Sanders of the Laboratory of Animal Sciences Program. Extramural collaborators at AID Autoimmun Diagnostika GmbH, Dr. Rosemarie Preyer and Dr. Martin Schuster; SeroScience LCC, Dr. Janos Szebeni; Massachusetts Institute of Technology, Andrew Van Dusen; Agilent Technologies, Dr. Jenny Zhang; and SN Bioscience, Dr. Young Hwan Park, are also owed thanks and recognition for contributing their expertise to several protocols for this edition. Finally, we would like to express our sincere gratitude to the hundreds of global biotech and academic collaborators we have had the privilege of working with over the last 20 years, whose insight and novel developments helped make all of this work possible.

Frederick, MD, USA

Jeffrey D. Clogston
Rachael M. Crist
Marina A. Dobrovolskaia
Stephan T. Stern

Contents

Contributors

DONNA BUTCHER • *Molecular Histopathology Laboratory, Laboratory of Animal Sciences Program, Frederick National Laboratory for Cancer Research, Frederick, MD, USA*
EDWARD CEDRONE • *Nanotechnology Characterization Laboratory, Cancer Research Technology Program, Frederick National Laboratory for Cancer Research, Frederick, MD, USA*
JEFFREY D. CLOGSTON • *Nanotechnology Characterization Laboratory, Cancer Research Technology Program, Frederick National Laboratory for Cancer Research, Frederick, MD, USA*
RACHAEL M. CRIST • *Nanotechnology Characterization Laboratory, Cancer Research Technology Program, Frederick National Laboratory for Cancer Research, Frederick, MD, USA*
SIMONE DIFILIPPANTONIO • *Animal Research Technical Support, Laboratory of Animal Sciences Program, Frederick National Laboratory for Cancer Research, Frederick, MD, USA*
MARINA A. DOBROVOLSKAIA • *Nanotechnology Characterization Laboratory, Cancer Research Technology Program, Frederick National Laboratory for Cancer Research, Frederick, MD, USA*
ANDREW VAN DUSEN • *Massachusetts Institute of Technology, Cambridge, MA, USA*
ELIJAH EDMONDSON • *Molecular Histopathology Laboratory, Laboratory of Animal Sciences Program, Frederick National Laboratory for Cancer Research, Frederick, MD, USA*
MATTHEW HANSEN • *Nanotechnology Characterization Laboratory, Cancer Research Technology Program, Frederick National Laboratory for Cancer Research, Frederick, MD, USA*
ENPING HONG • *Nanotechnology Characterization Laboratory, Cancer Research Technology Program, Frederick National Laboratory for Cancer Research, Frederick, MD, USA*
YINGWEN HU • *Nanotechnology Characterization Laboratory, Cancer Research Technology Program, Frederick National Laboratory for Cancer Research, Frederick, MD, USA*
ANNA ILINSKAYA • *Nanotechnology Characterization Laboratory, Cancer Research Technology Program, Frederick National Laboratory for Cancer Research, Frederick, MD, USA*
KRISHNA KATTEL • *Nanotechnology Characterization Laboratory, Cancer Research Technology Program, Frederick National Laboratory for Cancer Research, Frederick, MD, USA*
BARRY W. NEUN • *Nanotechnology Characterization Laboratory, Cancer Research Technology Program, Frederick National Laboratory for Cancer Research, Frederick, MD, USA*
HANNAH S. NEWTON • *Nanotechnology Characterization Laboratory, Cancer Research Technology Program, Frederick National Laboratory for Cancer Research, Frederick, MD, USA*
YOUNG HWAN PARK • *SN BioScience, Inc., #Na-420, Biz Growth Center, Seongnam-si, Gyeonggi-do, Republic of Korea*

Timothy M. Potter • *Nanotechnology Characterization Laboratory, Cancer Research Technology Program, Frederick National Laboratory for Cancer Research, Frederick, MD, USA*
Rosemarie Preyer • *AID Autoimmun Diagnostika GmbH, Straßberg, Germany; Rancho Santa Margarita, CA, USA*
Christina Robinson • *Animal Research Technical Support, Laboratory of Animal Sciences Program, Frederick National Laboratory for Cancer Research, Frederick, MD, USA*
Chelsea Sanders • *Animal Research Technical Support, Laboratory of Animal Sciences Program, Frederick National Laboratory for Cancer Research, Frederick, MD, USA*
Martin Schuster • *AID Autoimmun Diagnostika GmbH, Straßberg, Germany; Rancho Santa Margarita, CA, USA*
Ankit Shah • *Nanotechnology Characterization Laboratory, Cancer Research Technology Program, Frederick National Laboratory for Cancer Research, Frederick, MD, USA*
Sarah L. Skoczen • *Nanotechnology Characterization Laboratory, Cancer Research Technology Program, Frederick National Laboratory for Cancer Research, Frederick, MD, USA*
Stephan T. Stern • *Nanotechnology Characterization Laboratory, Cancer Research Technology Program, Frederick National Laboratory for Cancer Research, Frederick, MD, USA*
Janos Szebeni • *Nanomedicine Research and Education Center, Department of Translational Medicine, Semmelweis University, Budapest, Hungary; SeroScience LCC, Budapest, Hungary; Translational Nanobioscience Research Center, Sungkyunkwan University, Suwon, Korea*
Alison Vermilya • *Nanotechnology Characterization Laboratory, Cancer Research Technology Program, Frederick National Laboratory for Cancer Research, Frederick, MD, USA*
Teagan D. Ware • *Nanotechnology Characterization Laboratory, Cancer Research Technology Program, Frederick National Laboratory for Cancer Research, Frederick, MD, USA*
Jie Xu • *Nanotechnology Characterization Laboratory, Cancer Research Technology Program, Frederick National Laboratory for Cancer Research, Frederick, MD, USA*
Jenny Zhang • *Agilent Technologies, Santa Clara, CA, USA*

Part I

Introduction

Chapter 1

Advancements in Nanoparticle Characterization

Rachael M. Crist, Jeffrey D. Clogston, Stephan T. Stern, and Marina A. Dobrovolskaia

Abstract

Nanotechnology for drug delivery has made significant advancements over the last two decades. Innovations have been made in cancer research and development, including chemotherapies, imaging agents, and vaccine strategies, as well as other therapeutic areas, e.g., the recent commercialization of mRNA lipid nanoparticles as vaccines against the SARS-CoV-2 virus. The field has also seen technological advancements to aid in addressing the complex questions posed by these novel therapies. In this latest edition of protocols and methods for nanoparticle characterization, we highlight both old and new methodologies for defining physicochemical properties, present both in vitro and in vivo methods to test for a variety of immunotoxicities, and describe assays used for pharmacological studies to assess drug release and tissue distribution.

Key words Nanomedicine, Cancer, Oncology, Physicochemical characterization, Immunotoxicity, Pharmacology

1 Introduction

Since the creation and founding of the Nanotechnology Characterization Laboratory (NCL) 20 years ago (2004), the nanomedicine field has seen incredible advancements. There have been many new nanomedicine products brought to market in the field of cancer, including concepts such as liposomal irinotecan (Onivyde®) and daunorubicin plus cytarabine (Vyxeos®) for chemotherapy, iron oxide nanoparticles (Nanotherm®) for magnetic hyperthermia therapy, and hafnium oxide nanoparticles (Hensify®) for enhancement of radiotherapy (Table 1). Further, nanotechnology has also brought new concepts into other therapeutic areas, such as analgesia (liposomal bupivacaine, Exparel®), anemia (iron oxide, Feraheme®), and transthyretin-mediated amyloidosis (lipid nanoparticle with siRNA, Onpattro®) to name just a few. In total, spanning all applications and indications, there are about 100 nanomedicine products on the commercial market [1]. Owing to the

Jeffrey D. Clogston et al. (eds.), *Characterization of Nanoparticles Intended for Drug Delivery*, Methods in Molecular Biology, vol. 2789, https://doi.org/10.1007/978-1-0716-3786-9_1,

Table 1
Nanoformulations approved for cancer treatment since 2004 are listed and were compiled from the following references (1, 57, 58)

Formulation type	Commercial name	Indication	Year of approval
Albumin-bound paclitaxel	Abraxane®	Breast, pancreatic, and non-small cell lung cancers	2005 (FDA)
Non-PEGylated liposomal paclitaxel	Lipusu®	Breast, ovarian, and non-small cell lung cancers	2006 (China)
PEG-PLA polymeric micelle with paclitaxel	Genexol-PM®	Breast, lung, and ovarian cancers	2007 (South Korea)
Liposomal mifamurtide	Mepact®	Osteosarcoma	2009 (EMA)
Iron oxide	Nanotherm®	Glioblastoma imaging	2010 (EMA)
Liposomal vincristine	Marqibo®	Acute lymphoblastic leukemia	2012 (FDA; withdrawn in 2021)
Liposomal irinotecan	Onivyde®	Pancreatic cancer	2015 (FDA)
Liposomal daunorubicin and cytarabine	Vyxeos®	Acute myeloid leukemia	2017 (FDA)
Micellar paclitaxel	Apelea®	Ovarian cancer	2018 (EMA)
Hafnium oxide	Hensify®	Locally advanced soft tissue sarcoma	2019 (EMA)
Albumin-bound sirolimus	Fyarro®	Perivascular epithelioid cell tumors	2021 (FDA)

sustained interest and advancements in nanotechnology, in 2022, the Food and Drug Administration (FDA) finalized the guidance document, "Drug Products, Including Biological Products, that Contain Nanomaterials" [2]. They have also put significant efforts into updating several other nanotechnology-relevant documents such as "Liposome Drug Products: Chemistry, Manufacturing, and Controls; Human Pharmacokinetics and Bioavailability; and Labeling Documentation [3]" (finalized in 2018) and "Draft Guidance on Doxorubicin Hydrochloride [4]" (updated in 2022). Standards organizations such as ASTM International and the International Organization of Standards (ISO) have also been very active in developing, testing, and releasing a plethora of standards for the characterization of both physicochemical and biological properties of nanotechnology-based products. The ASTM E56 Committee on nanotechnology has more than 30 active standards [5], and the ISO Technical Committee 229 on nanotechnologies has more than 100 active standards documents on everything from nomenclature to proper handling to characterization techniques [6].

The field has also seen the rise of novel platforms and uses of the technology. Of recent prominence, Pfizer and BioNTech's Comirnaty® and Moderna's Spikevax® lipid nanoparticle mRNA vaccines quickly came to market as the stars of the COVID-19 pandemic. The significance of this achievement is highlighted by the recent Nobel Prize in Science and Medicine awarded to Drs. Katalin Karikó and Drew Weissman for research into nucleic acid base modifications which decreased immunotoxicity and allowed for mRNA lipid nanoparticle vaccine therapy [7]. Since their market approval, there has been an explosion of research in the lipid nanoparticle space for concepts extending to anticancer therapeutics and vaccine strategies, and there are already many new clinical trials evaluating various nucleic acid-encapsulated lipid nanoparticle strategies for solid tumors. Moderna, for example, has a clinical trial to study lipid nanoparticle–encapsulated mRNA encoding human OX40L, IL-23, and IL-36γ in advanced solid tumors (NCT03739931) [8, 9].

There has also been a large uptick in the combination of nanotechnology, including lipid nanoparticles, with immunotherapy strategies such as immune checkpoint inhibitors like PD-1 (pembrolizumab, nivolumab, and cemiplimab), PD-L1 (atezolizumab, avelumab, and durvalumab), CTLA-4 (ipilimumab), and anti-LAG-3 (relatlimab). Within the last decade, immunotherapy—named "Breakthrough of the Year" by *Science* magazine in 2013 and "Advance of the Year" by the American Society of Clinical Oncology (ASCO) in 2016 and 2017, and earning a Nobel Prize in 2018—has had a tremendous impact on survivorship for many cancer patients, including those with melanoma, breast, lung, bladder, kidney, prostate, and colorectal cancers. In fact, Merck's Keytruda® (pembrolizumab) is the first therapy to be granted FDA approval for any cancer type with a common biomarker, namely microsatellite instability-high (MSI-H) or mismatch repair deficient (dMMR). Famously, former US President Jimmy Carter was successfully treated with pembrolizumab in combination with radiotherapy for melanoma with metastases to the brain, and today, immunotherapy is a significant part of current US President Joseph Biden's "Cancer Moonshot" initiative. Immunotherapy works to restore T cell-mediated defense against malignancies. The Moderna mRNA lipid nanoparticle trial (NCT03739931), in fact, is exploring the effectiveness of their strategy in combination with the PD-L1 checkpoint inhibitor durvalumab [8, 9]. The OX40L is a T cell stimulant, while IL-23 and IL-36γ are pro-inflammatory cytokines. Early results have shown significant increases in PD-L1 expression up to 29 days after treatment, and of the several dozen patients treated in the immunotherapy combination arm, several patients have observed a partial response with tumor shrinkage.

Finally, in addition to the advancements of these new technologies and strategies, nanomedicine has entered a new phase wherein generic and biosimilar versions are now being explored. According to the FDA's Orange Book, in the last decade, several generic formulations of the liposomal doxorubicin formulation, Doxil®, have been approved by the FDA (Ayana Pharma, LTD [2021]; Celerity Pharmaceuticals, LLC [2022]; Dr. Reddys Laboratories, LTD [2017]; Sun Pharmaceutical Industries, LTD [2013]; and Zydus Worldwide, DMCC [2020]) [10]. Factors such as changes in manufacturing conditions, scale-up processes, and alternative characterization methods and assays versus what is used for the reference listed drug can make it difficult to ensure the bioequivalence of complex nanomedicine products. For a review of some of the discussions surrounding follow-on nanomedicine products and the regulatory challenges they face, *see* references [11–13].

With these advancements comes the need for improved and more advanced characterization methodologies. The aspects of a nanomedicine's physicochemical and biological properties, including immunology, pharmacology, and toxicology, are becoming increasingly more complex due in part to the novel and more sophisticated formulations, as well as increased understanding of their intricate mechanisms of action. The NCL has been developing novel methods to evaluate these properties for two decades, many of which are published in two earlier editions of this book series and on our website [14–16]. This third edition serves to add to the nanotechnology characterization space by providing both updated versions of previously published chapters as well as new protocols and techniques developed within the last several years to accommodate the recent trends and needs in nanotechnology development.

2 Physicochemical Characterization

Physicochemical characterization, which essentially covers the chemistry, manufacturing, and controls (CMC) section of a regulatory application, is typically where researchers begin the complex characterization process. Size/size distribution, zeta potential, morphology, and compositional analysis comprise the majority of physicochemical characterization on nanomedicine products, though other factors should also be explored, including stability under a variety of conditions, drug release, and others specific to each formulation. Many of these assays have been previously described in editions one and two of this book series [15, 16], including the use of techniques such as dynamic light scattering (DLS), atomic force microscopy (AFM), and transmission electron microscopy (TEM) for size analyses—even TEM and scanning electron microscopy (SEM) for biological and cell culture suspensions—and techniques for compositional analysis such as thin-layer chromatography (TLC), thermogravimetric analysis (TGA), reversed-phase high-performance liquid chromatography (RP-HPLC), inductively coupled plasma mass spectrometry (ICP-MS), and matrix-assisted laser desorption ionization time-of-flight mass spectrometry (MALDI TOF MS).

For this third volume, we are including several new techniques. Size distribution is further explored using the versatile asymmetric-flow field-flow fractionation (AF4) technique. Coupled with in-line detectors such as UV–vis, refractive index (RI), multi-angle light scattering (MALS), and DLS, as well as off-line analyses such as RP-HPLC, ICP-MS, or TEM, the AF4 provides a much more detailed analysis and identification of the size populations present in a nanoformulation. The first of the two AF4 chapters herein describes size-based separation of the nanoformulation, collection of each size population, and analysis by RP-HPLC to ascertain drug content in each size population (Chapter 2) [17, 18]. This not only affords an estimation of the drug loading in each population but also provides insight into the drug's stability within the nanoparticle. The AF4 is passivated with bovine serum albumin (BSA). As such, samples can either retain the drug after flowing through the passivated membrane (drug stability), or the drug can be stripped from the formulation (drug instability), binding to the BSA. This could indicate a fast-releasing formulation upon systemic administration.

The AF4 technique is also used to evaluate protein binding to the formulation. In this assay, the nanoformulation is incubated in human plasma prior to analysis to allow sufficient time for protein binding to the nanoparticle (Chapter 3) [17, 18]. Following this incubation period, the formulation is diluted and injected into the AF4 system. Analyses include a measurement of the hydrodynamic radius and radius of gyration for both the plasma-incubated sample as well as a non-incubated sample as the control. The shape factor, which is the ratio of the radius of gyration to the hydrodynamic radius, or $\rho = Rg/Rh$, of each sample can be compared and used as a good indicator of protein binding to the nanoparticle and expectation of potential protein binding in vivo [17, 18]. A change in the shape factor for the plasma-incubated sample versus that of the non-plasma-incubated sample usually indicates nanoparticle protein binding. While this technique is merely qualitative, it is much more facile and rapid than other techniques to assess protein binding, e.g., mass spectrometry [19].

The Spectradyne instrument also analyzes nanoparticle size but is combined with the analysis of nanoparticle concentration (Chapter 5) [20]. This is a resistive pulse sensing technology that has seen increased utility over the last few years. A single run can afford both size and particle concentration information; though, the instrument does have certain limitations. For example, the technology is not appropriate for nanoparticles smaller than approximately 50 nm. Furthermore, microfluidic cartridges used in the process are single-use consumables and are limited to measuring a specified, limited nanoparticle size range. The Spectradyne technology has been shown to produce similar results to more recent light scattering instruments that now also offer particle concentration analysis (NCL data; not published). Alternatively, single-particle ICP-MS can also be used for both size and concentration analysis (Chapter 6) [21, 22]. This technique is primarily applicable to metallic nanoparticles and can be more labor-intensive than alternative size and concentration instruments but has superior sensitivity in the pg/g range.

Finally, several protocols are described for the quantitation of various components of a nanoformulation. First, quantitation of the active pharmaceutical ingredient (API) in a polymeric formulation is described in Chapter 4. This assay brings to the forefront an old technique that is often overshadowed in most modern-day labs—the carbon, hydrogen, nitrogen, and sulfur elemental analyzer. An instrument most of us became familiar with in undergraduate laboratory classes, the elemental analyzer is particularly useful for quantitating drugs that have nitrogen and/or sulfur in them. Chapter 7 describes an RP-HPLC technique for quantitation of ion components in liposomal formulations. Liposomes are often loaded using an ion gradient. Combining RP-HPLC analysis with simple centrifugal filtration techniques can afford the total, internal, and external ion concentrations for a liposome—often a requirement for a complete CMC package. Lastly, gas chromatography is described as an ideal technique for quantitating residual organic solvents in a formulation (Chapter 8). Precise quantitation of these impurities is critical, as many can be toxic or potentially lethal when dosed too high in humans. The protocol describes the quantitation of many of the most commonly used organic solvents, including methanol, ethanol, acetone, diethyl ether, 2-propanol, acetonitrile, 1-propanol, ethyl acetate, tetrahydrofuran, dichloromethane, chloroform, 1-butanol, and pyridine. The protocol also references the concentration limits for these formulations as set forth by the International Council for Harmonisation (ICH) [23].

3 Immunology

Immunotoxicity is one of the leading causes of drug products' clinical failure [24]. For this reason, early testing of potential immunotoxicity concerns is critical. This fact, combined with the recent trend in immunotherapy combination treatment, pushes immunological testing even closer to the forefront of a thorough preclinical characterization scheme. Other novel approaches to combating cancer through the use of nanotechnology, such as vaccines and personalized therapies, also require a thorough immunological characterization to ensure patient safety and optimal efficacy, as well as to provide an understanding of the mechanism of action.

The NCL has dozens of in vitro and in vivo immunotoxicity assays at its disposal. Included are assays evaluating hematotoxicity, immune cell functions, cytokine expression, autoimmunity, immune-modulating impurities, and more. The first two editions of this book focused largely on in vitro assays for hematotoxicity, e.g., hemolysis, platelet aggregation, plasma coagulation and complement activation, and select immunotoxicity assays, e.g., phagocytosis and chemotaxis. For this third edition, we began the immunological discussion with two protocols focusing on the biological contaminants endotoxin (Chapter 9) and beta-glucans (Chapter 10). These innate immune response modulating impurities are commonplace in many laboratories. While endotoxin testing is a regulatory requirement for all drugs and biologics due to its pyrogenic property, beta-glucans are less pyrogenic. As such, beta-glucan testing is not formally required for regulatory submissions, but it is recommended that levels be monitored, especially in the context of immunogenicity of biotherapeutic formulations [25]. For more detailed information on the regulatory requirements for bacterial endotoxins, *see* reference FDA guidance documents [26–32] and US Pharmacopeia test methods [33, 34]; thorough discussion, including several case studies, of immunological impurities and their contribution to the immunotoxicity of drug products is also available elsewhere [35].

Several chapters describe protocols that can be used to further investigate the mechanism of action for certain drugs and formulations. Chapter 11 describes the detection of intracellular complement activation. Most studies of complement activation are focused on plasma complement. However, in recent years, intracellular complement activation has been the focus of growing attention following the discovery of the complement C3 protein in many cell types, including cancer cells [36]. The described protocol specifically looks at complement in T lymphocytes, as this is expected to play a role in hypersensitivity reactions induced by many nanoformulations, including PEGylated liposomes [37]. Several other mechanistic immunotoxicology protocols look at aspects such as total oxidative stress (Chapter 14), mitochondrial oxidative stress (Chapter 15), and changes in mitochondrial membrane potential (Chapter 16).

Hypersensitivity is further discussed in Chapter 20, with a focus on both in vitro and in vivo methods for evaluating delayed-type hypersensitivity reactions. Traditionally, the local lymph node assay (LLNA) and local lymph node proliferation (LLNP) assays have been used to predict a drug's potential to induce delayed-type hypersensitivity. More recently, two in vitro tests have been developed, the human cell line activation test and myeloid U937 skin sensitization test. The in vitro assays were developed in response to calls for decreased animal testing in drug development, and although they have better throughput than in vivo assays, it is still recommended that positive findings be verified using an in vivo method. Continuing the hypersensitivity discussion, Chapter 23 describes a porcine model for the assessment of complement activation-related pseudoallergy (CARPA). Formulations such as Taxol and Doxil are known to trigger CARPA, an infusion reaction causing a cascade of hemodynamic changes, and many other experimental nanoformulations have shown the same trend. While there are in vitro methods for the assessment of complement activation, which are good indicators of potential in vivo immune responses, the in vivo porcine model has been shown to be highly predictive of the complex human reactions, enabling researchers to take preventative measures prior to the administration of these formulations.

In addition to the LLNA, LLNP, and pig CARPA model, other in vivo methods are presented for the analysis of nanoparticle adjuvant properties (Chapter 21) and immunosuppressive properties (Chapter 22). Nanoparticles are increasingly being studied as potential vaccines for the targeted delivery of antigens to dendritic and other immune cells. Chapter 21 describes considerations for designing in vivo studies which will evaluate the potential adjuvanticity of these formulations, including discussions pertaining to the route of administration and dosing regimen. Immunosuppression is a common adverse effect of nanomedicines which could halt their clinical translation. Chapter 22 describes two protocols for assessing such potential: the in vivo T cell dependent antibody response (TDAR) and in vitro human lymphocyte activation (HuLA) assays. The in vitro HuLA alternative has been shown to be a good preliminary assessment of immunosuppression; however, it is not a full substitute for the TDAR assay, which incorporates in vivo aspects such as target organ biodistribution and toxicity effects.

Finally, two protocols are described for immunophenotyping of human peripheral blood mononuclear cells (PBMC) (Chapters 24 and 25). Immunophenotyping provides a way to probe the viability and activation status of cells. As the intent of many nanoparticle vaccine and cancer therapy strategies is to alter an immune cell's response to trigger a reaction against cancer cells, the immunophenotyping study can provide important mechanistic information relevant to the formulation's mode of action.

4 Pharmacology and Toxicology

Pharmacology and toxicology studies are most often thought of in an in vivo context. However, there are many assays that can be conducted in vitro which would inform the in vivo analyses of formulations—and help assess the in vivo success or failure of a formulation. Many of these can be found in the earlier editions of this book series, such as cytotoxicity, autophagy induction, inflammasome activation, reactive oxygen species generation, lipid peroxidation, glutathione homeostasis, and lysosomal activity. These all provide data on the possible toxicity of a formulation and insights into potential mechanisms of toxicity.

There are also in vitro assays to assess the pharmacological attributes of a formulation. One example is the stable isotope tracer ultrafiltration assay (SITUA), published in the second edition of this book series [38, 39]. This assay affords quantitation of nanomedicine drug populations, e.g., nanomedicine-bound drug, free and unencapsulated drug, as well as protein-bound drug fractions. This method has found tremendous utility in the evaluation of generic and follow-on formulations, especially as the industry is rapidly pursuing these for doxorubicin and paclitaxel-containing products. This assay can be used both in vitro to assess drug release properties and for the analysis of pharmacokinetic samples from an in vivo study, with good correlation between the two. This was recently demonstrated by comparing the controlled-release innovator Doxil formulation to a follow-on liposomal doxorubicin product, as well as with the solubilizing innovator Abraxane formulation to Genexol-PM and the classic equilibrium-binding cremophor-micelle formulation, Taxol [40].

The technique has since been adapted to evaluate release from the entire dosage vial, satisfying regulatory questions surrounding drug release, and as a discriminatory assay for lot release of certain formulations. Discussed in Chapter 27, temperature-dependent drug release uses a stable isotope of the drug to determine encapsulated, unencapsulated, and protein-bound drug fractions in a physiologically relevant human albumin matrix. Given the temperature-dependent relationship of drug release, the assay can readily identify differences in lots, flagging those that fail to meet the desired release acceptance criteria.

Finally, Chapter 28 describes a tissue distribution assay using Alexa-488 and polyethylene glycol (PEG) immunohistochemistry. Most tissue distribution studies are adept at identifying drug distribution. However, few studies are designed to look at the distribution of the nanoparticle carrier. By using Alexa-488 labeling or PEG immunohistochemistry, the nanoparticle platform can also be readily tracked, and since PEG is used in a vast majority of drug delivery vehicles being studied, the method is widely applicable.

5 The Future of Characterization

There have been significant advancements in nanoparticle development and techniques to evaluate the fast-paced landscape of nanomedicine. This third-edition book summarizes the most recent advancements in the field of nanoparticle characterization. As some technologies mature and new ones emerge, the future of nanoparticle characterization will improve with the implementation of these new technologies. We foresee future physicochemical characterization will include more methodologies for the separation and analysis of single particles. Several analytical techniques for measuring size and charge on a per-particle basis (and hence concentration) are currently available. However, we envision these techniques will evolve to also measure drug loading on a per-particle basis, providing drug distribution data as a function of size. Researchers have long asked questions such as what percentage of nanoparticles lack drug, what percentage of nanoparticles contain more drug than expected, and do these different nanoparticle populations affect the overall safety and efficacy of the formulation. With improved single-particle characterization methodologies, we may begin to address these pertinent questions.

Pharmacology and toxicology are rapidly evolving with the use of in silico techniques to screen for receptor interaction, metabolism, and toxicity profiles [41]. Thousands of compounds can be screened in this way without the use of any biological system, dramatically reducing time and costs. It is anticipated that these in silico techniques will also find application in the field of nanomedicine, with a recent example being the use of artificial intelligence to identify drugs with ideal physicochemical properties for liposome remote loading [42]. Similarly, advanced pharmacokinetic–pharmacodynamic, physiologically based pharmacokinetic, and population modeling techniques are finding application in nanomedicine development [43, 44].

We envision immunological characterization will broaden to include more high content analysis of single immune cell transcriptomes and proteomes; involve genomics, epigenomics, and metabolomics profiling of immune cells from primary sources such as whole blood; employ ex vivo cultures and co-culture models (e.g., Skimune® [45]); and leverage engineered tissues (e.g., artificial skin, blood vessels, bone marrow, and lymph nodes), laminar flow, and microfluidic devices (e.g., μLYMPH [46] and NutriChip [47]), bioreactors, and 3D hydrogels [48]. Some recent developments in this direction and associated challenges have recently been discussed elsewhere [49].

Included among other new developments in the field are reconsidering approaches and optimizing methodologies aimed at reducing animal use and expanding on the current understanding of molecular and cellular mechanisms underlying nanoparticle toxicity. For example, the recent passage of the FDA Modernization Act 2.0 offers insight into one next possible phase of characterization development—animal-free test methods—and follows similar recommendations from the European Union (EU) Reference Laboratory for Alternatives to Animal Testing. Many organizations, including the FDA and various EU organizations, have already begun to implement current knowledge of in vitro–in vivo correlations of traditional toxicology methods to reasonably reduce animal usage. For instance, the FDA guidance for nonclinical evaluation of immunotoxicity of novel pharmaceuticals released in June of 2023 suggests a comprehensive step-by-step approach to using the results of in vitro immunological studies for hazard identification [50]. Significant attention in this revised and updated approach is paid to the mechanism of action of the drug substance, highlighting the importance of future studies and improved and novel models to understand mechanisms of drug toxicity, taking toxicology to a deeper molecular level. Techniques such as computer modeling, organ-on-a-chip, and three-dimensional organoids have been studied for years. Organ-on-a-chip and organoid models aim to recreate the cellular microenvironment of human organs and serve as a convenient in vitro system to bridge the gap between laboratory animal testing and human patient experiences. Despite being around for many years, these methods have not become mainstream assays for many drug developers. The primary reason is that while important data can be gleaned from these humanized models, they do not fully capture the complex biological responses of targeting and immunological cascades like an in vivo system can.

Human-based computer modeling can be used to help predict some of these responses. Various software, programs, and databases are being used to help predict toxicities and side effects, such as the University of Oxford's program that predicts drug-induced heart arrhythmias [51], and machine learning strategies are being used to extrapolate immune responses. In fact, some would argue that the predictions of machine learning outperform the data gleaned from traditional rodent models. A small US-based biotechnology company, EpiVax, Inc., developed several algorithms and created in silico platforms to predict protein immunogenicity; these platforms include the so-called Interactive Screening and Protein Reengineering Interface (ISPRI), Epimatrix, and Pre-Deimmunization of a Functional Therapeutic (PreDeFT) Report (https://epivax.com/). In addition, the National Center for Advancing Translational Sciences at the National Institutes of Health (NIH) has developed a quantitative structure–activity relationship model and artificial immune cell platform (https://aicell.ncats.io/) to assist researchers in predicting type I and type III interferon responses to nucleic acid nanoparticles [52, 53].

To stimulate transparency, improve the reproducibility of basic research studies, encourage collaborations, and support data mining and machine learning, the NIH implemented a data-sharing policy in January 2023 [54]. As part of this effort, NIH encourages investigators to deposit scientific data into data repositories. One such data repository specific to nanomedicines is the caNanoLab (https://cananolab.cancer.gov/#/). Metrics of this platform and comparisons to the projects submitted for characterization to the Nanotechnology Characterization Lab, including the most commonly studied nanoparticle platforms, API, and cancer indications, have been recently described elsewhere [55]. Warren Chan and his team at the University of Toronto created a similar database summarizing published pharmacokinetic studies of nanomedicines [56]. This database was recently transferred to caNanoLab and is now available to researchers via this online platform. One goal of the caNanoLab database is to afford a vast array of nanoparticle characterization data to support artificial intelligence and machine learning efforts aimed at informing improved nanoparticle drug development.

Nanomedicines are complex, and understanding their efficacy and safety will continue employing in vivo models to answer questions, accompanied by in vitro, ex vivo, and in silico models. We envision in vivo animal testing will likely remain an integral part of preclinical studies. Many approaches (ex vivo, in vitro, and in silico) already exist to complement traditional in vivo studies; using these models alone or in combination is reasonable as long as they answer the specific scientific question(s) being asked and are scientifically justified. As more and more researchers and companies integrate these new methods and technologies into their developmental pathways, insight into those best suited for predictability will be realized, paving the way for the next leap in nanomedicine characterization.

Acknowledgments

This project has been funded in whole or in part with federal funds from the National Cancer Institute, National Institutes of Health, under contract 75N91010D00024. The content of this publication does not necessarily reflect the views or policies of the Department of Health and Human Services, nor does the mention of trade names, commercial products, or organizations imply endorsement by the US Government.

References

1. Thapa RK, Kim JO (2023) Nanomedicine-based commercial formulations: current developments and future prospects. J Pharm Investig 53(1):19–33
2. Food and Drug Administration (2022) Final guidance for industry: drug products, including biological products, that contain nanomaterials, April. https://www.fda.gov/media/157812/download. Accessed 21 Aug 2023
3. Food and Drug Administration (2018) Liposome drug products: chemistry, manufacturing, and controls; human pharmacokinetics and bioavailability; and labeling documentation, April. https://www.fda.gov/media/70837/download. Accessed 21 Aug 2023
4. Food and Drug Administration (2022) Draft guidance on doxorubicin hydrochloride, May. https://www.accessdata.fda.gov/drugsatfda_

docs/psg/PSG_050718.pdf. Accessed 21 Aug 2023

5. Committee E56 Subcommittees and Standards. https://www.astm.org/get-involved/technical-committees/committee-e56/subcommittee-e56. Accessed 1 Sept 2023
6. Standards by ISO/TC 229. https://www.iso.org/committee/381983/x/catalogue/p/1/u/0/w/0/d/0. Accessed 1 Sept 2023
7. Meet the authors: Katalin Kariko and Drew Weissman (2021). Immunity 54 (12): 2673–2675
8. Patel M, Jimeno A, Wang D, Stemmer S, Bauer T, Sweis R, Geva R, Kummar S, Reagan P, Perets R, LoRusso P, Gupta S, Zacharek S, Laino A, Milberg O, Frederick J, Chen S, Pascarella S, Randolph W, Aanur P, Johansen L, Do K, Meehan R, Sullivan R (2021) 539 Phase 1 study of mRNA-2752, a lipid nanoparticle encapsulating mRNAs encoding human OX40L/IL-23/IL-36γ, for intratumoral (ITu) injection +/− durvalumab in advanced solid tumors and lymphoma. J Immunother Cancer 9(Suppl 2):A569–A569
9. Patel MR, Bauer TM, Jimeno A, Wang D, LoRusso P, Do KT, Stemmer SM, Maurice-Dror C, Geva R, Zacharek S, Laino AS, Sun J, Frederick J, Zhou H, Randolph W, Cohen PS, Meehan RS, Sullivan RJ (2020) A phase I study of mRNA-2752, a lipid nanoparticle encapsulating mRNAs encoding human OX40L, IL-23, and IL-36γ, for intratumoral (iTu) injection alone and in combination with durvalumab. J Clin Oncol 38(15_suppl):3092
10. FDA Orange Book. https://www.accessdata.fda.gov/scripts/cder/ob/index.cfm. Accessed 21 Aug 2023
11. Hertig JB, Shah VP, Fluhmann B, Muhlebach S, Stemer G, Surugue J, Moss R, Di Francesco T (2021) Tackling the challenges of nanomedicines: are we ready? Am J Health Syst Pharm 78(12):1047–1056
12. Muhlebach S (2018) Regulatory challenges of nanomedicines and their follow-on versions: a generic or similar approach? Adv Drug Deliv Rev 131:122–131
13. Soares S, Sousa J, Pais A, Vitorino C (2018) Nanomedicine: principles, properties, and regulatory issues. Front Chem 6:360
14. https://www.cancer.gov/nano/research/ncl. Accessed 29 Aug 2023
15. McNeil SE (ed) (2011) Characterization of nanoparticles intended for drug delivery, Methods in molecular biology, vol 697. Humana Press, New York. isbn:978-1-60327-197-4
16. McNeil SE (ed) (2018) Characterization of nanoparticles intended for drug delivery, Methods in molecular biology, vol 1628. Humana Press. isbn:978-1-4939-7350-7
17. Caputo F, Mehn D, Clogston JD, Rosslein M, Prina-Mello A, Borgos SE, Gioria S, Calzolai L (1635) Asymmetric-flow field-flow fractionation for measuring particle size, drug loading and (in)stability of nanopharmaceuticals. The joint view of European Union Nanomedicine Characterization Laboratory and National Cancer Institute – Nanotechnology Characterization Laboratory. J Chromatogr A 2021: 461767
18. Hu Y, Crist RM, Clogston JD (2020) The utility of asymmetric flow field-flow fractionation for preclinical characterization of nanomedicines. Anal Bioanal Chem 412(2): 425–438
19. Dobrovolskaia MA, Neun BW, Man S, Ye X, Hansen M, Patri AK, Crist RM, McNeil SE (2014) Protein corona composition does not accurately predict hematocompatibility of colloidal gold nanoparticles. Nanomedicine 10(7):1453–1463
20. Spectradyne: nCS1 Particle Analyzer. https://nanoparticleanalyzer.com/ncs1/. Accessed 29 Aug 2023
21. Laborda F, Bolea E, Jimenez-Lamana J (2014) Single particle inductively coupled plasma mass spectrometry: a powerful tool for nanoanalysis. Anal Chem 86(5):2270–2278
22. Mozhayeva D, Engelhard C (2020) A critical review of single particle inductively coupled plasma mass spectrometry – a step towards an ideal method for nanomaterial characterization. J Anal Atom Spectrom 35(9):1740–1783
23. International Council for Harmonisation (2021) Impurities: guideline for residual solvents Q3C(R8), April. https://database.ich.org/sites/default/files/ICH_Q3C-R8_Guideline_Step4_2021_0422_1.pdf. Accessed 29 Aug 2023
24. Craveiro NS, Lopes BS, Tomas L, Almeida SF (2020) Drug withdrawal due to safety: a review of the data supporting withdrawal decision. Curr Drug Saf 15(1):4–12
25. Food and Drug Administration (2014) Guidance for industry: immunogenicity assessment for therapeutic protein products, August. https://www.fda.gov/media/85017/download. Accessed 27 Oct 2023
26. Food and Drug Administration (2008) Guidance for industry: CGMP for Phase 1 investigational drugs, July. https://www.fda.gov/media/70975/download. Accessed 1 Sept 2023

27. Food and Drug Administration (2020) Draft guidance for industry: setting endotoxin limits during development of investigational oncology drugs and biological products, July. https://www.fda.gov/media/140410/download. Accessed 1 Sept 2023
28. Food and Drug Administration (2012) Guidance for industry: pyrogen and endotoxins testings: questions and answers, June. https://www.fda.gov/media/83477/download. Accessed 1 Sept 2023
29. Food and Drug Administration (1995) Guidance for industry: content and format of investigational new drug applications (INDs) for Phase 1 studies of drugs, including well-characterized, therapeutic, biotechnology-derived products. https://www.fda.gov/media/71203/download. Accessed 1 Sept 2023
30. Food and Drug Administration (2014) Bacterial endotoxins/pyrogens, November. https://www.fda.gov/inspections-compliance-enforcement-and-criminal-investigations/inspection-technical-guides/bacterial-endotoxinspyrogens. Accessed 1 Sept 2023
31. Food and Drug Administration (2004) Guidance for industry: sterile drug products produced by aseptic processing—current good manufacturing practice, September. https://www.fda.gov/media/71026/download. Accessed 1 Sept 2023
32. Food and Drug Administration (2021) Guidance for industry: questions and answers on quality related controlled correspondence, September. https://www.fda.gov/media/152281/download. Accessed 1 Sept 2023
33. United States Pharmacopeia and National Formulary (2012) General Chapter <85> bacterial endotoxins test. https://www.usp.org/sites/default/files/usp/document/harmonization/gen-method/q06_current_webpage_stage_6_monograph_23_nov_2011.pdf. Accessed 1 Sept 2023
34. United States Pharmacopeia and National Formulary (2023) General Chapter <1085> guidelines on the endotoxin test. https://doi.usp.org/USPNF/USPNF_M2245_03_01.html. Accessed 1 Sept 2023
35. Reijers JAA, Malone KE, Bajramovic JJ, Verbeek R, Burggraaf J, Moerland M (2019) Adverse immunostimulation caused by impurities: the dark side of biopharmaceuticals. Br J Clin Pharmacol 85(7):1418–1426
36. Liszewski MK, Kolev M, Le Friec G, Leung M, Bertram PG, Fara AF, Subias M, Pickering MC, Drouet C, Meri S, Arstila TP, Pekkarinen PT, Ma M, Cope A, Reinheckel T, Rodriguez de Cordoba S, Afzali B, Atkinson JP, Kemper C (2013) Intracellular complement activation sustains T cell homeostasis and mediates effector differentiation. Immunity 39(6): 1143–1157
37. Pavlos R, Mallal S, Ostrov D, Buus S, Metushi I, Peters B, Phillips E (2015) T cell-mediated hypersensitivity reactions to drugs. Annu Rev Med 66:439–454
38. Skoczen S, McNeil SE, Stern ST (2015) Stable isotope method to measure drug release from nanomedicines. J Control Release 220 (Pt A):169–174
39. Skoczen SL, Stern ST (2018) Improved ultrafiltration method to measure drug release from nanomedicines utilizing a stable isotope tracer. Methods Mol Biol 1682:223–239
40. Skoczen SL, Snapp KS, Crist RM, Kozak D, Jiang X, Liu H, Stern ST (2020) Distinguishing pharmacokinetics of marketed Nanomedicine formulations using a stable isotope tracer Assay. ACS Pharmacol Transl Sci 3(3): 547–558
41. Shaker B, Ahmad S, Lee J, Jung C, Na D (2021) In silico methods and tools for drug discovery. Comput Biol Med 137:104851
42. Zucker D, Marcus D, Barenholz Y, Goldblum A (2009) Liposome drugs' loading efficiency: a working model based on loading conditions and drug's physicochemical properties. J Control Release 139(1):73–80
43. Moss DM, Siccardi M (2014) Optimizing nanomedicine pharmacokinetics using physiologically based pharmacokinetics modelling. Br J Pharmacol 171(17):3963–3979
44. Villa Nova M, Lin TP, Shanehsazzadeh S, Jain K, Ng SCY, Wacker R, Chichakly K, Wacker MG (2022) Nanomedicine ex machina: between model-informed development and artificial intelligence. Front Digit Health 4:799341
45. Ahmed S, Dickinson A (2021) Detecting immunotoxicity-clinically relevant studies in the COVID-19 era. J Pharmacol Toxicol Methods 111:107000
46. Gong MM, Lugo-Cintron KM, White BR, Kerr SC, Harari PM, Beebe DJ (2019) Human organotypic lymphatic vessel model elucidates microenvironment-dependent signaling and barrier function. Biomaterials 214: 119225
47. Ramadan Q, Jafarpoorchekab H, Huang C, Silacci P, Carrara S, Koklu G, Ghaye J, Ramsden J, Ruffert C, Vergeres G, Gijs MA (2013) NutriChip: nutrition analysis meets microfluidics. Lab Chip 13(2):196–203
48. Roh TT, Chen Y, Paul HT, Guo C, Kaplan DL (2019) 3D bioengineered tissue model of the

large intestine to study inflammatory bowel disease. Biomaterials 225:119517

49. Hammel JH, Cook SR, Belanger MC, Munson JM, Pompano RR (2021) Modeling immunity in vitro: slices, chips, and engineered tissues. Annu Rev Biomed Eng 23:461–491
50. Food and Drug Administration (2023) Non-clinical evaluation of the immunotoxic potential of pharmaceuticals, June. https://www.fda.gov/media/169117/download. Accessed 27 Oct 2023
51. Virtual assay: drug safety and efficacy prediction software. https://www.cs.ox.ac.uk/ccs/virtual-assay/. Accessed 27 Oct 2023
52. Chandler M, Jain S, Halman J, Hong E, Dobrovolskaia MA, Zakharov AV, Afonin KA (2022) Artificial immune cell, AI-cell, a new tool to predict interferon production by peripheral blood monocytes in response to nucleic acid nanoparticles. Small 18(46): e2204941
53. Johnson MB, Halman JR, Satterwhite E, Zakharov AV, Bui MN, Benkato K, Goldsworthy V, Kim T, Hong E, Dobrovolskaia MA, Khisamutdinov EF, Marriott I, Afonin KA (2017) Programmable nucleic acid based polygons with controlled neuroimmunomodulatory properties for predictive QSAR modeling. Small 13(42)
54. 2023 NIH Data Management and Sharing Policy. https://oir.nih.gov/sourcebook/intramural-program-oversight/intramural-data-sharing/2023-nih-data-management-sharing-policy. Accessed 27 Oct 2023
55. Ke W, Crist RM, Clogston JD, Stern ST, Dobrovolskaia MA, Grodzinski P, Jensen MA (2022) Trends and patterns in cancer nanotechnology research: a survey of NCI's caNanoLab and nanotechnology characterization laboratory. Adv Drug Deliv Rev 191:114591
56. Wilhelm S, Tavares AJ, Dai Q, Ohta S, Audet J, Dvorak HF, Chan WCW (2016) Analysis of nanoparticle delivery to tumours. Nat Rev Mater 1(5)
57. Patra JK, Das G, Fraceto LF, Campos EVR, Rodriguez-Torres MDP, Acosta-Torres LS, Diaz-Torres LA, Grillo R, Swamy MK, Sharma S, Habtemariam S, Shin HS (2018) Nano based drug delivery systems: recent developments and future prospects. J Nanobiotechnol 16(1):71
58. https://www.cancer.gov/nano/cancer-nanotechnology/current-treatments. Accessed 21 Aug 2023

Part II

Physicochemical Characterization

Chapter 2

Nanoparticle Size Distribution and Stability Assessment Using Asymmetric-Flow Field-Flow Fractionation

Matthew Hansen and Jeffrey D. Clogston

Abstract

Nanomaterials are inherently polydisperse. Traditional techniques, such as the widely used batch-mode dynamic light-scattering (DLS) analysis, are not ideal nor thoroughly descriptive enough to define the full complexity of these materials. Asymmetric-flow field-flow fractionation (AF4) with various in-line detectors, such as ultraviolet–visible (UV–vis), multi-angle light scattering (MALS), refractive index (RI), and DLS, is an alternative technique that can provide flow-mode analysis of not only size distribution but also shape, drug release/stability, and protein binding.

Key words AF4, MALS, DLS, Separation, Fractionation, Nanomedicine

1 Introduction

Field-flow fractionation (FFF) is a separation technique first introduced by J. Calvin Giddings in 1966 [1] and has become an increasingly important tool for analyzing pharmaceutical materials [2–5]. The technique takes advantage of the principles of lamellar flow for separating particles of different sizes. The flow of liquid through the separation channel is parabolic; the slowest speeds reside near the depletion and accumulation walls of the channel while the fastest speeds are in the center [1]. By applying a perpendicular force to the flow of the channel, samples can be striated to these different flows based on the particle's size/mass in solution, with the smaller particles residing in faster flows near the midline of the channel and hence will elute at shorter retention times while the larger particles are driven closer to the accumulation wall and will move slower through the channel [1]. This protocol deals with a flow-based perpendicular force, but other forces may be applied depending on sample types and characteristics. These additional forces include centrifugal (sedimentation FFF) [6], thermal (thermal FFF) [7], or electrical (electrical FFF) [8].

Jeffrey D. Clogston et al. (eds.), *Characterization of Nanoparticles Intended for Drug Delivery*, Methods in Molecular Biology, vol. 2789, https://doi.org/10.1007/978-1-0716-3786-9_2,

However, this technique is not without its disadvantages. To maximize the resolution through the separating channel, samples often need to go through a lengthy focusing step. Overall, this focusing step is beneficial to the separation process as it allows for the concentration of the sample into a compact area to reduce the spread of the sample while simultaneously allowing the sample to relax or settle into the proper flows, thus increasing resolution. During focusing, flow through the channel is stopped, and only the cross-flow is active. This can lead to a significant dilution of the sample, which can be detrimental if samples are not dilution stable. To overcome this limitation, frit-inlet flow channels have been developed [9]. These channels still use cross-flow to focus the samples, but channel flow remains active during the focusing step, so the sample never stops moving through the channel. Instead, the focusing step happens as the sample passes underneath a permeable frit through which the cross-flow is introduced into the channel. As the sample passes under the frit, it experiences a stronger force from the introduction of the cross-flow leading to hydrodynamic relaxation [9].

2 Materials

1. Wyatt AF4 system or other suitable AF4 systems with MALS, UV/Vis, and RI detectors.
2. Channel spacer (350 μm).
3. Membrane (regenerated cellulose, 10 kDa molecular weight cutoff [MWCO]).
4. Bovine serum albumin monomer (BSA).
5. Phosphate-buffered saline (PBS, pH 7.4).
6. Double-distilled water (DDI H_2O, 18 MΩ cm resistance).

3 Methods

3.1 AF4 Method Optimization

1. Determine the type of channel you will use for your separation. For most applications, we find the long channel most useful (*see* **Note 1**).
2. Determine an appropriate channel spacer for your sample. A good starting spacer thickness is 350 μm (*see* **Note 2**).
3. Determine the optimal cross-flow and cross-flow profile for the sample and spacer chosen in **step 1** (*see* **Note 3**).
4. Determine the optimal focus flow and focus time for the sample and AF4 channel setup. If you are using a frit-inlet channel, skip this step since there will be no focusing step (*see* **Note 4**).

5. Choose a membrane for the experiment. The important parameters to consider are membrane material and molecular weight cutoff. Typically, we use a regenerated cellulose membrane with a 10 kDa molecular weight cutoff (*see* **Note 5**).
6. Determine sample loading (*see* **Note 6**).

3.2 System Setup

1. Launch the AF4 software.
2. Prepare a fresh membrane (determined in Subheading 3.1) for the AF4 flow channel by either rinsing or soaking a new membrane in DDI water to allow it to swell and remove any residual coatings present from manufacturing.
3. Open the AF4 channel and disassemble the parts.
4. Rinse the parts with DDI water to ensure they are clean.
5. Reassemble the AF4 channel using the optimal components selected in Subheading 3.1.
6. Reconnect the channel to the Eclipse chassis, following the user manual for the instrument.
7. Adjust the detector flow rate to 2–3 mL/min to push the air out of the channel, and reduce the flow rate to 1 mL/min once the channel pressure begins to increase.
8. Purge the system by opening the high-performance liquid chromatography (HPLC) pump to the purge setting and increasing the channel flow to 3 mL/min, and allow the pump to purge for at least 3 min.
9. Once the pump has been purged, switch it back over to normal operating mode, set the flow rate to 1 mL/min, and wait until channel pressure and cross-flow pressure are constant.

3.3 Conditioning the New Membrane

1. Prepare 500 μL of a 5 mg/mL solution of BSA monomer in PBS by weighing out 2.5 mg of BSA monomer (*see* **Note 7**).
2. Filter 1 mL of PBS solution through a 0.2 μm filter.
3. Dissolve the BSA in 500 μL of the filtered PBS.
4. Vortex the BSA solution until all the BSA has dissolved and no solids are visible in solution.
5. Add the solution to an HPLC vial and place it in the autosampler (if using a manual injector, you may skip to **step 7**).
6. Set up your AF4 method using a flow rate of 1 mL/min, an injection rate of 0.2 mL/min, and a focus flow rate of 3 mL/min. A detailed table containing the elution profile can be found in **Note 8**.
7. For conditioning the membrane, a minimum of three 100 μL injections of the 5 mg/mL BSA solution is recommended.

8. Use the third data set to determine the necessary system parameters in the analysis software, including band broadening, alignment, and normalization (refer to the user guide for your instrument).
9. Check the calculated molecular weight for the BSA monomer to assess the accuracy/suitability of the AF4 system (*see* **Note 9**).

3.4 Sample Preparation

1. Using the instrument parameters for alignment, broadening, and detector normalization determined from the BSA samples, prepare a method for running your samples (*see* **Note 10**).
2. Dilute samples to an appropriate dilution for running AF4 (*see* **Note 11**).
3. Run the sample multiple times (*see* **Note 12**).
4. Run the sample with no focusing or cross-flow to determine recovery (*see* **Note 13**).

3.5 Data Analysis

1. Data analysis should be conducted according to your manufacturer and software guides.
2. If all the information in Subheading 3.4, **step 1**, was entered into the method before running your samples, skip to **step 3**. Otherwise, enter the parameters for detector alignment and band broadening and import the detector normalization from the BSA run.
3. Determine the baselines for the data and perform the baseline subtraction.
4. Identify and mark the peaks to be analyzed. Determine whether the full peak, full-width, half-max peak, or peak center should be analyzed.
5. Choose the proper fitting type for your data (*see* **Note 14**).
6. Determine which MALS detectors to use (*see* **Note 15**).

4 Notes

1. For less stable or easily aggregating samples, we recommend using a frit-inlet channel to avoid the focusing step by the conventional AF4 technique. Frit-inlet AF4 takes advantage of the frit flow to promote hydrodynamic relaxation as the sample passes under the frit inlet. Using this technique, higher cross-flow may be used to increase the efficiency of hydrodynamic relaxation. Additionally, samples tend to benefit from lower channel flows to increase the dwell time under the frit and allow for better relaxation of the sample. Like other parameters, however, these parameters will be sample dependent and will need to be optimized on a per-sample basis.

2. According to the working principle of AF4 (Eq. 1), the channel's geometry, especially its height, is critical to high-quality fractionation. The channel height is generally the thickness of the channel, determined by the spacer utilized in between the upper wall and the bottom accumulation membrane. A series of spacers with different thicknesses (190, 250, 350, 490, and 600 μm) are generally provided with the instrument. Different channel heights give different parabolic laminar flow rate profiles as well as different system pressures and thus affect separation resolution. Also, a thicker channel allows the analysis of larger sample amounts. In this protocol, to recover a sufficient amount of sample for downstream analysis, we used a spacer with a thickness of 350 μm.

3. Based on AF4 theory, cross-flow is the driving force counteracting the Brownian motion of particles to separate particles with different hydrodynamic sizes at different channel flow laminae at steady state (Eq. 1) [10]. Thus, cross-flow is a defining factor in AF4 fractionation quality. To determine the optimal cross-flow for sample fractionation, it is necessary to evaluate various cross-flow rate ($\dot{V}_{\text{cross}}$)/channel flow rate ($\dot{V}_{\text{out}}$) settings. We advise starting with varying the $\dot{V}_{\text{cross}}/\dot{V}_{\text{out}}$ ratio in a constant gradient based on the particle sizes and channel pressure. A rule of thumb is to start by analyzing an unknown sample using a gradient elution profile from a cross-flow which allows sufficient retention of early components of interest (usually from literature), and then gradually decrease it to 0 mL/min [10].

$$t_{\text{r}} = \frac{\pi \eta d w^2}{2 k_{\text{B}} T} \cdot \ln\left(1 + \frac{\dot{V}_{\text{cross}}}{\dot{V}_{\text{out}}}\right) \tag{1}$$

where t_{r} = retention time, w = channel thickness, $\dot{V}_{\text{cross}}$ = cross-flow rate, $\dot{V}_{\text{out}}$ = channel flow rate, d = Stokes diameter, η = solvent viscosity, k_B = Boltzmann's constant, T = absolute temperature (in Kelvin).

4. The focus point can directly affect the total sample inject amount. We used a 100 μL sample loop in our instrument for sample loading. During the focus step, a flow opposing the channel forward flow was introduced from the outlet, and, together with the channel flow, it focuses the sample into a narrow band close to the injection port. The focusing flow rate and focusing time determine focusing efficiency. It is recommended to perform focus position tests, using methods listed by your manufacturer. The channel spacer also comes with wide and narrow taper widths to be used at different focus points.

5. In AF4, a semi-permeable membrane is placed at the bottom of the channel with a specific cutoff size, allowing the penetration of solvent and small molecules below the cutoff size but

retaining larger sample components. Because the sample fractionation is performed close to the membrane, the sample may nonspecifically bind to the membrane surface and cause sample loss during the separation. Thus, it is important to choose a compatible membrane material with the specific sample. A smaller MWCO will affect overall system pressure, which means the cross-flow range is less and therefore may lead to poor resolution.

6. Once the key fractionation parameters are determined, the next step is to choose the right amount of sample to be injected into the AF4 system. The minimal amount of material required for AF4 is mainly determined by the sensitivity limit of the in-line detectors, i.e., the UV, MALS, DLS, and RI signals, and the fraction concentration needed for downstream analysis. The signal/noise (S/N) ratio must be adequate for accurate data collection and interpretation. It is also important to avoid overloading the channel. The maximal amount of material is determined by the required resolution of fractionation, which depends on the purpose of the experiment and the complexity of the sample. At very high concentrations, the mutual interference between nanoparticles near the accumulation wall may result in steric effects, reversing the fractionation order and perturbing the shape (broadening/widening of the peaks) of the fractogram.
7. To minimize nonspecific binding onto the membrane surface and improve mass recovery, conditioned injections are made such that the sample (or a sample of similar chemical composition) is overloaded onto the channel in order to saturate any active sites on the new membrane surface with a defined covering layer. For an aqueous system, it is recommended to use BSA with a mobile phase solution of PBS.
8. The recommended method for the separation/conditioning of a new membrane using BSA with a standard long channel and a 350 μm spacer is given in Table 1.
9. Occasionally, we find membranes that, for whatever reason, are not suitable for use in an AF4 measurement. This can usually be determined during the conditioning step of the membrane. Either the fractogram for the BSA monomer will look inconsistent after multiple runs and/or the calculated molecular weight will be wrong. If either or both of these occur, we advise changing to a new membrane and starting over.
10. Certain instrument parameters need to be determined to have accurate results from AF4 and MALS. These parameters should be determined prior to running the sample and done using a monodisperse, isotropic scattering sample. We suggest using BSA monomer for this in aqueous solutions or latex

Table 1
Recommended eclipse method timetable for BSA injection (long channel)

Start time (min)	End time (min)	Duration (min)	Flow mode	Vx Start (mL/min)	Vx End (mL/min)
0.00	2.00	2.00	Elution	3.00	3.00
2.00	4.00	2.00	Focus	–	–
4.00	9.00	5.00	Focus + inject	–	–
9.00	11.00	2.00	Focus	–	–
11.00	26.00	15.00	Elution	3.00	3.00
26.00	28.00	2.00	Elution	3.00	0.00
28.00	31.00	3.00	Elution	0.00	0.00

nanoparticles smaller than 20 nm for other media. Of these parameters, the most important are the detector alignment and the detector normalization. The detector alignment is used to align the light-scattering detector with the UV/Vis and RI signals. Once this factor is determined, it should be the same for all subsequent runs assuming there has been no changing in tubing/volume delays between the detectors. The other important factor is detector calibration. This parameter uses the scattering from an isotropic scatterer to determine the response of the various MALS detectors. It is important to use an isotropic scatterer since these particles have no angular dependence on their scattering, allowing for the correct normalization of all the MALS detectors.

11. Sample dilutions will depend on the many factors/characteristics of the nanoparticle solutions, including sample concentration, size, and stability. Smaller particles and less stable particles will need less dilution to see appropriate levels of signal, while larger samples can be diluted more due to their increased scattering. There is no hard rule on what starting dilution factor is needed, but starting between 10- and 100-fold is generally a good starting point.

12. We generally suggest running the sample multiple times, depending on how much sample is available. Often, the first time running the sample will lead to some interaction of the sample with the membrane even after it has been passivated and will display an atypical fractogram. We suggest running the sample at least two times and discarding the first run. If ample sample is available, four runs while discarding the first run are preferred.

13. Running the sample with no focusing or cross-flow will allow the sample to run through the system with the least amount of interaction or loss. This is a good method for accessing sample stability and recovery. A stable sample with respect to the elution profile should yield >70% recovery [3]. If the recovery is lower than this value, it may indicate that the sample is unstable or interacting with/sticking to the membrane. If this is the case, the separation method will need to be adjusted to reduce the intensity of the focusing/cross-flow. If adjusting neither of these increases the recovery, then switching to a frit-AF4 channel may be necessary.
14. For most cases, Zimm treatment of the scattering data is recommended. Other useful treatments include the Debye treatment (useful for larger particles) or spherical treatment (for spherical nanoparticles).
15. For detector selection, if your particle is not an isotropic scatterer, there should be a linear fit to the detector data. It is important to remove any detectors that do not fall along the linear regression fit and not include them in the calculations.

Acknowledgments

This project has been funded in whole or in part with federal funds from the National Cancer Institute, National Institutes of Health, under contract 75N91019D00024. The content of this publication does not necessarily reflect the views or policies of the Department of Health and Human Services, nor does the mention of trade names, commercial products, or organizations imply endorsement by the US Government.

References

1. Giddings JC (1966) A new separation concept based on a coupling of concentration and flow nonuniformities. Sep Sci 1(1):123–125
2. Ansar SM, Mudalige T (2020) Characterization of doxorubicin liposomal formulations for size-based distribution of drug and excipients using asymmetric-flow field-flow fractionation (AF4) and liquid chromatography-mass spectrometry (LC-MS). Int J Pharm 574: 118906
3. Caputo F, Arnould A, Bacia M, Ling WL, Rustique E, Texier I, Mello AP, Couffin AC (2019) Measuring particle size distribution by asymmetric flow field flow fractionation: a powerful method for the preclinical characterization of lipid-based nanoparticles. Mol Pharm 16(2):756–767
4. Fraunhofer W, Winter G (2004) The use of asymmetrical flow field-flow fractionation in pharmaceutics and biopharmaceutics. Eur J Pharm Biopharm 58(2):369–383
5. Yohannes G, Jussila M, Hartonen K, Riekkola ML (2011) Asymmetrical flow field-flow fractionation technique for separation and characterization of biopolymers and bioparticles. J Chromatogr A 1218(27):4104–4116
6. Yang FJF, Myers MN, Giddings JC (1974) Programmed sedimentation field-flow fractionation. Anal Chem 46(13):1924–1930

7. Giddings JC, Martin M, Myers MN (1978) High-speed polymer separations by thermal field-flow fractionation. J Chromatogr 158 (October):419–435
8. Caldwell KD, Gao YS (1993) Electrical field-flow fractionation in particle separation. 1. Monodisperse standards. Anal Chem 65(13):1764–1772
9. Giddings JC (1990) Hydrodynamic relaxation and sample concentration in field-flow fractionation using permeable wall elements. Anal Chem 62(21):2306–2312
10. Wahlund KG, Giddings JC (1987) Properties of an asymmetrical flow field-flow fractionation channel having one permeable wall. Anal Chem 59(9):1332–1339

Chapter 3

Assessment of Protein Binding Using Asymmetric-Flow Field-Flow Fractionation Combined with Multi-angle Light Scattering and Dynamic Light Scattering

Matthew Hansen and Jeffrey D. Clogston

Abstract

Asymmetric-flow field-flow fractionation (AF4) is a valuable tool to separate and assess different size populations in nanotherapeutics. When coupled with both static light scattering and dynamic light scattering, it can be used to qualitatively assess protein binding to nanoparticles by comparing the shape factors for both non-plasma-incubated samples and plasma-incubated samples. The shape factor is defined as the ratio of the derived root mean square radius (by static light scattering) to the measured hydrodynamic radius (by dynamic light scattering). The shape factor gives an idea of where the center of mass lies in a nanoparticle, and any shift in the shape factor to larger values is indicative of a mass addition to the periphery of the nanoparticle and suggests the presence of protein binding. This protocol will discuss how to set up an experiment to assess protein binding in nanoparticles using AF4, multi-angle light scattering (MALS), and dynamic light scattering (DLS).

Key words AF4, MALS, DLS, Separation, Nanomedicine, Protein corona

1 Introduction

The physicochemical properties of nanoparticles are important parameters in understanding the fate of nanotherapeutics once injected into a biological system. Both size and surface charge are important parameters affecting nanoparticle circulation and clearance [1]. Reducing the interaction of nanoparticles with serum proteins is an important focus for nanoformulations and has led to extensive studies looking at how proteins form around nanoparticles and ways to reduce these interactions, including understanding how surface charge and "stealth" coatings can affect these interactions [1, 2].

Asymmetric-flow field-flow fractionation (AF4), as discussed in Chap. 2, is an important tool in separating and evaluating nanotherapeutics due to its gentler separation of these materials

Jeffrey D. Clogston et al. (eds.), *Characterization of Nanoparticles Intended for Drug Delivery*, Methods in Molecular Biology, vol. 2789, https://doi.org/10.1007/978-1-0716-3786-9_3,

and its ability to work with a wide variety of biologically relevant media [3–6]. AF4 can be used to separate free serum proteins from nanoparticles when there is an appreciable difference in size between them. When combined with light scattering, it can become a useful tool for qualitatively assessing protein binding to nanoparticles. To assess protein binding, both static light scattering (multi-angle light scattering or MALS) and dynamic light scattering (DLS) must be used. Each of these gives a measure of size populations in the nanoparticle solution, but the sizes they measure are different. MALS uses the angular dependence of the scattered light along with fitting by an appropriate model (Zimm, Debye, spherical, etc.) to convert this angular dependence into the root mean squared radius (R_{rms}) [7]. This value can be thought of as the radius at which the center of mass for the particle resides. DLS measures the Brownian motion of particles and gives the measurement of the hydrodynamic radius (R_h), which can be thought of as the hydrated or solvated size of the nanoparticle [8].

By dividing the R_{rms} by the R_h, a qualitative parameter known as the shape factor can be calculated. As an example of how this factor can assess shape, consider a solid sphere and a hollow sphere (i.e., an empty liposome). For the solid sphere, the center of mass will be somewhere between the center of the sphere and the outer edge of the sphere and will be smaller than the hydrodynamic radius. When the shape factor is determined for an ideal sphere, the shape factor has a value of approximately $R * \sqrt{\frac{3}{5}}$, or ~ 0.77 [9]. Whereas for a hollow sphere, most of the mass is located on the periphery, and R_{rms} and R_h will be nearly identical, yielding a shape factor between 0.93 and 1 [9]. This shape factor can be leveraged to assess protein binding to nanoparticles. Since protein absorption takes place at the surface, any mass added to the nanoparticle will be added to the exterior which will cause a shift in the shape factor to larger values. By comparing the shape factor of a sample fractionated without plasma incubation to one that has been incubated with plasma, an estimation of the amount of protein binding to a nanoparticle can be assessed [10]. This protocol deals with the steps necessary to conduct a protein binding experiment and how to analyze and assess the results.

2 Materials

1. Wyatt AF4 system, or other suitable AF4 systems with MALS, UV/Vis, and Refractive Index (RI) detectors.
2. Channel spacer (350 μm).
3. Membrane (regenerated cellulose, 10 kDa molecular weight cutoff [MWCO]).
4. Bovine serum albumin monomer (BSA).

5. Phosphate-buffered saline (PBS, pH 7.4).
6. Double-distilled water (DDI H_2O, 18 MΩ cm resistance).
7. Human plasma.

3 Methods

3.1 AF4 Method Optimization, System Setup, and Membrane Conditioning

1. See the previous Chap. 2 for steps on AF4 optimization, AF4 system setup, and steps to condition new membranes.

3.2 Sample Preparation

1. Determine the desired total dilution for the sample. The following procedure assumes a final dilution of 1:100. The total plasma concentration should not exceed 10% (*see* **Note 1**).
2. Combine 10 μL of the sample with 100 μL of human plasma.
3. Incubate at 37 °C while shaking at 150 rpm for 2 h.
4. Dilute the sample by adding 890 μL of PBS.
5. Prepare a sample without plasma incubation at the same total dilution as the plasma-incubated sample.
6. Run both samples using the same elution profile (*see* **Note 2**).

3.3 Data Analysis

1. See the previous Chap. 2 for a discussion of data analysis.
2. Align the MALS and DLS signals.
3. Determine the peak area of interest for the MALS and DLS traces using the scattering and concentration detectors.
4. Determine the root mean squared radius (R_{rms}) from the MALS data for both samples.
5. Determine the hydrodynamic radius (R_h) using the DLS data for both samples.
6. Determine the shape factor for each run by dividing the root mean squared radius by the hydrodynamic radius.
7. Compare and evaluate the shape factors for both the normal sample and the plasma-incubated sample (*see* **Note 3**).

4 Notes

1. We suggest keeping the concentration of human plasma below 10% in the final sample dilution. This is to ensure that the AF4 membrane is not saturated/fouled with too many proteins since this can have detrimental effects on separation.

2. It is important that both the sample and the plasma-incubated sample be run with the same elution profile at the same dilution so that a proper comparison between the shape factor of both samples can be accomplished.
3. By comparing the shape factor between the sample with and without plasma incubation, a qualitative assessment of the amount of protein binding can be determined. If the shape factor for each sample is identical, then little or no protein binding may be present. If the shape factor increases, this suggests there is protein binding. Minor increases suggest minimal protein binding, whereas large increases suggest significant protein binding.

Acknowledgments

This project has been funded in whole or in part with federal funds from the National Cancer Institute, National Institutes of Health, under contract 75N91019D00024. The content of this publication does not necessarily reflect the views or policies of the Department of Health and Human Services, nor does the mention of trade names, commercial products, or organizations imply endorsement by the US Government.

References

1. Albanese A, Tang PS, Chan WCW (2012) The effect of nanoparticle size, shape, and surface chemistry on biological systems. Annu Rev Biomed Eng 14:1–16
2. Dobrovolskaia MA, Neun BW, Man S, Ye XY, Hansen M, Patri AK, Crist RM, McNeil SE (2014) Protein corona composition does not accurately predict hematocompatibility of colloidal gold nanoparticles. Nanomed Nanotechnol 10(7):1453–1463
3. Ansar SM, Mudalige T (2020) Characterization of doxorubicin liposomal formulations for size-based distribution of drug and excipients using asymmetric-flow field-flow fractionation (AF4) and liquid chromatography-mass spectrometry (LC-MS). Int J Pharm 574: 118906
4. Caputo F, Arnould A, Bacia M, Ling WL, Rustique E, Texier I, Mello AP, Couffin AC (2019) Measuring particle size distribution by asymmetric flow field flow fractionation: a powerful method for the preclinical characterization of lipid-based nanoparticles. Mol Pharm 16(2):756–767
5. Fraunhofer W, Winter G (2004) The use of asymmetrical flow field-flow fractionation in pharmaceutics and biopharmaceutics. Eur J Pharm Biopharm 58(2):369–383
6. Yohannes G, Jussila M, Hartonen K, Riekkola ML (2011) Asymmetrical flow field-flow fractionation technique for separation and characterization of biopolymers and bioparticles. J Chromatogr A 1218(27):4104–4116
7. Wyatt PJ (1993) Light scattering and the absolute characterization of macromolecules. Anal Chim Acta 272(1):1–40
8. Hackley VA, Clogston JD (2011) Measuring the hydrodynamic size of nanoparticles in aqueous media using batch-mode dynamic light scattering. In: McNeil SE (ed) Characterization of nanoparticles intended for drug delivery. Humana Press, Totowa, pp 35–52. https://doi.org/10.1007/978-1-60327-198-1_4
9. Some D, Johann C. WP2611: Characterization of nanopharmaceuticals with field-flow fractionation and light scattering (FFF-MALS-DLS)
10. Hu YW, Crist RM, Clogston JD (2020) The utility of asymmetric flow field-flow fractionation for preclinical characterization of nanomedicines. Anal Bioanal Chem 412(2): 425–438

Chapter 4

Quantitation of Active Pharmaceutical Ingredients in Polymeric Drug Products Using Elemental Analysis

Jie Xu, Yingwen Hu, and Jeffrey D. Clogston

Abstract

Polymeric prodrugs have gained significant popularity as a strategy to enhance the bioavailability and improve the pharmacokinetic properties of active pharmaceutical ingredients (API). Since the amount of the API in a polymeric prodrug product directly impacts both safety and efficacy, there is a pressing need for robust and accurate analytical methods to quantify the API in these formulations. Presently, drug quantification methods include reversed-phase high-performance liquid chromatography (RP-HPLC) and size exclusion chromatography (SEC)-based molecular weight determination. Even though these methods are highly precise and reproducible, a deep understanding of chromatography is required for complex method development, including optimization of the elution profile and selecting the appropriate column and mobile phase. In this chapter, we introduce the automated elemental analyzer for drug quantification, which is simple to use and does not require special method development.

Key words Drug loading, Polymer–drug conjugates, Prodrug, Nanoparticle, Nanomedicine, CHNOS compositional analysis

1 Introduction

Polymeric prodrugs, also known as polymer–drug conjugates, employ covalent conjugation to link active pharmaceutical ingredients (API) to the polymer backbone. This approach offers numerous benefits, including improved aqueous solubility, targeted delivery, bioavailability, and improved pharmacokinetics [1]. Moreover, through customization of the chemistry utilized to incorporate the API, it becomes possible to optimize the drug loading ratio and subsequent drug release rates [2]. Since the amount of the API in a polymeric prodrug product directly impacts both safety and efficacy, it is crucial to establish a reliable method to quantify the API loading in these prodrug products during the drug development process.

Jeffrey D. Clogston et al. (eds.), *Characterization of Nanoparticles Intended for Drug Delivery*, Methods in Molecular Biology, vol. 2789, https://doi.org/10.1007/978-1-0716-3786-9_4,

Here, we present a protocol for quantifying drugs in polymeric prodrugs utilizing an automated elemental analyzer (PE 2400 Series II CHNS/O, Perkin Elmer). Briefly, samples containing carbon (C), hydrogen (H), nitrogen (N), and/or sulfur (S) are oxidized in a pure oxygen environment at high temperatures. This combustion process generates gases such as CO_2, H_2O, N_2, and/or SO_2. The gases are then passed through an internal gas chromatography (GC) column, where they are separated in a stepwise steady-state manner. The separated gases are measured by a thermal conductivity detector, and the generated signals are converted to weight percentages of each element in the original sample [3]. To demonstrate the utility of this technique, we analyzed a polymeric prodrug prepared with poly(L-lysine succinylated) (PLS) and lamivudine (LAM) using the elemental analyzer.

2 Materials

1. Elemental analyzer (2400 CHNS/O Series II System, PerkinElmer).
2. AD6000 Autobalance with ±6 μg accuracy (PerkinElmer).
3. Standards: acetanilide or cystine.
4. Precleaned tin capsules, 5 mm × 8 mm.
5. Tweezers for holding and folding the capsules.
6. CHN mode: CHN combustion tubes and reduction tubes.
7. CHNS mode: CHNS combustion tubes and reduction tubes.
8. Samples: In the example presented herein, the following samples were analyzed: free drug, lamivudine; free polymer backbone, poly(L-lysine succinylated); polymeric PLS-LAM prodrug.

3 Methods

3.1 Instrument Preparation

1. This method is specific to the Perkin Elmer 2400 CHNS/O Series II System, so other instruments may vary slightly in operation.
2. Gas requirements: He > 200 psi in the tank; O_2 > 200 psi in the tank; N_2 regulator set at 60 psi. The purity of all gases should be >99.995%.
3. Select instrument mode: Press PARAMETER and enter 6. Enter 1 to select CHN mode or 2 to select CHNS mode.
4. Verify run capacity: Press PARAMETER and enter 4. Check the sample runs remaining for the tubes and vial receptacle.

Table 1
Theoretical weight percentages of carbon, hydrogen, nitrogen, and sulfur in acetanilide and cystine

	%C	%H	%N	%S
Acetanilide	71.09	6.71	10.36	
Cystine	29.99	5.03	11.66	26.69

Ensure that all values surpass the anticipated total count of runs, including blanks, conditioners, calibrants (K-factor), and samples.

5. Check for leaks: Press DIAGNOSTICS and enter 2 for gas diagnostics. Enter 1 for the leak test. Enter 2 to select the combustion zone. The leak test will start automatically. If the leak test is passed, the display will show "PASS."
6. Verify combustion settings: Press PARAMETER and enter 9 to verify combustion settings. For CHN mode, choose CONSTANT and verify the settings: 3 for OXYFILL; 5 for COMB; 3 for OXB1; and 0 for OXB2. For CHNS mode, choose CONSTANT and verify the settings: 1 for OXYFILL; 1 for COMB; 1 for OXB1; and 0 for OXB2.
7. Purge the system: Purge with He for 300 s and O_2 for 150 s whenever a tube or gas cylinder is replaced or before a run sequence is initiated.
8. Enter theoretical weight percentage: Press PARAMETER and enter 3 to select Theories of Standards. Enter 1 to select S1. Input the weight percentage of carbon, hydrogen, nitrogen, and/or sulfur in the standards sample. The theoretical weight percentages of each element in acetanilide for CHN mode and cystine for CHNS mode are listed in Table 1.

3.2 Preparation of Standards and Sample

1. Calibrate the autobalance.
2. Tare the weight of the capsule on the autobalance.
3. Add 1.5–2.5 mg of sample/standard to the capsule (*see* **Notes 1** and **2**).
4. Check the weight; adjust if necessary. When the capsule contains the appropriate amount of sample/standard, fold the capsule.
5. Re-weigh and record the weight.
6. Prepare at least six standards and three samples by repeating **steps 2–5**. Place the folded capsules in a clean well storage tray and record tray positions.

Table 2
Acceptable result ranges of carbon, hydrogen, nitrogen, and sulfur for the blank

	Carbon	Hydrogen	Nitrogen	Sulfur
CHN mode	±30	±100	±16	
CHNS mode	±30	±100	± 16	±20

Table 3
The range of K-factor values (carrier gas: helium)

	Carbon	Hydrogen	Nitrogen	Sulfur
Acetanilide	16.5 ± 3.5	50.0 ± 20.0	6.0 ± 3.0	
Cystine	16.5 ± 3.5	50.0 ± 20.0	6.0 ± 3.0	6.0 ± 3.0

3.3 Stabilization and Calibration

1. Blank run: Press the SINGLE RUN key. Press 1 for BK (blank). Enter 3 (minimum) for the number of runs. The system will switch to the "STANDBY" mode. Fold blank tin capsules and load them onto the auto-injector. Press START to start the series of blank runs. Ensure that the results of blanks are reproducible within the specified limits in Table 2. If the results are not within the limits, run several additional blanks for further evaluation.
2. Conditioner run: Press the SINGLE RUN key. Press 3 for Sample. Enter the standard capsule name and the weight in mg. Drop the standard capsule into sample drop opening at the top of auto-injector. Press START.
3. Additional runs: Perform the additional runs in the following order: blank run, conditioner run, blank run, conditioner run, and blank run.
4. K-factor run: Press the SINGLE RUN key and press 2 for K-factor. Press 1 for Standard S1 (acetanilide for CHN mode and cystine for CHNS mode). Enter the weight of the standard in mg. Drop the prepared standard into the funnel in the sample drop opening at the top of the auto-injector. Press START. Repeat this step (**step 4**) three times (minimal) to calibrate the system. The K-factor should be within the range detailed in Table 3 and should reproduce from the mean value to within the limits in Table 4. The recommended run sequence for stabilization and calibration is listed in Table 5.

3.4 Sample Analysis

1. Samples can be analyzed in either Single Run or Auto Run modes. Procedures for running samples in each of these modes are outlined below.

Table 4
Acceptable variation range of K-factor values

	Carbon	Hydrogen	Nitrogen	Sulfur
Acetanilide	±0.15	±3.75	±0.16	
Cystine	±0.15	±3.75	±0.16	±0.15

Table 5
Recommended run sequence for stabilization and calibration

Run #	Run type
1–3	Blank run
4	Conditioner run
5	Blank run
6	Conditioner run
7	Blank run
8	Conditioner run
9	Blank run
10–12	K-factor run

2. Single run mode: Press the SINGLE RUN key. Press 3 for the sample. Enter the sample name and the weight in mg. Drop the sample into the sample drop opening at the top of the auto-injector. Press START. Repeat this step (**step 2**) for the remaining samples.
3. Auto run mode: Adjust the autosampler clockwise to line up position #1 with the sample drop opening. Press AUTO RUN and then press 4 to select RP (Reset/Print). Press 1 to clear the old run sequence from memory and reset the run sequence to #1. Press 3 for the sample. Enter the sample name and the weight in mg. Place the sample in the appropriate position on the autosampler. Record the auto run # and autosampler position of the sample on a separate data sheet. Repeat this step (**step 3**) to load all the prepared samples on the autosampler. Press START to start the auto run.

3.5 Shut Down

1. Bleed valves: Press the DIAGNOSTICS key and press 2 to select GAS. Press 2 to select VALVE. Enter the valve code number 4 and press 1 to select ON. Enter valve code number 5 and press 1 to select ON. Wait for 15 seconds and then press the DIAGNOSTICS key again. The opened valves are closed automatically.

2. Gas saver mode: Press the PARAMETER key and enter 22. Press 1 to turn on the gas saver mode. Enter the date and time for the gas saver mode to turn on. Enter the date and time for the gas saver to turn off. Press the PARAMETER key to exit to the STANDBY mode.

3.6 Data Analysis

The macromolecular carrier, poly(L-lysine succinylated) (PLS), is a negatively charged synthetic polyamino acid. It possesses approximately 250 pendant carboxylic acid groups that can react to conjugate drugs through ester linkages. The small molecular drug, lamivudine (LAM), is an antiretroviral drug commonly used for treating hepatitis B and HIV infections. In this protocol, LAM, conjugated to PLS, can be quantitated using elemental analysis due to its small molecular size and its nitrogen and sulfur content [4].

The PLS-LAM prodrug was synthesized using DMAP-catalyzed carbodiimide esterification to link LAM to the pendant carboxyl groups of PLS, as illustrated in Fig. 1 [4]. The quantification of LAM content in the PLS-LAM prodrug was accomplished using an automated elemental analyzer. However, since impurities such as residual solvents, moisture, and buffers can contribute to the carbon and hydrogen content, it is advisable to employ nitrogen or sulfur to calculate the drug loading for a more accurate and reliable quantification. In the chemical structure presented in Fig. 1, nitrogen is found in both the polymer backbone and the drug component, while sulfur is exclusively present in the drug moiety.

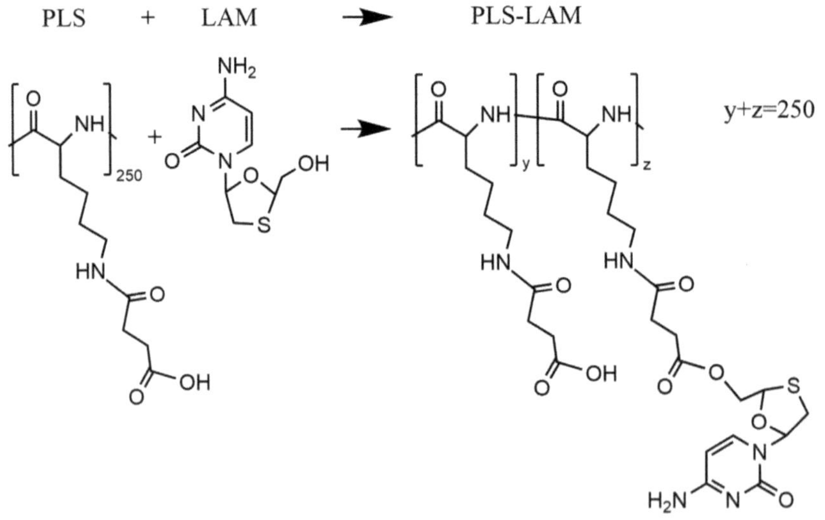

Fig. 1 Synthesis of poly(L-lysine succinylated) lamivudine prodrug (PLS-LAM) via carbodiimide esterification. *PLS* poly(L-lysine succinylated), *LAM* lamivudine

Table 6
CHN and CHNS mode elemental analysis

	CHN mode			CHNS mode			
	% C	% H	% N	% C	% H	% N	%S
PLS	52.0 ± 0.4	7.3 ± 0.1	12.2 ± 0.1	52.5 ± 1.4	7.4 ± 0.1	12.3 ± 0.9	0.5 ± 0.5
LAM	42.0 ± 0.1	4.9 ± 0.03	18.4 ± 0.1	42.8 ± 0.2	4.9 ± 0.03	18.9 ± 0.1	14.2 ± 0.6
PLS-LAM	50.5 ± 0.1	7.1 ± 0.1	12.6 ± 0.1	51.4 ± 0.1	7.2 ± 0.1	12.7 ± 0.04	1.4 ± 0.04

The average weight percentages of carbon (%C), hydrogen (%H), nitrogen (%N), and sulfur (%S) in the polymer (PLS), drug (LAM), and polymeric prodrug (PLS-LAM) obtained using the CHN and CHNS modes of the automated elemental analyzer [4]

Nitrogen and sulfur measurements are used to determine the drug loading percentage ($\%\mathrm{wt}_{\mathrm{drug}}$) by employing elemental mass balance using the equation provided below:

$$\underbrace{\%X_{\mathrm{drug}} \times \%\mathrm{wt}_{\mathrm{drug}}}_{\text{Element} \times \text{From drug}} + \underbrace{\%X_{\mathrm{polymer}} \times \left(1 - \%\mathrm{wt}_{\mathrm{drug}}\right)}_{\text{Element} \times \text{From polymer}} = \%X_{\mathrm{prodrug}} \qquad (1)$$

where %X represents the percentage of nitrogen (%N) or sulfur (%S) in the drug, polymer, or prodrug, and $\%\mathrm{wt}_{\mathrm{drug}}$ represents the drug weight percentage in the prodrug.

A simplified equation can be derived from Eq. 1 as follows:

$$\%\mathrm{wt}_{\mathrm{drug}} = \frac{\%X_{\mathrm{prodrug}} - \%X_{\mathrm{polymer}}}{\%X_{\mathrm{drug}} - \%X_{\mathrm{polymer}}} \times 100\% \qquad (2)$$

The average weight percentages of carbon (%C), hydrogen (%H), nitrogen (%N), and sulfur (%S) of the polymer (PLS), drug (LAM), and polymeric prodrug (PLS-LAM) obtained using the CHN mode and CHNS mode are presented in Table 6. In both the CHN and CHNS modes, the elemental analysis method provided consistent results across all three common elements: carbon, hydrogen, and nitrogen.

The %N value from the CHN mode and %S value from the CHNS mode were used to determine the drug loading in the prodrug using Eq. 2. The calculated drug loadings of LAM in the PLS-LAM prodrug were 6.5 ± 0.9% from the CHN mode and 6.6 ± 0.4% from the CHNS mode. For comparison, the drug weight percentage is 7.4 ± 0.1% obtained using RP-HPLC and 5.7 ± 3.8% obtained using SEC-MALS [4]. This remarkable agreement with other methods underscores the effectiveness of the elemental analyzer as a simple and accurate analytical method for quantifying drug loading in polymeric prodrug products.

In this case, sulfur is exclusively found in the drug (LAM) and does not exist in the polymer backbone (PLS), so, theoretically, it should be sufficient to only analyze the drug using the CHNS mode. However, as shown in Table 6, the polymer showed a minor contribution of 0.5% sulfur from an unidentified impurity. Although this percentage is low, failure to account for it would skew the final weight percentage of the drug (%wt_{drug}) to 9.8%. Thus, in practical terms, it is still recommended to analyze the polymer, even if it does not contain sulfur. Furthermore, hydrogen and carbon can also be used to calculate the weight percentage of the drug loading. However, the results obtained from these calculations are not consistent due to the presence of various impurities, such as moisture, solvents, and salts which contain carbon and hydrogen. In this study, when using %C obtained from the CHN or CHNS mode to determine the drug loading, the results ranged from 11% to 15%. On the other hand, %H showed approximately 8% LAM for both the CHN and CHNS modes [4].

4 Notes

1. It is best to use the free drug and free polymer backbone from the same lot that was used to prepare the prodrug. The impurities in the free drug or free polymer backbone could influence the final data.
2. A good analysis highly relies on accurate sample weight. Minimize exposure of standard/sample to light, air, and moisture. Use a brush to wipe the outside of the tin capsule. Avoid cross-contamination of samples/standards by cleaning all tools between groups. It is recommended to lyophilize the sample/standard to remove moisture, if possible. Use a homogeneous sample. If needed, pulverize the sample into fine particles. An insufficient surface area and/or inhomogeneity of the sample could lead to incomplete combustion.

Acknowledgments

This project has been funded in whole or in part with federal funds from the National Cancer Institute, National Institutes of Health, under contract 75N91019D00024. The content of this publication does not necessarily reflect the views or policies of the Department of Health and Human Services, nor does the mention of trade names, commercial products, or organizations imply endorsement by the US Government.

References

1. Ekladious I, Colson YL, Grinstaff MW (2019) Polymer-drug conjugate therapeutics: advances, insights and prospects. Nat Rev Drug Discov 18(4):273–294
2. Pang X, Jiang Y, Xiao Q, Leung AW, Hua H, Xu C (2016) pH-responsive polymer-drug conjugates: design and progress. J Control Release 222:116–129
3. Perkin Elmer 2400 series II CHNS/O elemental analysis application note. https://resources.perkinelmer.com/lab-solutions/resources/docs/bro_2400_seriesii_chnso_elemental_analysis.pdf. Accessed 24 July 2023
4. Hu Y, Stevens DM, Man S, Crist RM, Clogston JD (2019) Total drug quantification in prodrugs using an automated elemental analyzer. Drug Deliv Transl Res 9(6):1057–1066

Chapter 5

Particle Size and Concentration Measurement Using the Spectradyne nCS1 Instrument

Alison Vermilya and Jeffrey D. Clogston

Abstract

This protocol describes the use of the Spectradyne nCS1 instrument to measure the particles per mL concentration and size of nanoparticles. The Spectradyne nCS1 is a particle-analyzing instrument that uses microfluidic resistive pulse sensing, rather than optical measurements, to determine the size and concentration of samples. The size and concentration of a sample are determined by measuring the changes in voltage as particles travel through a nano-constriction in the microfluidic cartridge. This method also has the advantage over optical techniques in that measurements are not dependent on the type of material being measured (e.g., refractive index of the sample itself is not needed for accurate analysis), and only microliters (typically 5 μL) of a sample are needed for analysis.

Key words Per particle sizing, Particle counting, Microfluidic device, Resistive pulse sensing

1 Introduction

The Spectradyne nCS1 instrument relies on resistive pulse sensing (RPS), which is based on the Coulter principle [1]. A more detailed discussion of the theoretical basis of the Spectradyne nCS1 instrument is discussed elsewhere [2]. Here is a brief summary of instrument use and operation based on software Version 2.5.0.261.

The instrument uses microfluidic cartridges with varying size ranges for particle analysis. The cartridges have a lower size limit of 50 nm and an upper size limit of 10,000 nm. Within the microfluidic cartridge, there is a nano-constriction across which a voltage is applied. The sample reservoir is on one side of this constriction, and the running buffer (typically phosphate-buffered saline [PBS] + 1% Tween) flows on the opposite side of the constriction. As particles migrate through this constriction, the measured volt-

Jeffrey D. Clogston et al. (eds.), *Characterization of Nanoparticles Intended for Drug Delivery*, Methods in Molecular Biology, vol. 2789, https://doi.org/10.1007/978-1-0716-3786-9_5,

age across the nano-constriction will be disrupted. The magnitude of this disruption correlates with the size of the particle traveling through the nano-pore. The frequency at which these voltage disruptions occur correlates with the concentration of the sample [2].

The microfluidic cartridges are disposable, and users should select an appropriate size-range cartridge for each nanoparticle analyzed based on prior knowledge of the particle's size. If no such information is available, a wider size range can be run initially, followed by a smaller, more appropriate size-range cartridge. The commonly used cartridges are C-300 and C-400, whose ranges are ~50–300 nm and ~70–400 nm, respectively. External calibration of the cartridges is not necessary; however, particles of a known size can be added to the sample of interest as an internal standard.

2 Materials

1. Running buffer: phosphate-buffered saline (PBS) with 1% Tween 20 (*see* **Notes 1–4**).
2. Distilled water (DI) H_2O.
3. Spectradyne nCS1 instrument and software. The instrument will also need a nitrogen source with 15–30 psi.
4. Spectradyne nCS1 sample cartridges. The example uses a C-300 cartridge (50–300 nm range). Various sizes are available through the company; choose the size that best suits your sample of interest (*see* **Notes 5** and **6**).

3 Methods

3.1 Sample Measurement

1. Power on the instrument before loading the nCS1 software. Turn on the nitrogen gas connection and check that the power supply, gas supply, and USB connection are in proper working order.
2. Fill buffer bottles with at least 50 mL of running buffer (*see* **Notes 1–4**) and ensure the waste bottles are empty. All five bottles should be tightly attached via both the bottle lid and the quick connects.
3. Run a cleaning cartridge cycle prior to running samples (*see* **Note** 7).
4. Prepare the sample for loading by either spiking Tween 20 into the sample or diluting the sample in running buffer (*see* **Notes 8–11**).

5. Allow the diluted sample to sit at room temperature for approximately 15 min to equilibrate in the running buffer solution.
6. Add 3–5 μL of the prepared sample to the selected microfluidic cartridge (*see* **Notes 5**, **6**, **12**, and **13**). Ensure there are no bubbles in the sample chamber by gently tapping. Be cautious not to break the glass cartridge base.
7. Load the cartridge into the front of the instrument with the glass lip edge first. When the cartridge is loaded properly, the indicator light on the front panel will light.
8. Press the front button to close the instrument and raise the cartridge stage into position for sample analysis. The white light will light up to indicate that the sample cartridge has been loaded correctly, and sample analysis can begin.
9. Fill in the various "run details" required by the software, e.g., Mold ID, Save location, Data file prefix (*see* **Notes 14** and **15**).
10. Click "GO" to begin the cartridge priming process (*see* **Notes 16–18**).
11. The instrument will begin collecting data immediately after the priming is complete (*see* **Notes 19** and **20**).
12. To end a run, click "End Run." This will begin the instrument flush procedures to clean the fluid lines and prepare the instrument for another sample run. The instrument will indicate when it is ready to have you lower the stage and remove the cartridge.
13. Dispose of the cartridge in an appropriate refuse container (*see* **Note 21**).

3.2 Shutdown and Cleaning Procedure

1. Click "Instrument Shutdown" to begin. Follow the prompts on the software to progress through the shutdown procedure. Two instrument-compatible bottles of DI water and the cleaning cartridge will be required for shutdown and cleaning.
2. The instrument will note when it is safe to turn the instrument off. Once completed, the running buffer should be placed in the refrigerator, the waste bottles emptied, the air supply turned off, the power supply turned off, and the computer software closed.

3.3 Data Analysis

1. All data analysis functions can be completed using the data analysis software, including (but not limited to) data processing, data filtering, peak fitting, size and concentration determination, and graph creation (*see* **Note 22**). A detailed explanation of all software functions will not be given here; please refer to the Spectradyne nCS1 manual and website for more thorough explanations [3].

2. Using the file explorer, select the raw data files to be processed; files can be added one at a time or in a batch.
3. Once the files to be processed are added to the queue, select the desired processing settings. Click "Process All" to process the raw data files into analyzed data files (*see* **Note 23**).
4. Once the raw data files are processed, data files containing the concentration spectral density (CSD) data can be viewed, combined, and analyzed. There are three options to view CSD data—single, multiple, or combined. The single option will only pull up the CSD file highlighted. The multiple option will pull up all loaded CSD files and overlay them. The combined option will combine all loaded CSD files into a single trace.
5. Review each CSD file to determine whether there are runs without particles detected; these files should be removed and not used in the determination of particle size nor concentration. The multiple view is helpful in identifying if any runs are outliers or if the sample is degrading over time.
6. Once data is reviewed, the individual run files can be combined into a single data file. Using the software, click "Combine All Loaded" to create a single file from the individual CSD data files loaded on the software. This new file can be saved with the file name of your choice.
7. Using the combined final data file, use the software tools to determine particle size and concentration; this is found in the "quantification" tab. The simplest manner to determine size is to select the particle peak using the software and use the "Gaussian fit" function. This will fit a simple Gaussian function to the graphed data and is used to determine the average particle size and coefficient of variation of the data. The Gaussian fit will also report a particle concentration and standard deviation; this concentration corresponds to the Gaussian peak fit, not the integrated concentration under the graphed data. The integrated concentration is given at the right of the graph. For samples with a single particle population, the integrated concentration is typically very close to the Gaussian fit concentration.
8. For an example of the final processed data, *see* Fig. 1, which was generated using the commercial Doxil formulation.

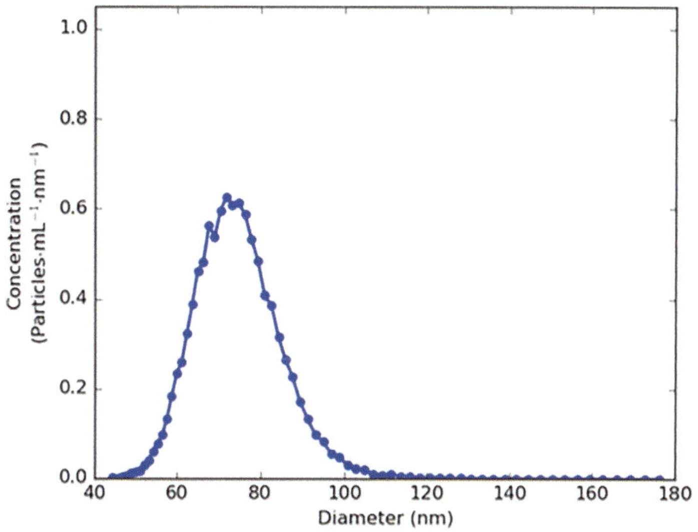

Fig. 1 Doxil data, collected and analyzed using the nCS1. Doxil was diluted 100-fold in running buffer (PBS with 1% Tween 20), and the instrument used a C-300 cartridge (50–300 nm range). Doxil had an average diameter of 73.3 ± 13.5 nm and a concentration of 1.5 × 10^{13} particles per mL

4 Notes

1. Spectradyne recommends using PBS with 1% Tween 20 as the running buffer and sample diluent. The running buffer should be filtered through a 0.2 μm filter prior to use. Store the running buffer at 4 °C and discard it after 2 weeks.
2. Some nanoparticle test samples may not be compatible with the Tween 20 component of the buffer. According to Spectradyne, BSA can also be used in place of Tween 20, though this has never been tested in our lab.
3. To test sample compatibility with Tween 20, prepare the sample dilution in the recommended PBS + 1% Tween 20 running buffer and measure using dynamic light scattering (DLS). Compare the DLS size distribution versus a test sample without the presence of 1% Tween 20. Ensure the same dilution is measured for both samples.
4. Given the viscous nature of Tween 20, it is paramount to ensure the Tween is sufficiently dissolved in the PBS. Use a disposable plastic syringe or positive-displacement style pipette to measure the necessary volume of Tween 20. Once the PBS and Tween 20 are combined, the solution should be stirred for 30–60 min to ensure proper mixing; use a low stir speed to avoid foaming of the Tween 20. We typically prepare running buffer batches in 200 mL increments; for example, 198 mL of PBS and 2 mL of Tween 20.

5. Spectradyne currently offers five cartridge sizes: C-300, 50–300 nm; C-400, 75–400 nm; C-900, 130–900 nm; C-2000, 250–2000 nm; and C-10 k, 1.5–10 μm. It is recommended that your particle of interest have a nominal diameter approximately 20% smaller than the maximum size for the cartridge.
6. On occasion, a cartridge may function improperly. For example, a cartridge may fail to fully prime for measurement after being inserted into the instrument. If this occurs, simply discard the cartridge and replace it with a new cartridge. As another example, no particles may be detected following the measurement if the cartridge does not allow the particles to travel within the channels. This could be a result of a clog in the channel. Running the "clear cartridge" function will help clear a suspected clog. If running this method does not result in improved measurements, simply discard the cartridge and replace it with a new cartridge.
7. Cleaning cartridges are used to clean the instrument between sample runs and in the shutdown process. Running a cleaning cartridge cycle before sample analysis can reduce sample cartridge priming time by ensuring that fluids are flushed throughout the instrument.
8. Spectradyne recommends a particle concentration of approximately 10^8–10^9 particles/mL. We have found that a 100-fold dilution is often a good starting dilution for nCS1 analysis.
9. A sample that is not concentrated enough will result in very few or no particles passing through the pore of the cartridge; this will appear as very few or no changes in voltage during the measurement.
10. A sample that is too concentrated will often result in clogs in the pore of the cartridge. The nCS1 software will indicate suspected clogs in the run data as it is collected; this can be indicated by a dramatic change in voltage or particle peaks completely disappearing.
11. For samples that are stable in the running buffer, samples should be diluted from stock in the running buffer. For dilute samples that are compatible with the running buffer, the samples can be spiked with Tween. For example, add 0.5 μL of Tween to a 50 μL aliquot of the sample.
12. Pipetting the sample into the chamber at an angle often reduces bubbles.
13. Optionally, a nanoparticle of a known size can be spiked into the sample as an internal standard to calibrate the cartridge measurement. It is recommended that the size of the internal standard not overlap with the size(s) of the test sample(s).

14. Many of these details are required entries before the software will permit the sample to be analyzed. "Mold ID" identifies the size and mold of the cartridge. Not all cartridges of the same size (e.g., TS-300) will have the same mold ID. "Save location" is where the files will be saved on your computer. "Data file prefix" is the name of your data file. "Optional-description" can include any other information desired. "Optional-calibration size and concentration" is used when using an internal standard spike.
15. Acquisition time can be chosen in the drop-down menu, with the option of selecting "single" or "continuous" data acquisition. Typically, we collect 60 s runs continuously. Each run will be numbered sequentially beginning with "prefix_date_00000.h5." Partial acquisitions will not be saved.
16. The instrument manual notes a typical priming time of 1–2 min; however, we have found that priming time is usually closer to 5 min. Running a cleaning cartridge prior to a sample cartridge typically encourages faster priming.
17. You can watch the four fill monitors on the instrument panel (sliders with red/yellow/green background) to monitor the progress of the priming.
18. If the priming does not complete properly, you can first try to end the priming and restart it. If this does not solve the issue, discard the cartridge and replace it with a new cartridge.
19. The voltage should be between ±0.8 V and preferably between ±0.2 V. The voltage balance can often indicate the presence of a clog in the chamber.
20. Should the voltage balance be out of range or sample traces appear to have multiple clogs or no sample, the clear constriction function can be used. If running continuous runs, stop collecting runs and then click "clear constriction." The clear constriction program will run to flush the cartridge of any clogs. Once the program is complete, allow the voltage to return to the normal range and resume collecting runs.
21. The cartridge contains glass; a sharps container or other glassware disposal container may be better suited for the disposal of nCS1 cartridges. The disposal method should also be appropriate for nanomaterial waste.
22. The Spectradyne instrument control software and data analysis software are two different programs; it is recommended that you close the instrument control software before running the data analysis software.
23. This step can take a significant amount of time; it can be set up to run overnight. The processed data files will be saved in the same location as the raw data files.

Acknowledgments

This project has been funded in whole or in part with federal funds from the National Cancer Institute, National Institutes of Health, under contract 75N91019D00024. The content of this publication does not necessarily reflect the views or policies of the Department of Health and Human Services, nor does the mention of trade names, commercial products, or organizations imply endorsement by the US Government.

References

1. Coulter WH (1953) Means for counting particles suspended in a fluid. US Patent 2,656,508
2. Fraikin JL, Teesalu T, McKenney CM, Ruoslahti E, Cleland AN (2011) A high-throughput label-free nanoparticle analyser. Nat Nanotechnol 6(5):308–313
3. Spectradyne nCS1 user manual, Revision 2.5.3 (2020) https://www.nanoparticleanalyzer.com/pdf/nCS1operate250r3.pdf?_gl=1*5dy6ft*_ga*MTUwODI1ODg2Mi4xNjk3ODA2NjM1*_ga_BFTLJFR01N*MTY5NzgwNjYzNS4xLjEuMTY5NzgwNzY2Mi4wLjAuMA. Accessed 20 Oct 2023

Chapter 6

Measuring Size and Number Concentration of Metallic Nanoparticles Using spICP-MS

Matthew Hansen and Jeffrey D. Clogston

Abstract

This protocol describes how to measure the size and concentration of individual metallic nanoparticles using inductively coupled plasma-mass spectrometry (ICP-MS) in single-particle (sp) mode. Accurately determining the size of individual nanoparticles on a per-particle basis, both quickly and accurately, is an ever-increasing need within nanoparticle characterization. ICP-MS is capable of measuring a broad range of metallic nanoparticle sizes with high resolution, thus allowing the measurement of multiple particle populations for the quality assessment of nanoformulations. Additionally, spICP-MS can accurately determine particle concentrations without the need for concentration standards.

Key words Single-particle sizing, Mass spectrometry, Elemental analysis, Nanomedicine, Single-particle concentration, Nanoformulation

1 Introduction

Inductively coupled plasma-mass spectrometry (ICP-MS) is an ultrasensitive technique that can determine the concentration of trace elements in solution down to the pg/g level. Delguard et al. first demonstrated in 2006 that ICP-MS could accurately measure the mass of individual metallic nanoparticles in what has come to be known as spICP-MS [1]. By introducing a nanoparticle solution into the ICP, the individual nanoparticles are atomized into an ion cloud, which is then carried as an intact unit to the detector. After a brief rapid expansion of the ion cloud, the duration of this cloud, defined as the time it first hits the detector to the time the last ion hits, is approximately 100–500 μs [2]. This allows for two methods for measuring nanoparticles in sp mode, both of which will be outlined in the protocols below.

In method 1, a large integration time (10 ms) with respect to the duration of the ion cloud is used in combination with a sufficiently dilute solution to measure individual nanoparticles. Since

Jeffrey D. Clogston et al. (eds.), *Characterization of Nanoparticles Intended for Drug Delivery*, Methods in Molecular Biology, vol. 2789, https://doi.org/10.1007/978-1-0716-3786-9_6,

the integration time is much larger than the particle cloud duration, this method measures the entire particle cloud as a single event, and as a consequence, it is extremely important that the nanoparticle solution is sufficiently dilute to prevent multiple particles from hitting the detector during the integration window. This method has the advantage of being usable for many modern ICP-MS instruments that do not have a dedicated single-particle counting mode, since most of these instruments are capable of integration times as low as 10 ms. Additionally, the long integration times allow for a longer period for the detector to recover after a nanoparticle event, leading to a lower occurrence of detector saturation and small particle events.

Method 2 uses a shorter integration time than the duration of the ion cloud, allowing for the measurement of nanoparticle events by rastering multiple points across the detection event. The 50 μs integration time is generally available in instruments with a dedicated single-particle mode and provides a few advantages over 10 ms integration measurements. The faster integration time allows for the collection of more information than the 10 ms integration, which is limited to the total or summed intensity of the nanoparticle event. Three types of information are available due to the ability to raster across the peak. Like 10 ms integrations, the total or summed intensity is available; however, also available are the max height intensity of the peak and the duration of the ion cloud. Due to the faster integration time, samples about 10 times higher in concentration than the 10 ms method can be accurately analyzed, simplifying the sample preparation process.

2 Materials

1. A quadrupole ICP-MS (Perkin Elmer Nexion 200B) equipped with a peristaltic pump, MicroMist nebulizer, cyclonic spray chamber, and peristaltic pump tubing with an inner diameter of 0.76 mm were used to develop this protocol; however, this method should be applicable for any ICP-MS system, and users should consult their manufacturer's documentation for adapting this protocol for their system. For measurement of the nanoparticle samples, all samples should be diluted in buffers prepared from ultrapure water, 18 MΩ cm resistivity (*see* **Note 1**).
2. Dissolved gold standards buffer: 4% hydrochloric acid (HCl) and 1.5% nitric acid (HNO_3) v/v.
3. NIST SRM 3121 (10 mg/g Au) or NIST traceable RM dissolved gold standard (*see* **Note 2**).
4. Citrate-stabilized gold nanoparticle solutions with diameters of 30, 40, 50, 60, 80, 100, and 150 nm.

5. Citrate-stabilized silver nanoparticle solutions with diameters of 30, 50, 70, and 100 nm.
6. Plastic bottles, preferred materials include high-density polyethylene (HDPE), low-density polyethylene (LDPE), or polypropylene.

3 Methods

3.1 Measuring the Flow Rate of the Peristaltic Pump

1. Move the sampling probe to a vial containing the running buffer, in this case, 18 MΩ cm water, and allow the sampling tube to fill with water until all air or residual solvent has been evacuated from the sample tubing.
2. Collect and label four microcentrifuge tubes or other small plastic vessels. Tube #1 will be discarded after use and serves as a sacrificial run to ensure that the tubing is adequately wetted and no air is remaining in the tubing. The remaining tubes (#2–4) will be used to calculate the flow rate.
3. Weigh and record the weights of all four empty 2 mL microcentrifuge tubes.
4. Turn off the peristaltic pump and disconnect the sampling tube from the nebulizer and place the end of the tube into tube #1.
5. Set the speed of the peristaltic pump to the speed used during single-particle measurements (*see* **Note 3**).
6. Start the peristaltic pump while simultaneously starting a timer.
7. Collect water for 1–2 min, stopping the pump when the timer completes.
8. Cap the tube and set aside.
9. Repeat **steps 4–8** three additional times using tubes #2–4, respectively.
10. Weigh and record the weight of all four tubes.
11. Subtract the weight of the empty tube from the weight of the tube plus solvent to get the total amount of liquid collected. Multiply this weight by the density of the liquid (*see* **Note 4**). Divide the calculated volume by the total time in minutes to get the flow rate for the subsequent set of measurements. Discard the flow rate for tube #1 and average the results for tubes #2–4 together to get the final flow rate (*see* **Note 4**).
12. Reconnect the tubing to the peristaltic pump.

3.2 Preparation of Dissolved Elemental Standards (Au Standards)

1. Prepare the dissolved standards buffer by adding 871 mL of water to a 1 L plastic bottle. To this, add 21 mL of concentrated HNO_3 (70% w/w, trace metal grade or higher) and 108 mL of concentrated HCl (37% w/w, trace metal grade or higher) to the plastic bottle and mix.

2. Use serial dilutions to prepare the dissolved elemental standards (*see* **Note 5**).
3. Prepare a 50 parts per million (ppm, μg/g) solution of NIST SRM 3121 by weighing an empty 30 mL LDPE sample bottle. Record the weight. Add 150 μL of NIST SRM 3121 to the pre-weighed bottle and record the weight. Dilute to a total volume of 30 mL using the 1.5% HNO_3: 4% HCl solution.
4. Prepare a 1 ppm (μg/g) solution of SRM 3121 by weighing a 60 mL LDPE sample bottle. Add 1 mL of the 50 ppm Au solution prepared in **step 3** to the bottle and record the weight. Dilute the solution to a total volume of 50 mL using the 1.5% HNO_3: 4% HCl solution and record the weight.
5. Prepare the dissolved parts per billion (ppb) Au standards from the 1 ppm SRM solution prepared in **step 4**. For example, to make a 10 ppb solution of SRM 3121, first weigh an empty 60 mL LDPE sample bottle. Next, add 500 μL of the 1 ppm solution of SRM 3121 prepared in **step 4** and weigh the sample bottle. Finally, dilute to a total volume of 50 mL using the 1.5% HNO_3: 4% HCl solution and record the final weight. Repeat until a minimum of five standards are prepared across the concentration ranges of 10–100 parts per billion (ppb, ng/g) for 50 μs measurements or 0–10 ppb for 10 ms measurements.
6. Determine the exact concentration in each of the ppb standards (*see* **Note 6**).

3.3 Prepare Nanoparticle Standard Solutions and Unknown Samples

1. Determine the ideal nanoparticle flux (*see* **Note** 7) for the integration time and total sampling time for your experiments.
2. Determine the number concentration (particles/mL) in the undiluted nanoparticle sample (*see* **Note 8**).
3. Using the equation in **Note 6** and estimating a transport efficiency, determine the number concentration necessary to minimize using the nanoparticle flux, the estimated nebulization efficiency, and the flow rate measured in Subheading 3.1.
4. Dilute the sample using serial dilution to that number concentration using water or buffer (*see* **Notes 9** and **10**).
5. To dilute the samples, weigh an empty sample vial and record the weight.
6. Add the sample to the vial and record the weight of the vial plus the sample.
7. Dilute the sample with water or buffer and record the total weight.
8. Repeat **steps 5**–7 until all the samples have been diluted to the proper concentration for analysis.

3.4 Sample Measurement

1. Load the samples into the autosampler for your instrument.
2. Turn on the instrument and allow it to warm up for 15 min (*see* **Note 11**).
3. Tune the instrument according to the manufacturer's specifications (*see* **Note 12**).
4. Open the single-particle application/module (*see* **Note 13**).
5. Determine the integration and total run time for your sample (*see* **Note 14**).
6. Set up the method for measuring the nanoparticles and select the proper mass to analyze.
7. Create a sequence for running your samples (*see* **Note 15**).
8. Run your samples.

3.5 Data Analysis

1. Perform data analysis using your system's software (*see* **Note 16**).
2. Determine the average signal intensity of each dissolved standard (*see* **Note 17**).
3. Construct the dissolved standard calibration curve by graphing the mean dissolved Au counts as a function of the mass of analyte reaching the detector (*see* **Note 18**).
4. Determine the slope of the linear fit of this graph.
5. Determine the average of your buffer blanks counts by averaging the counts in entire blank datasets (*see* **Note 19**).
6. Determine the Lc (Poisson critical limit/limit of detection) and Ld (Poisson limit of detection/limit of quantitation) for the measurements defined as $2.326 \times \sqrt{\mu_B + 0.5}$ and $2.706 + 4.65\sqrt{\mu_B + 0.5}$ the standard deviation of the blanks, respectively [3] (*see* **Note 20**).
7. Filter all nanoparticle datasets by the Ld.
8. Determine the nanoparticle events and isolate them by considering any remaining nonzero intensity to be a peak, and any consecutive nonzero intensities as a single peak.
9. The duration of the peak can be determined by counting the consecutive nonzero values and multiplying by the integration time.
10. Determine the peak area/summed intensity of each nanoparticle event.
11. Create a histogram of the peak area/summed intensity.
12. Fit the histogram and determine the peak area/summed intensity at the peak maximum.
13. Construct a nanoparticle standard calibration plot by graphing the area/summed peak intensity maximum determined in **step**

11 for the nanoparticle standards as a function of the mass of the nanoparticles.

14. Determine the slope of the linear fit from this plot.
15. Calculate the transport efficiency for the analysis by dividing the dissolved calibration slope by the nanoparticle calibration slope (*see* **Note 21**).

3.5.1 Processes for Determining the Sizes for the Unknown Particles, Method 1: Nanoparticle Calibration Curve (See ***Note 22****)*

1. Using the nanoparticle calibration curve linear regression formula, convert nanoparticle signals to size.
2. Construct a histogram for the frequency of each size.
3. Fit the histogram and determine the size at the peak maximum and determine the standard deviation.

3.5.2 Processes for Determining the Sizes for the Unknown Particles, Method 2: Using the Dissolved Standards Curve and the Density of the Nanoparticle

1. Convert the summed nanoparticle signal to mass per measured event using the dissolved standard calibration curve linear regression formula.
2. Multiply this mass by the transport efficiency.
3. Using the density of the nanoparticle, convert the mass in **step 2** per event to a spherical radius using the density of the nanoparticle by first converting mass to a total volume.
4. Assuming a spherical geometry, convert the volume of the nanoparticle to a spherical diameter.
5. Construct a histogram of the frequency of the nanoparticle diameters in nm.
6. Fit the histogram and determine the size at the peak maximum, and determine the standard deviation of the nanoparticle sample.

3.5.3 Determine the Particle Concentration of the Unknown Nanoparticle Solution

1. Count the total number of nanoparticle events during the measurement.
2. Convert the counted number of particles to a particle concentration using the following equation:

$$\mathrm{CNP} = \frac{n_{\mathrm{NP}} \times 60}{t \times \eta_{\mathrm{NEB}} \times q_{\mathrm{LIQ}}} \times \text{Dilution factor} \quad (1)$$

where concentration of nanoparticles (CNP) is the undiluted particle concentration in particles per mL, n_{NP} is the number of particles counted, t is the total time in seconds, q_{LIQ} is the flow rate in mL/min, and η_{NEB} is the nebulization efficiency measured.

4 Notes

1. For the nanoparticles used in this protocol, the dilution buffers are water only, but other surfactants/stabilizers may be chosen depending on the sample. Diluted samples can be stored and used at room temperature. Samples should be freshly prepared prior to running to minimize any destabilizing effects caused by the large dilutions required to run the samples.
2. For this protocol, Au is used as the calibration standard; however, any element can be used for calibration if Standard Reference Material (SRM)/Reference Material (RMs) are available or a well-defined set of nanoparticles with known sizes and concentrations are available.
3. Choosing a peristaltic pump speed is system dependent and depends on the sample introduction, including the type of peristaltic pump, spray chamber, nebulizer, and peristaltic pump tubing. For our instrument setup, the speed is set to 18 rpm.
4. Prior to running the samples, the flow rate needs to be determined for the specific peristaltic pump, tubing, and the solvent system being used in spICP-MS. Due to the deterioration/stretching of the tubing over time, the flow rate should be measured just prior to each single particle (SP) sample run. For our system, the flow rate is usually around 0.350 mL/min. If your system uses a syringe pump or another sample introduction method that allows for the precise control of the sample flow, you can skip ahead to Subheading 3.2.
5. National Institute of Standards and Technology (NIST) SRM 3121 has an initial concentration of 10 mg/g (10 parts per 1000), so the dissolved standards need to be prepared by serial dilutions to achieve running concentrations of 10–100 parts per billion (ppb, ng/g) for 50 μs measurements or 0–10 ppb for 10 ms measurements.
6. The exact concentration (M) of each dissolved standard in ppb (*ng/g*) can be calculated using Eq. 2 and solving for M2 and using the weights (W) of the solutions added.

$$M_{\mathrm{STD1}} W_{\mathrm{STD1}} = M_{\mathrm{STD2}} W_{\mathrm{Total}} \tag{2}$$

7. To prepare the nanoparticle calibration standards, the optimal nanoparticle flux, or the rate at which the nanoparticles hit the detector, needs to be determined. Two sources of instrumental uncertainty depend on the nanoparticle flux. Bias is introduced when the sample flux is too high due to the increased frequency of multiple particles hitting the detector simultaneously. When this happens, it is difficult to distinguish between single and

multiple particle events. Bias is related to the nanoparticle flux (f_{NP}) and the dwell time (t_{dwell}) by Eq. 3 [4, 5].

$$\text{Bias} = \frac{t_{\text{dwell}}}{2} \times f_{\text{NP}} \tag{3}$$

On the other hand, the relative standard deviation (RSD) decreases with increasing f_{NP} due to the larger total number of nanoparticles counted. Since the total nanoparticles counted is proportional to the total time (t), their relationship with RSD is given in Eq. 4 [4, 5].

$$\text{RSD}_{\text{NP}} = \frac{1}{\sqrt{t \times f_{\text{NP}}}} \tag{4}$$

The target nanoparticle flux is determined from the point where the RSD is equal to the bias and is given by Eq. 5 [5].

$$f_{\text{NP}} = \sqrt[3]{\frac{4}{t \times t_{\text{dwell}}^2}} \tag{5}$$

8. The number concentration and total dilution of the nanoparticle samples can be determined for each of the integration times according to Eq. 6 [5].

$$N_{\text{NP}} = \frac{f_{NP} \times 60}{q_{\text{LIQ}} \times \eta_{\text{n}}/100} \tag{6}$$

Two additional factors must be determined before the number concentration in Eq. 6 can be used to estimate the particle numbers: the flow rate (q_{liq}) and the nebulization efficiency (η_n). The flow rate is easy to manually measure as described in Subheading 3.1. The nebulization efficiency, however, must be determined experimentally after nanoparticle standards have been analyzed. Without prior knowledge or an estimate of the nebulization efficiency, it may be necessary to run a single nanoparticle standard at a few dilutions to determine its response and use the response of a single dissolved standard of similar signal to obtain an estimate of the nebulization efficiency. The estimated particle flux and number concentration for particles in solution are given for 10 ms and 50 μs and a total acquisition time of 100 s in Table 1. A nebulization efficiency of 0.1 and a flow rate of 0.285 mL/min were used to calculate the number concentrations in Table 1.

Table 2 gives a list of various sizes of Au nanoparticles and the approximate total dilution necessary to achieve the number concentrations reported in Table 1 (based on the manufacturer-supplied gold content of 0.01% gold chloride).

9. Serial dilutions should be utilized to make the proper run dilutions for each nanoparticle solution. The following dilutions will be made in 18 MΩ cm water, with the total solution

Table 1
Particle flux for spICP-MS for dwell times of 10 ms and 50 μs for a 100 s acquisition

t_{dwell}	f_{NP}	N_{NP} (part./mL)
10 ms	7.37	1.5×10^4
50 us	251	5.3×10^5

Table 2
Approximate dilution factors necessary for various-sized gold nanoparticles

Au nanoparticle diameter (nm)	Dilution factor, 50 μs	Dilution factor, 10 ms
150	3,000	110,000
100	10,000	340,000
80	21,000	730,000
60	49,000	1,700,000
50	83,000	2,800,000
40	150,000	5,200,000
30	390,000	13,000,000

volumes equaling 10 mL. The weights of all solutions will be used in place of volume, with conversion to volume units assuming a 1 mg/mL density of each solution. We typically find weights to be more accurate due to variation with pipetting and pipetting technique from sample to sample. Additionally, by using weight in combination with knowing the volume pipetted, there is a built-in redundancy for double-checking the sample preparations. Typical dilutions will take place by first weighing an empty conical centrifuge tube. Then the sample solution is added to the tube, and the weight of the tube and the sample is recorded. Finally, the diluent is added to the tube, and the total weight is recorded. The dilution factor and concentration of each sample can be determined similarly to the procedure described in **Note 3**. It is recommended that the minimum amount of sample to weigh out is 100 mg, or 100 μL of solution, to ensure an accurate weight is obtained. Tables 3 and 4 give a summary of the serial dilutions targeting different dilution factors presented in Table 2. In the tables, the amounts for each solution are given in mL since it is easier to add amounts based on volume, but the weight of each solution should be recorded to ensure accuracy.

Table 3
Serial dilutions to achieve necessary dilution factors for spICP-MS for 50 μs integration

Solution ID (Sample amount)	Diluent amount (mL)	Dilution factor (Au NP solution ID)
Stock (0.10 mL)	9.90 mL	1:100 (all)
1:100 solution (0.10 mL)	9.90 mL	1:10 k (all except 150 nm; final dilution for 100 nm)
1:100 solution (0.33 mL)	9.67 mL	1:3 k (150 nm)
1:10 k solution (4.76 mL)	5.24 mL	1:21 k (80 nm)
1:10 k solution (2.04 mL)	7.96 mL	1:49 k (60 nm)
1:10 k solution (1.20 mL)	8.80 mL	1:83 k (50 nm)
1:10 k solution (0.67 mL)	9.33 mL	1:150 k (40 nm)
1:10 k solution (0.26 mL)	9.74 mL	1:390 k (30 nm)

Table 4
Serial dilutions to achieve necessary dilution factors for spICP-MS for 10 ms integration

Solution ID (Sample amount)	Diluent amount (mL)	Dilution factor (Au nanoparticle solution ID)
Stock (0.10 mL)	9.90 mL	1:100 (all)
1:100 solution (0.10 mL)	9.90 mL	1:10 k (all)
1:10 k solution (0.87 mL)	9.13 mL	1:110 k (150 nm)
1:10 k solution (0.28 mL)	9.72 mL	1:340 k (100 nm)
1:10 k solution (0.14 mL)	9.86 mL	1:730 k (80 nm)
1:10 k solution (0.10 mL)	9.90 mL	1:1 M (30, 40, 50, 60 nm)
1:1 M solution (5.78 mL)	4.22 mL	1:1.7 M (60 nm)
1:1 M solution (3.57 mL)	6.43 mL	1:2.8 M (50 nm)
1:1 M solution (1.92 mL)	8.08 mL	1:5.2 M (40 nm)
1:1 M solution (0.77 mL)	9.23 mL	1:13 M (30 nm)

10. Unknown samples should be diluted as outlined in Subheading 3.3, **step 3**. To determine the amount of dilution for each sample, the particle concentration can be estimated using the size of the nanoparticle from a different method (DLS, TEM, etc.) and the total ionic concentration of the analyte in solution according to Eq. 7.

$$N_{\mathrm{NP,estimate}} = \frac{3 \times [\text{Analyte Ion}]}{\rho_{\mathrm{NP}} \times \pi \times r^3 \times 4 \times 1x10^{-21}} \tag{7}$$

where ρ_{NP} is the density and r is the radius of the nanoparticle. Once the particle concentration has been estimated, the necessary dilution can be calculated using Eq. 6, and the solutions should be made as outlined in **Note 9**.

11. A good time to measure the flow rate for the instrument/experimental run is during the instrument warm-up period.
12. If extra sensitivity is needed for an analyte, move the sample tubing to a vial containing a dilute solution of the analyte (~1 to 10 ng/g) and tune the instrument to maximize the response of the analyte.
13. If your ICP-MS does not have a dedicated spICP-MS mode but has a chromatography mode (time-resolved analysis), you can make single-particle measurements using this mode. Additionally, if you have an older instrument that cannot do microsecond dwell times, your instrument is still likely to make measurements down to 10 ms, which is ample to run single-particle ICP-MS; however, you will not be able to raster across the single nanoparticle peaks.
14. The calculations presented in **Notes 8** and **9** were done using a run time of 100 ms. If you choose to use a different run time, the flux, nanoparticle run concentration, and total dilutions will need to be recalculated using the formulas found in **Notes** 7 and **8**.
15. Run a minimum of three buffer blanks before any samples have been run to get a stable background signal and make sure there is no residual analyte signal. Additionally, while setting up the sequence, it is advisable to run at least three buffer blanks after each of the calibrations (dissolved and nanoparticle) in order to remove any analyte carryover. Run at least two buffer blanks between each sample. These buffer blanks can then be used later to calculate the Lc and Ld for the samples.
16. You may perform the data analysis using either the built-in spICP-MS program or using your own analysis template. The following data analysis description will outline manual data processing. If your instrument does not have a dedicated single-particle mode to analyze your data, your only option will be to manually analyze the raw data.
17. A typical spectrum of a dissolved standard is given in Fig. 1. Because the signal of the dissolved standards is constant, there will be no peaks like in a discrete event. As the concentration of the dissolved standard increases, so does the height of the signal (both the max and min signal). By taking the average of this signal, indicated by the red line, the dissolved calibration curve can be constructed as the average signal as a function of the mass of the calibrant (g Au) hitting the detector during the integration window.

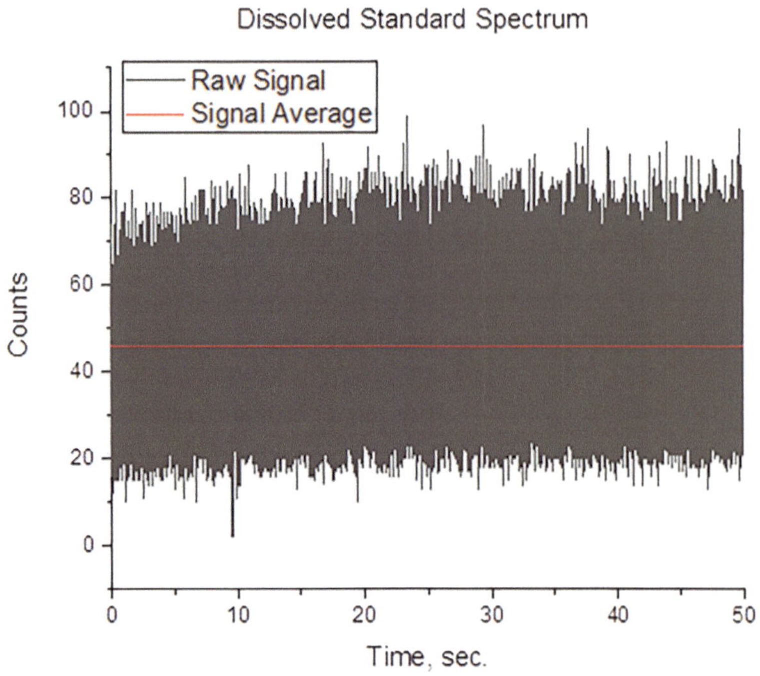

Fig. 1 Dissolved standard spectrum for a 50 ppb dissolved gold standard to illustrate how the counts are determined for calibration

18. The mass hitting the detector can be calculated by converting the concentration to mass per second and multiplying by the integration time. Equation 8 illustrates the calculation.

$$\frac{\text{Au ng}}{\text{sol g}} \times \frac{1x10^{-9}\text{g Au}}{1\text{ ng Au}} \times \frac{\rho_{\text{sol}}\text{g}}{\text{mL}} \times \frac{q_{\text{LIQ}}\text{mL}}{\text{min}} \times \frac{1\text{ min}}{60\text{ s}} \times 5x10^{5}\text{ s} \quad (8)$$

19. An example of the signal collected for a buffer (blank) sample is given in Fig. 2 for the analysis of Au. In a sufficiently clean system with low background, the background counts appear as in the insert in Fig. 2 with the vast majority of the background signal consisting of single-point "blips" of 1–2 total counts. Both the counts and duration are significantly lower than what is expected from the actual nanoparticle samples.
20. The equations calculating the Lc and Ld are based on Poisson statistics for rare events. In each calculation, an adjustment factor is given in each equation (+0.5 within the square root) and is needed when the mean of the background counts is below 5. This number can be adjusted from 0 to 1 if necessary, and more information on this can be found in [3].
21. There is a second method using particle concentration to determine the transport efficiency. This method requires a well-characterized sample in which the size and total concentration of the analyte in solution are known. The size and

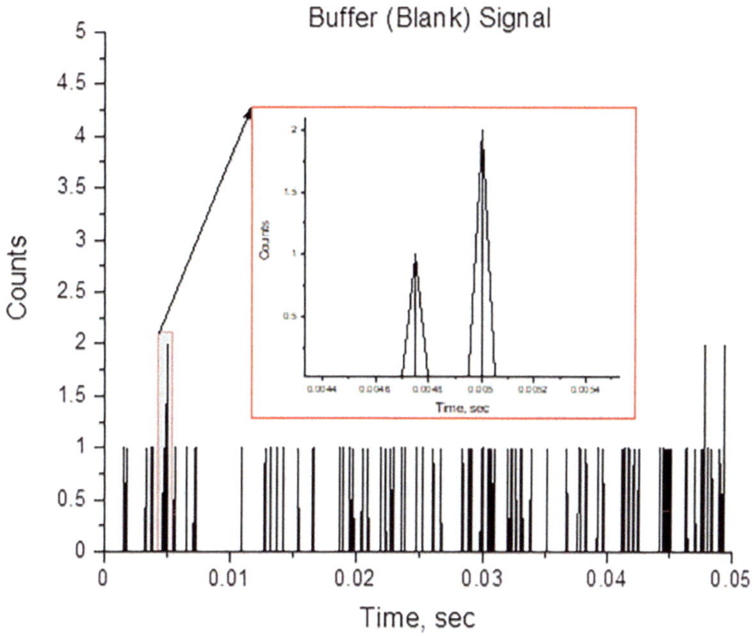

Fig. 2 Buffer (blank) signal obtained during a 50 μs integration spICP-MS acquisition. The inset shows a blown-up portion of two peaks to show the width of the duration of the spikes in the background signal

concentration are then used to determine the particle concentration in solution. To determine the transport efficiency, first convert the total number of particles measured during a run to particles per mL using the flow rate and the total acquisition time. Next, using the dilution factor, calculate the theoretical particle concentration of the measured sample. The transport efficiency is then the measured particle concentration in particles/mL divided by the theoretical particle concentration at that run dilution in particles/mL [6].

22. This method is most useful when a nanoparticle standard is available for the element being measured. When there is no nanoparticle standard available, using a dissolved standard plot constructed from the element being measured is preferred. A third method for determining the size of the nanoparticles, when no nanoparticle standards are available, is to determine a response factor between the calibration element and the unknown element. Once this response factor is known, you can adjust the unknown counts by that factor and use the nanoparticle calibration curve to determine size. Table 5 demonstrates this using 100 nm Ag nanoparticles. The actual measured diameter is 97 nm, and the particles were measured versus both calibrations from dissolved standards and nanoparticle standards and for both Au and Ag.

Table 5
Nanoparticle sizes for 100 nm silver nanoparticles determined from both dissolved calibration standards and nanoparticle standards for both Au and Ag

	Calibrant	Size from dissolved standards	Size from nanoparticle standards
100 nm Ag nanoparticle	Au	112.9 nm	107.0 nm
100 nm Ag nanoparticle	Ag	97.6 nm	97.9 nm

Acknowledgments

This project has been funded in whole or in part with federal funds from the National Cancer Institute, National Institutes of Health, under contract 75N91019D00024. The content of this publication does not necessarily reflect the views or policies of the Department of Health and Human Services, nor does the mention of trade names, commercial products, or organizations imply endorsement by the US Government.

References

1. Degueldre C, Favarger PY, Wold S (2006) Gold colloid analysis by inductively coupled plasma-mass spectrometry in a single particle mode. Anal Chim Acta 555:263–268
2. Montaño MD, Olesik JW, Barber AG, Challis K, Ranville JF (2016) Single particle ICP-MS: advances toward routine analysis of nanomaterials. Anal Bioanal Chem 408:5053–5074
3. Currie LA (2008) On the detection of rare, and moderately rare, nuclear events. J Radioanal Nucl Chem 276:285–297
4. Laborda F, Jiménez-Lamana J, Bolea E, Castillo JR (2013) Critical considerations for the determination of nanoparticle number concentrations, size and number size distributions by single particle ICP-MS. J Anal At Spectrom 28:1220
5. Murphy KE, Liu J, Montoro Bustos AR, Johnson ME, Winchester MR (2016) Characterization of nanoparticle suspensions using single particle inductively coupled plasma mass spectrometry. National Institute of Standards and Technology, Gaithersburg. Report No.: NIST SP 1200-21
6. Geiss O, Bianchi I, Bucher G, Verleysen E, Brassinne F, Mast J et al (2022) Determination of the transport efficiency in spICP-MS analysis using conventional sample introduction systems: an Interlaboratory comparison study. Nanomaterials 12:725

Chapter 7

Ion Quantitation in Liposomal Products Using RP-HPLC with Charged Aerosol Detection

Jie Xu and Jeffrey D. Clogston

Abstract

Ion concentration in liposomal drugs is important for drug stability and drug release profile. However, quantifying ion concentration in liposomal drugs is challenging due to the absence of chromophores or fluorophores of ions and the efficiency of their release from the liposome structure. To address these issues, a method based on reversed-phase high-performance liquid chromatography (RP-HPLC) coupled with a charged aerosol detector (CAD) has been developed to determine total, internal, and external ions in drug-loaded liposomal products. In this protocol, we focused on the quantitation of ammonium and sulfate ions in liposomal products, using generic PEGylated liposomal doxorubicin as an example. This method can be extended to calcium, acetate, and other ions in different liposomal formulations with slight modifications.

Key words Ion concentration, Ammonium, Sulfate, Nanoparticle, PEGylated liposomal doxorubicin, Nanomedicine

1 Introduction

The charged aerosol detector (CAD) is a universal detector that can measure all non-volatile and many semi-volatile analytes with a uniform response. In CAD, the elution from reversed-phase high-performance liquid chromatography (RP-HPLC) is nebulized into droplets, and the droplets are subsequently dried into analyte particles. The ionized nitrogen gas charges the analyte particles in a mixing chamber, and then a sensitive electrometer is used to measure the aggregate charge of the analyte particles. Thus, the detector's response is directly related to the amount of the analyte entering the detector [1]. This mechanism grants CAD a distinctive capability in analyzing analytes that lack chromophores or fluorophores. Thanks to this advantage, CAD is commonly used in the pharmaceutical and biopharmaceutical industries for the analysis of organic and inorganic ions, surfactants, lipids, and more.

Jeffrey D. Clogston et al. (eds.), *Characterization of Nanoparticles Intended for Drug Delivery*, Methods in Molecular Biology, vol. 2789, https://doi.org/10.1007/978-1-0716-3786-9_7,

Ion concentrations in liposomal formulations play an important role in ensuring an efficient drug encapsulation process. Furthermore, inorganic ion pairs in liposomal drugs can be released into the body, potentially influencing the metabolism of related elements [2]. This protocol outlines a simple method that employs liquid chromatography coupled with a CAD to determine the total, internal, and external ions in liposomal drug products. To measure the total ion concentration, the liposome is disrupted through lyophilization to release all components, and ion concentrations are measured after reconstitution. External ion concentrations are measured by conducting membrane centrifugation without disrupting the liposome structure and subsequently analyzing the filtrate. The internal ion concentrations are calculated from the mass balance of the total and external ions. To demonstrate the utility of this method, we measured the total and external concentrations of ammonium and sulfate ions in a generic PEGylated liposomal doxorubicin formulation.

2 Materials

1. Thermo Fisher Vanquish HPLC equipped with a quaternary pump (VF-P20-A) and an autosampler (VF-A10-A-02).
2. Charged aerosol detector (Thermo Fisher VH-D20-A).
3. Labconco FreeZone Plus 2.5 L cascade console freeze-dry system.
4. Generic PEGylated liposomal doxorubicin (native solvent: 10% (w/v) sucrose and 10 mM L-histidine).
5. Acetonitrile (ACN), HPLC grade.
6. Trifluoroacetic acid (TFA), HPLC grade.
7. Water, HPLC grade.
8. Ammonium sulfate ($\geq$99.0%).
9. L-Histidine ($\geq$99.0%).
10. Sucrose ($\geq$99.5%).
11. Formic acid, HPLC grade.
12. Ammonium formate ($\geq$99.0%).
13. SIELC Primesep 100 column (4.6 $\times$ 150 mm, 5 μm particle size, SIELC Technologies).
14. Thermo Fisher acclaim trinity P2 column (3.0 $\times$ 100 mm, 3 μm particle size, Thermo Fisher).
15. Microcon DNA Fast Flow centrifugal device (100 kDa molecular weight cutoff (MWCO)).

3 Methods

3.1 Total Ammonium Ion Concentration

1. Add the liposomal sample, in this case, 10 μL of generic PEGylated liposomal doxorubicin, to a 1.5 mL tube. Seal the tube using parafilm and puncture three holes using a needle. Do not close the cap.
2. Freeze the sample in dry ice and lyophilize it in Labconco FreeZone Plus overnight at −81 °C under high vacuum (<0.01 mbar).
3. Reconstitute the lyophilized sample with 100 μL of 25% ACN with 0.2% (v/v) TFA.
4. Vortex and sonicate the sample to make sure the sample is completely dissolved.
5. Prepare ammonium ion standards at concentrations ranging from 1 to 8 mM by dissolving ammonium sulfate in 25% ACN containing 0.2% (v/v) TFA, 1 mM L-histidine, and 1% (w/v) sucrose (*see* **Note 1**).
6. Set up RP-HPLC with the following conditions:
   (a) SIELC Primesep 100 column, with a column temperature of 35 °C.
   (b) CAD evaporation temperature of 35 °C.
   (c) Solvent system A comprised of water with 0.2% (v/v) TFA. Solvent system B comprised of ACN with 0.2% (v/v) TFA.
   (d) An isocratic elution profile of 70% B for 30 min.
   (e) Injection volume of 10 μL and a flow rate of 0.4 mL/min.
7. Run each calibration and nanoparticle sample (*see* **Note 2**). The ammonium ion elutes at 7.1 min (Fig. 1).

3.2 External Ammonium Ion Concentration

1. Centrifuge the liposomal sample, 100 μL of generic PEGylated liposomal doxorubicin in this case, to dryness using the rinsed Microcon DNA Fast Flow centrifugal device at 14,000 rpm and 24 °C. The recommended duration is 1 h (*see* **Notes 3** and **4**).
2. Dilute the filtrate tenfold with water containing 0.2% (v/v) TFA.
3. Prepare ammonium ion standards at concentrations ranging from 5 to 40 mM by dissolving ammonium sulfate in water containing 10% sucrose with 10 mM L-histidine.
4. Centrifuge standards to dryness using a rinsed Microcon DNA Fast Flow centrifugal device at 14,000 rpm and 24 °C. The recommended duration is 1 h (*see* **Note 3**).

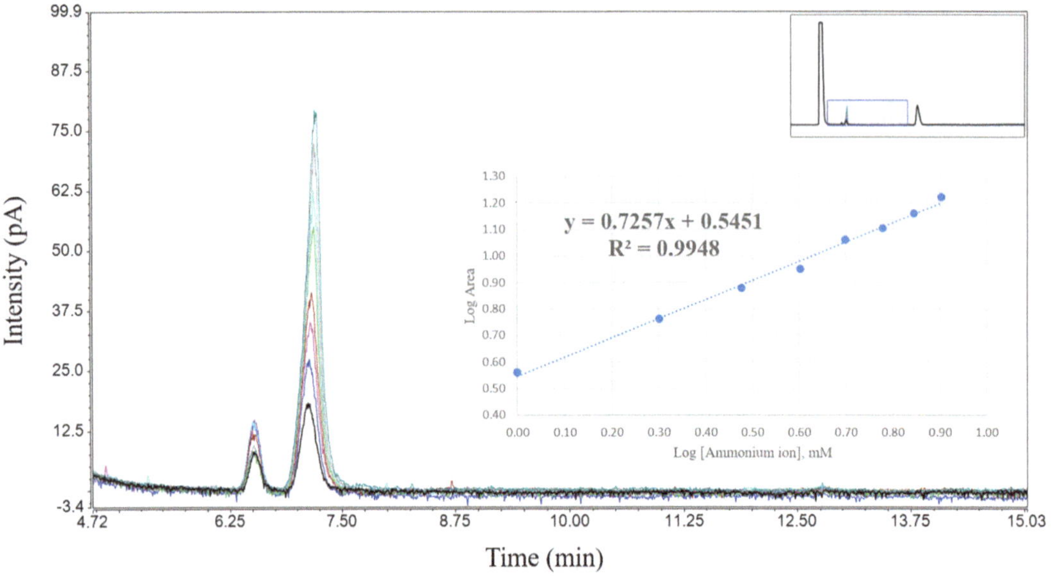

Fig. 1 HPLC chromatogram and calibration curve used for total ammonium ions concentration determination

5. Dilute centrifuged standards tenfold with water containing 0.2% (v/v) TFA to yield final concentrations ranging from 0.5 to 4 mM.
6. Run each calibration and nanoparticle sample using the same RP-HPLC conditions outlined in Subheading 3.1 (*see* **Note 2**). The ammonium ion elutes at 7.1 min.

3.3 Total Sulfate Ion Concentration

1. Add liposomal sample, in this case, 10 μL generic PEGylated liposomal doxorubicin, to a 1.5 mL tube. Seal the tube using parafilm and puncture three holes using a needle. Do not close the cap.
2. Freeze the sample in dry ice and lyophilize in Labconco Free-Zone Plus overnight at −81 °C under high vacuum (<0.01 mbar).
3. Reconstitute the lyophilized sample with 400 μL 25% ACN with 0.6% (v/v) formic acid.
4. Vortex and sonicate the sample to make sure the sample is completely dissolved.
5. Prepare sulfate ion standards at concentrations ranging from 0.1 to 1 mM by dissolving ammonium sulfate in 25% ACN containing 0.6% (v/v) formic acid (*see* **Note 1**).
6. Set up RP-HPLC with the following conditions:
   (a) Thermo Fisher acclaim trinity P2 column, with a column temperature of 40 °C.
   (b) CAD evaporation temperature of 35 °C.

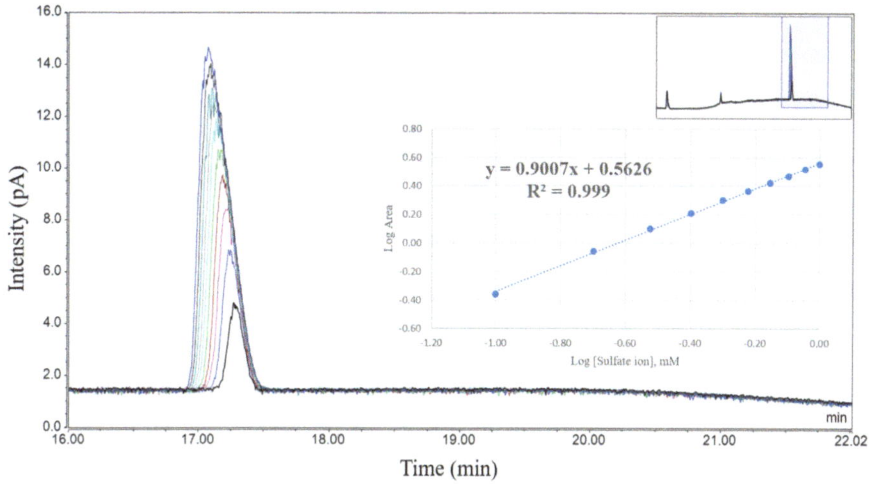

Fig. 2 HPLC chromatogram and calibration curve used for total sulfate ions concentration determination

(c) Solvent system A comprised of water. Solvent system B comprised of 100 mM ammonium formate with 0.3% (v/v) formic acid.

(d) A gradient elution profile of 10% B for 2 min, ramp to 100% B in 10 min, hold at 100% B for 5 min and ramp down to 10% B in 5 min, with a 3 min reconditioning time at 10% B between runs.

(e) Injection volume of 10 μL and a flow rate of 0.4 mL/min.

7. Run each calibration and nanoparticle sample (*see* **Note 2**). The sulfate ion elutes at 17.2 min (Fig. 2).

3.4 External Sulfate Ion Concentration

1. Centrifuge the liposomal sample, 100 μL of generic PEGylated liposomal doxorubicin in this case, to dryness using the rinsed Microcon DNA Fast Flow centrifugal device at 14,000 rpm and 24 °C. The recommended duration is 1 h (*see* **Notes 3** and **4**).
2. Dilute the filtrate 20-fold with water containing 10% sucrose and 10 mM L-histidine.
3. Prepare sulfate ion standards at concentrations ranging from 2 to 20 mM by dissolving ammonium sulfate in water containing 10% sucrose with 10 mM L-histidine.
4. Centrifuge standards (100 μL) to dryness using a rinsed Microcon DNA Fast Flow centrifugal device at 14,000 rpm and 24 °C. The recommended duration is 1 h (*see* **Note 3**).
5. Dilute centrifuged standards 20-fold with water containing 10% sucrose and 10 mM L-histidine to yield final concentrations ranging from 0.1 to 1 mM.

6. Run each calibration and nanoparticle sample using the same RP-HPLC conditions outlined in Subheading 3.3 (*see* **Note 2**). The sulfate ion elutes at 17.2 min.

3.5 Data Analysis

1. Open the elution profiles and integrate the peaks of ammonium ion or sulfate ion for each of the calibration and liposomal samples.
2. Plot the log of the peak area against the log of the ion concentration to generate the standard calibration curve and generate a linear fit equation.
3. Use the linear fit equation and the ion peak area of the test samples to calculate the ion concentrations in the liposomal sample.

4 Notes

1. The CAD response for ammonium ions is approximately 20% higher when the standards and samples are prepared with sucrose and L-histidine additives [2]. To eliminate this effect, it is crucial to ensure that both the standards and samples have the same concentrations of sucrose and L-histidine. For the total ion measurements, the 10 μL stock sample is reconstituted to a final volume of 100 μL after lyophilization. Consequently, the concentration of sucrose and L-histidine in the sample is diluted by tenfold, resulting in concentrations of 1% (w/v) for sucrose and 1 mM for L-histidine, respectively. Thus, the calibration standards should contain 1% (w/v) sucrose and 1 mM L-histidine. The same practice is applied to the method of the external ammonium ion measurement. The CAD response to sulfate ions is not affected by the presence of sucrose and L-histidine additives.
2. Prior to injecting the sample or standards, perform at least two injections of blank solvent. Ensure that the CAD baseline is reproducible and shows no noticeable drift.
3. To rinse a Microcon DNA Fast Flow centrifugal unit, centrifuge 300 μL of water containing 10% sucrose with 10 mM L-histidine three times at 14,000 rpm. For the first two cycles, centrifuge for 5 min, and for the third cycle, centrifuge until dryness. The recommended duration for the third cycle is 1 h.
4. After centrifugation, it is recommended to analyze the filtrate using dynamic light scattering (DLS) to ensure liposomal particles do not pass through the filter.

Acknowledgments

This project has been funded in whole or in part with federal funds from the National Cancer Institute, National Institutes of Health, under contract 75N91019D00024. The content of this publication does not necessarily reflect the views or policies of the Department of Health and Human Services, nor does the mention of trade names, commercial products, or organizations imply endorsement by the US Government.

References

1. Thermo Fisher technical guide: charged aerosol detection. See what other universal detectors are missing. https://assets.thermofisher.com/TFS-Assets/CMD/brochures/tg-70712-charged-aerosol-detection-tg70712-en.pdf. Accessed 11 Oct 2023
2. Wu J, Crist RM, McNeil SE, Clogston JD (2019) Ion quantification in liposomal drug products using high performance liquid chromatography. J Pharm Biomed Anal 165:41–46

Chapter 8

A Static Headspace Gas Chromatography Method for Quantitation of Residual Solvents in Nanoformulations

Krishna Kattel and Jeffrey D. Clogston

Abstract

Various organic solvents are widely used in the manufacturing, processing, and purification of drug substances, drug products, formulations, excipients, etc. These solvents must be removed to the lowest amount permitted, as they do not possess any therapeutic advantages and may cause undesirable toxicities. Therefore, a rapid and sensitive analytical method for the quantitation of residual solvents is needed. The following chapter presents a static headspace gas chromatographic (HSGC) method for determining the concentration of common residual solvents in various nanoformulations. An efficient and sensitive HSGC method has been developed using PerkinElmer's headspace autosampler/gas chromatographic system with a flame ionization detector (FID) and validated according to the International Conference for Harmonization (ICH) guideline Q3C. The method validation indicates that the method is specific, linear, accurate, precise, and sensitive for the analyzed solvents. The method is suitable for the analysis of 13 residual solvents (methanol, ethanol, acetone, diethyl ether, 2-propanol, acetonitrile, 1-propanol, ethyl acetate, tetrahydrofuran, dichloromethane, chloroform, 1-butanol, and pyridine) and utilizes an Elite 624 Crossbond 6% cyanopropylphenyl, 94% dimethylpolysiloxanes column with helium as a carrier gas.

Key words Volatile organic impurities, Gas chromatography, Headspace autosampler, Flame ionization detector, Nanomedicine, Ethanol, Liposomes

1 Introduction

Various organic solvents are used in the synthesis of complex drug products, such as nanomedicines, and in the manufacturing/purification of drug substances and other excipients. Residual solvents may originate from the synthesis, manufacturing process, and/or purification of drug materials, as well as from the cleaning/maintenance of equipment used to manufacture the drug products. These residual solvents are considered drug product impurities having no therapeutic benefit. Irregular and unregulated concentrations of these volatile residual impurities not only pose health risks for patients but also affect the product's quality. In addition, residual

Jeffrey D. Clogston et al. (eds.), *Characterization of Nanoparticles Intended for Drug Delivery*, Methods in Molecular Biology, vol. 2789, https://doi.org/10.1007/978-1-0716-3786-9_8,

solvents can also affect the physicochemical properties of therapeutics, such as particle size, dissolution, and wettability [1].

The amount of residual solvent tolerated in final drug products is described in pharmacopeias such as the European Pharmacopeia (EP) and the United States Pharmacopeia (USP) [2] and is closely monitored by regulatory agencies. The International Council for Harmonization of Technical Requirements for Registration of Pharmaceuticals for Human Use (ICH) and the Food and Drug Administration (FDA) have released guidelines for residual solvents based on their classification [3]. As per the guidance, residual solvents were evaluated for their possible risk to human health and were placed into one of three classes based on their toxicity data and environmental impact.

Class 1 solvents, such as benzene, 1,2-dichloroethane, and 1,1-dichloroethene, are highly suspected to be carcinogenic, and carbon tetrachloride and 1,1,1-trichloroethane pose serious environmental hazards. Class 1 solvents are recommended to be avoided in therapeutic preparations (USP limits for class 1 solvents in ppm: benzene: 2; 1,2-dichloroethane: 5; 1,1-dichloroethene: 8; carbon tetrachloride: 4; and 1,1,1-trichloroethane: 1500 [2]). Class 2 solvents, such as acetonitrile, chlorobenzene, chloroform, cyclohexane, hexane, methanol, etc., are non-genotoxic carcinogens and can induce neurotoxicity. Class 3 solvents, such as acetone, ethanol, ethyl acetate, formic acid, heptane, propanol, etc., have low toxic potential. Class 3 solvents are typically limited to 5000 ppm or 0.5% (w/w), whereas class 2 solvents have their own individual limits. The acceptable levels of residual solvents are specified by ICH Q3C(R8) [3].

Static headspace gas chromatography (HSGC) is the preferred technique for the analysis of residual solvents in nanoformulations/drug products because it offers several advantages over direct injection by GC. In static HS, the vapor in equilibrium with the sample is analyzed; thus, only volatile components are introduced into the GC system, resulting in an extended column lifetime and reduced instrument maintenance, providing superior sensitivity and reproducibility. Headspace sampling is conducted by placing a liquid or solid sample in a sealed vial until a thermodynamic equilibrium between the sample and gas phase is reached. A known aliquot of the gas phase analyte is then transferred to the gas chromatography for analysis.

The present chapter provides the details of a static HSGC method for the analysis of 13 common solvents and method performance data for the tested analytes.

2 Materials

1. Chromatographic system and operating conditions: A Clarus 690 gas chromatograph equipped with an FID detector and with a Turbomax 40 trap headspace autosampler (*see* **Note 1**). Total Chrom Navigator (TNav, PerkinElmer) software is used to control the instrument, data acquisition, and processing. The optimized headspace conditions are presented in Table 1, and GC/FID conditions are summarized in Table 2.
2. Elite 624 (Crossbond 6% cyanopropyl phenyl 94% dimethyl-polysiloxane) column, 0.32 mm ID × 30 m capillary column with a 1.8 μm layer.
3. Stock standard solution: All solvent analyte concentrations in the standards are based on a sample size of 200 mg nanoformulation in 1 mL dimethyl sulfoxide (DMSO). Transfer the required amount of residual solvent (approx. 150 mg methanol, 250 mg ethanol, 250 mg acetone, 250 mg diethyl ether, 250 mg 2-propanol, 20 mg acetonitrile, 250 mg 1-propanol, 250 mg ethyl acetate, 40 mg tetrahydrofuran, 30 mg dichloromethane, 3 mg chloroform, 250 mg 1-butanol, and 10 mg pyridine) into a 25 mL volumetric flask (VF) half-filled with DMSO. Dilute to volume with DMSO and vortex. The analytes have been prepared in two groups: set I (methanol, ethanol, acetone, 2-propanol, acetonitrile, 1-propanol, ethyl acetate, tetrahydrofuran, and 1-butanol) and set II (diethyl ether, dichloromethane, chloroform, and pyridine) prepared at concentrations within the calibration range (*see* **Note 2**). *See* Table 3 for the calibration ranges of the analytes.
4. Working standard solution: Transfer 2.0 mL of the stock standard solution into a 20 mL VF half-filled with DMSO. Dilute to volume with DMSO and vortex. For HSGC analysis, 1 mL of the standard solution was pipetted into a 22 mL headspace vial (20 mm, clear, PerkinElmer) and crimped immediately (*see* **Note 3**) with a Teflon-lined septum and aluminum crimp cap (PerkinElmer).
5. Analytical balance with 0.0001 g accuracy.
6. Parker hydrogen generator.
7. Headspace PTFE/SIL liner, GC vials, hand crimper.
8. Volumetric flasks (50 mL, 25 mL, 10 mL, A grade).
9. Glass pipettes/Pasteur pipettes/gas-tight syringes.
10. Analytical reference standards for the solvent, e.g., methanol, ethanol, acetone, diethyl ether, 2-propanol, acetonitrile, 1-propanol, ethyl acetate, tetrahydrofuran, dichloromethane, chloroform, 1-butanol, and pyridine. Standards should be GC grade (potency ≥99.99).

Table 1
Headspace conditions

Instrument	Turbomax 40 Trap, PerkinElmer
Oven temperature	120 °C
Needle temperature	115 °C
Carrier gas/pressure	Helium/15 psi
Thermostatting time	2 min
Pressurization time	0.5 min
Withdraw time	0.2 min
Operating mode	Constant
Transfer line temperature	120 °C
Injection time	0.04 min
GC cycle	2.5 min

Table 2
GC/FID conditions

Instrument	Clarus® 690, PerkinElmer
Column	Elite, 624 (Crossbond 6% cyanopropylphenyl-94% dimethylpolysiloxane), 30 m, 0.32 mm ID, 1.8 μm df
Detector temperature	250 °C
Carrier gas/pressure	Helium/10 psi
Detector gas flow	Hydrogen: 35 mL/min Air: 400 mL/min
Ramp	Initial temp: 40 °C and hold for 6 min, ramp to 200 °C @ 10 °C/min, hold for 3 min
Injector temperature	200 °C
Split ratio	None (splitless)
Hydrogen flow	35 mL/min
Air flow	400 mL/min
Run time	25 min

11. Dimethyl sulfoxide (DMSO), 99.99%.
12. Ultrapure helium (research grade, purity >99.999%).
13. Zero-grade air.
14. Ultrapure hydrogen (research grade, purity >99.999%).

3 Method

3.1 Method Validation

1. Analytical method validation: The analytical method validation is performed per ICH guidelines [3]. The validation parameters are system suitability, linearity, specificity, limit of detection (LOD), limit of quantitation (LOQ), and spike recovery/method precision.
2. System suitability: Inject six injections of each standard analyte to calculate the relative standard deviation (% RSD, acceptance criteria <5.0%).
3. Linearity: Prepare a set of calibration standards in DMSO from LOQ to 150% of the nominal concentration (USP limit). A minimum of seven standards is recommended. The summary of linearity data is presented in Table 3, and example chromatograms are presented in Fig. 1.
4. Specificity: Inject diluent (DMSO) and an appropriate matrix solution in DMSO and evaluate whether there is any interference on the retention time of the analyte.

Table 3
Linearity of class 2 and class 3 solvents

Analytes	ICH limit[a] (ppm)	Retention time (min)	Linearity range (μg/mL)	Regression coefficient (R^2)
Methanol	3000	2.92	15.01–900.51	0.9998
Ethanol	5000	3.32	26.76–1605.44	0.9998
Diethyl ether	5000	3.37	1.02–1524.5	0.9998
Acetone	5000	3.91	23.69–1421.40	0.9999
2-propanol	5000	4.19	25.53–1531.65	0.9999
Acetonitrile	410	4.53	2.46–147.59	0.9996
Dichloromethane	600	4.60	6.88–206.34	0.9991
1-propanol	5000	6.47	25.16–1509.30	0.9998
Ethyl acetate	5000	7.49	25.42–1525.34	0.9991
Tetrahydrofuran	720	7.80	4.31–258.56	0.9998
Chloroform	60	7.97	3.05–18.29	0.9991
1-butanol	5000	10.27	25.33–1519.94	0.9998
Pyridine	200	12.34	2.38–71.40	0.9978

Analytes are listed in the order of increasing retention time
[a] *See* ref. [2]

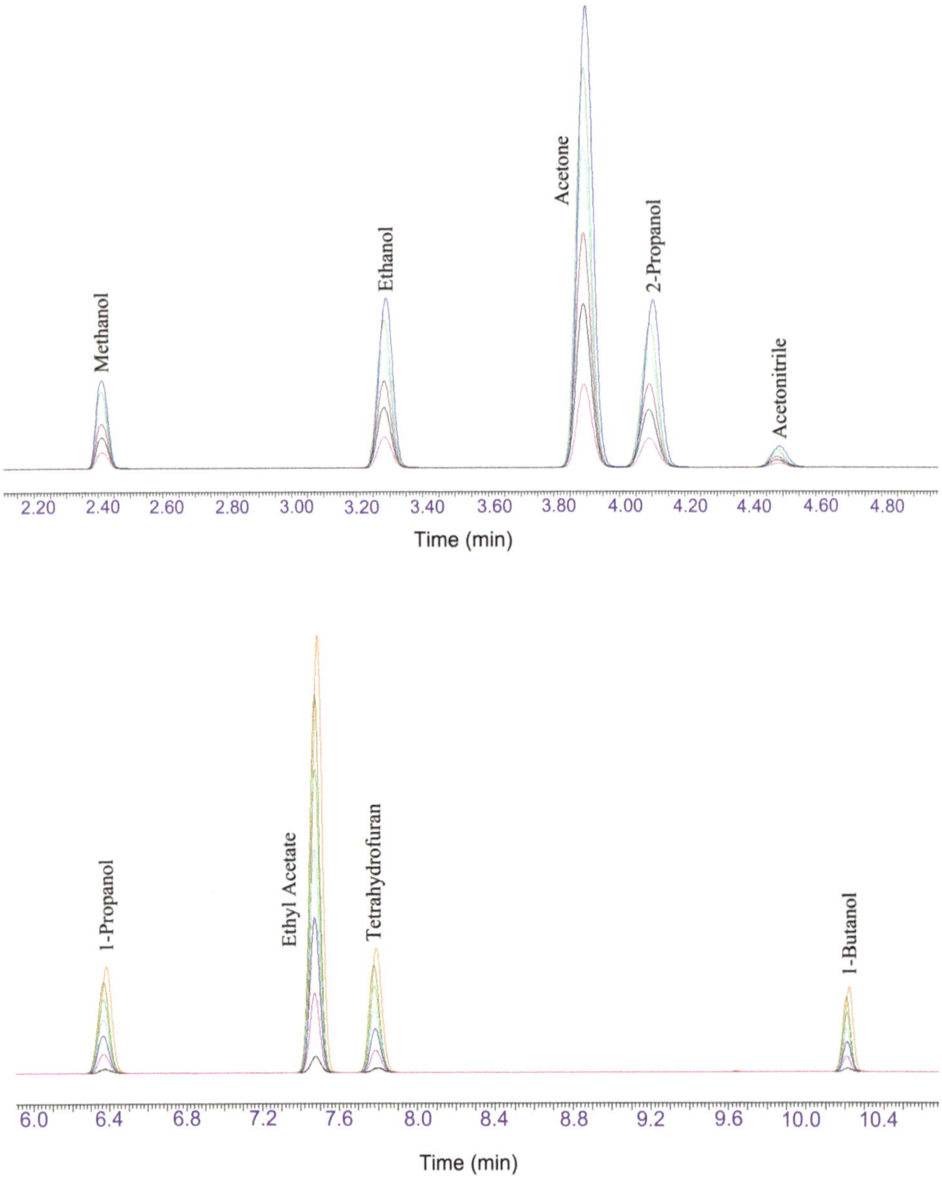

Fig. 1 Overlaid chromatograms of common residual solvents at various linearity levels

5. Method sensitivity: The sensitivity of the method is evaluated in terms of the limit of detection and the limit of quantitation by performing six injections of each analyte at LOD and LOQ levels. The LOQ of 13 analytes has been established between 5 and 34 ppm (equivalent to 0.0005–0.003% (v/v) of the sample concentration, as presented in Table 4).
6. Accuracy/spike recovery: Spike recovery of the method is determined by analyzing triplicate sample preparations at the practical limit of quantification (PLOQ), 50% of the limit,

Table 4
Limit of quantitation (LOQ) of class 2 and class 3 solvents

Analytes	LOQ concentration (μg/g)	LOQ with respect to sample conconcentration (ppm)	% RSD (six injections)
Methanol	2.101	11	2.9
Ethanol	3.746	19	2.6
Diethyl ether	1.016	5	1.1
Acetone	3.317	17	1.0
2-propanol	3.574	18	4.0
Acetonitrile	2.46	12	5.9
Dichloromethane	6.878	34	2.3
1-propanol	3.522	18	3.9
Ethyl acetate	3.559	18	1.7
Tetrahydrofuran	1.80	9	2.4
Chloroform	6.096	30	6.9
1-butanol	3.547	18	3.7
Pyridine	2.38	12	9.8

150% of the limit, and six preparations at 100% of the limit. The samples ($n = 6$) at 100% of the limit for each analyte also serve as method precision. The results are presented in Table 5. The acceptance criterion for accuracy/spike recovery is 80–120% per ICH specifications.

3.2 Calculations

Report the residual solvent content in the sample in terms of % (w/w) or ppm as:

$$\text{Result (\%)} = \frac{\text{Peak area (sample)}}{\text{Peak area (standard)}} \times \text{Std.conc (mg/mL)} \times \frac{\text{Dilution (mL)}}{\text{Sample weight (mg)}} \times 100$$

$$\text{Result (ppm)} = \frac{\text{Peak area (sample)}}{\text{Peak area (standard)}} \times \text{Std.conc (mg/mL)} \times \frac{\text{Dilution (mL)}}{\text{Sample weight (mg)}} \times 10^6$$

3.3 Procedure for Shutting Down the GC

1. Many parts of the GC operate at very high temperatures, such as the oven compartment, inlet area, and FID detector area. Therefore, special precautions should be taken by bringing the operating instrument down to normal temperature, especially when changing the GC column and shutting down the instrument after use. Never open the oven door during operation or immediately after the heating sequence (*see* **Note 4**).

Table 5
Spike recovery of class 2 and class 3 solvents

	Spike recovery (%)			
Analytes	**PLOQ ($n = 3$)**	**50% level ($n = 3$)**	**100% level ($n = 6$)**	**150% level ($n = 3$)**
Methanol	100.6 (RSD: 2.3)	102.6 (RSD: 0.5)	100.1 (RSD: 1.5)	100.3 (RSD: 2.1)
Ethanol	100.0 (RSD: 1.9)	103.4 (RSD: 0.4)	100.6 (RSD: 1.1)	100.5 (RSD: 1.8)
Diethyl ether	90.8 (RSD: 7.9)	93.3 (RSD: 5.9)	97.0 (RSD: 1.6)	100.1 (RSD: 0.8)
Acetone	101.9 (RSD: 0.4)	102.9 (RSD: 0.4)	100.4 (RSD: 0.7)	100.4 (RSD: 0.0)
2-propanol	95.7 (RSD: 0.2)	104.1 (RSD: 0.3)	100.6 (RSD: 1.0)	100.6 (RSD: 1.5)
Acetonitrile	95.2 (RSD: 5.7)	102.3 (RSD: 0.7)	98.3 (RSD: 2.6)	99.5 (RSD: 2.6)
Dichloromethane	99.1 (RSD: 2.5)	99.7 (RSD: 2.3)	100.7 (RSD: 0.6)	102.6 (RSD: 1.0)
1-propanol	93.1 (RSD: 2.8)	94.8 (RSD: 1.4)	100.4 (RSD: 1.8)	100.6 (RSD: 2.1)
Ethyl acetate	96.6 (RSD: 0.2)	95.3 (RSD: 0.4)	100.7 (RSD: 0.8)	100.8 (RSD: 1.5)
Tetrahydrofuran	98.6 (RSD: 2.6)	103.2 (RSD: 0.2)	100.1 (RSD: 1.0)	100.3 (RSD: 1.7)
Chloroform	96.8 (RSD: 5.8)	92.3 (RSD: 1.5)	101.0 (RSD: 3.8)	98.0 (RSD: 0.9)
1-butanol	98.1 (RSD: 2.8)	92.6 (RSD: 0.4)	100.1 (RSD: 1.5)	101.7 (RSD: 2.1)
Pyridine	83.3 (RSD: 4.1)	97.6 (RSD: 1.0)	100.9 (RSD: 1.4)	96.1 (RSD: 2.5)

2. Reduce the column temperature to <40 °C and the detector and inlet temperatures to <100 °C by keeping the carrier gas flow on (*see* **Note 5**). When the GC is cool, turn it off. Turn off all gas sources. Remove the column and cap its end to prevent contamination.

4 Notes

1. Turn on the supply gas first, and power on the GC system and headspace autosampler.
2. All preparations should be made in glass vials to avoid any interaction of organic solvents with plastics.
3. Caps of flasks and vials should be closed immediately after preparation due to the volatile nature of the analytes.
4. Vendor-certified regulators, hose connections, and tubing should be used. Compressed gas cylinders should be placed away from any types of heat sources.
5. Hydrogen gas may flow into the GC oven and create an explosion hazard when used as a carrier gas. Therefore, the GC gas

regulator should be turned off until all connections are made, and ensure that the inlet and detector column fittings are either connected to a column or always capped when hydrogen gas is supplied to the instrument.

Acknowledgments

This project has been funded with federal funds from the National Cancer Institute, National Institutes of Health, under contract 75N91010D00024. The content of this publication does not necessarily reflect the views or policies of the Department of Health and Human Services, nor does the mention of trade names, commercial products, or organizations imply endorsement by the US Government.

References

1. Witschi C, Doelker E (1997) Residual solvents in pharmaceutical products: acceptable limits, influences on physicochemical properties, analytical methods and documented values. Eur J Pharm Biopharm 43(3):215–242
2. <467> Residual solvents. United States Pharmacopeia-National Formulary. https://www.uspnf.com/sites/default/files/usp_pdf/EN/USPNF/generalChapter467Current.pdf. Accessed 19 July 2023
3. Impurities: Guideline for Residual Solvents, Q3C(R8). International Council for Harmonisation of Technical Requirements for Pharmaceuticals for Human Use. https://database.ich.org/sites/default/files/ICH_Q3C-R8_Guideline_Step4_2021_0422_1.pdf. Accessed 19 July 2023

Part III

Immunotoxicity

Chapter 9

Current Considerations and Practical Solutions for Overcoming Nanoparticle Interference with LAL Assays and Minimizing Endotoxin Contamination

Barry W. Neun and Marina A. Dobrovolskaia

Abstract

Monitoring endotoxin contamination in drugs and medical devices is required to avoid pyrogenic responses and septic shock in patients receiving these products. Endotoxin contamination of engineered nanomaterials and nanotechnology-based medical products represents a significant translational hurdle. Nanoparticles often interfere with an in vitro limulus amebocyte lysate (LAL) assay commonly used in the pharmaceutical industry for the detection and quantification of endotoxin. Such interference challenges the preclinical development of nanotechnology-formulated drugs and medical devices containing engineered nanomaterials. Protocols for the analysis of nanoparticles using LAL assays have been reported before. Here, we discuss considerations for selecting an LAL format and describe a few experimental approaches for overcoming nanoparticle interference with the LAL assays to obtain more accurate estimations of endotoxin contamination in nanotechnology-based products. The discussed approaches do not solve all types of nanoparticle interference with the LAL assays but could be used as a starting point to address the problem. This chapter also describes approaches to prevent endotoxin contamination in nanotechnology-formulated products.

Key words Endotoxin, LAL, Interference, Inhibition enhancement control

1 Introduction

Endotoxin is a building block of the gram-negative bacterial cell wall [1]. Many cell types have an endotoxin receptor complex and respond to this bacterial ligand. Mononuclear phagocytes are particularly sensitive and produce cytokines, chemokines, and other pro-inflammatory messengers in response to low (picogram) quantities of endotoxin [2]. These pro-inflammatory substances result in pathophysiological events leading to fever and hypotension and, at high levels, may also lead to severe tissue and organ damage [1]. Common life-threatening conditions observed in response to high endotoxin levels in humans are septic shock syndrome, toxic

Jeffrey D. Clogston et al. (eds.), *Characterization of Nanoparticles Intended for Drug Delivery*, Methods in Molecular Biology, vol. 2789, https://doi.org/10.1007/978-1-0716-3786-9_9,

anterior segment syndrome, multiple organ failures (mainly kidney, lungs, and liver), and disseminated intravascular coagulation. Endotoxin contamination is particularly problematic for pharmaceutical products and medical devices. If these products get contaminated with endotoxin, the results of preclinical toxicity and efficacy studies may be confounded and lead to wrong conclusions. In the clinical setting, they will lead to immunotoxicity in patients. Regulatory authorities recognize this problem and mandate endotoxin limits that depend on the type of product, the dose, and the route of administration of the given pharmaceutical product. The biological activity of endotoxin is measured in endotoxin units (EU). The endotoxin limit (EL), specific for each product, depends on the threshold pyrogenic dose (K) and the maximum dose (M) of the given product administered per kilogram of body weight per single hour. The EL for parenteral drugs is calculated according to the formula K/M. The K-value described in the United States Pharmacopeia is 5 EU/kg for all parenteral routes except for the intrathecal route, for which the threshold pyrogenic dose is 0.2 EU/kg [3]. The EL for radiopharmaceuticals depends on the volume of the product in milliliters (V) and is equal to 175/V and 14/V for most parenteral routes and the intrathecal route, respectively [3]. The allowable EL mandated for the surfaces of medical devices is 0.5 EU/mL (or 20 EU/device) and for products in contact with the cardiovascular and lymphatic systems and cerebrospinal fluid is 0.06 EU/mL (or 2.15 EU/device) [4]. According to the new FDA guideline, the EL for intraocular fluids is 0.2 EU/mL and that for solid ocular devices is 0.2 EU/device [5].

Due to its presence virtually everywhere (air, water, many common laboratory reagents, equipment, and surfaces) and its resistance to autoclaving, endotoxin is a common contaminant in engineered nanomaterials and nanotechnology-formulated drugs and devices produced in research laboratories [6, 7]. As much as one-third of nanotechnology formulations fail in the preclinical phase due to excessive endotoxin contamination [8].

In addition to the concerns regarding endotoxin presence in pharmaceutical products and devices discussed above, nanotechnology products experience an additional challenge. Recent reports from our laboratory [9] and other investigators [10–14] have demonstrated that some nanoparticles are not pro-inflammatory alone but exaggerate the inflammatory properties of low concentrations of endotoxin. This challenge complicates the estimation of the EL using the formulas we discussed above and, in the absence of a specific regulatory guideline or standard, suggests that nanotechnology products possessing the endotoxin-exaggerating property should be virtually free from endotoxin.

The in vitro method traditionally used in the pharmaceutical industry to detect and quantify endotoxin contamination is the limulus amebocyte lysate (LAL) test, which exists in three formats:

turbidity, chromogenic, and gel-clot [15]. Both turbidity and chromogenic assays can be performed in an end-point and kinetic mode. In addition, a newer version of an LAL derivative assay has been developed with recombinant factor C [16–18]. This assay utilizes a substrate that, upon cleavage by endotoxin-activated lysate, produces a fluorescent product and operates in the kinetic mode [16–18]. The kinetic assays, in general, have greater assay sensitivity (or lambda) than their end-point counterparts [15]. Understanding the difference in lambda of various LAL assays is necessary for both experimental design and inter-assay data comparison because this value is used to calculate the maximum valid dilution (MVD). Many chemicals may interfere with the LAL assays; however, due to their complex composition and unique physicochemical properties, nanotechnology-based products experience a greater spectrum of interference with the LAL [15, 19, 20].

In this chapter, we discuss the factors one may consider when selecting an LAL assay suitable for the given nanoformulation and suggest several practical approaches which may be used to overcome interference from certain nanomaterials. Although the proposed experimental steps do not completely solve the problem of nanoparticle interference with the LAL assays, they may serve as a starting point to analyze nanoparticles with physicochemical properties similar to those used in our case studies. We will also discuss approaches to prevent endotoxin contamination in nanotechnology-formulated products and to remove endotoxin from contaminated particles.

2 Choosing an LAL Format

Some of the nanoparticle interferences with the LAL can be predicted based on the nanomaterial's basic physicochemical properties such as the absorbance spectrum at or about 405 and 660 nm because chromogenic and turbidity LAL assays are performed at these wavelengths, respectively. The decision tree presented in Fig. 1 may be helpful in making a choice about the LAL format.

3 Overcoming False-Positive Results Due to Contamination with Beta-Glucans

Despite its high sensitivity to endotoxin, traditional LAL assays also detect beta-glucans. The LAL proteins activated by endotoxin and beta-glucans are different. Factor C is activated by endotoxin, while factor G is triggered by beta-glucans [15]. Despite the difference in the sensing protein and pathway, the result (e.g., gelation or turbidity) of the LAL reaction triggered by endotoxin and beta-glucans is the same and often leads to overestimation of endotoxin in the sample containing beta-glucans. While monitoring for the

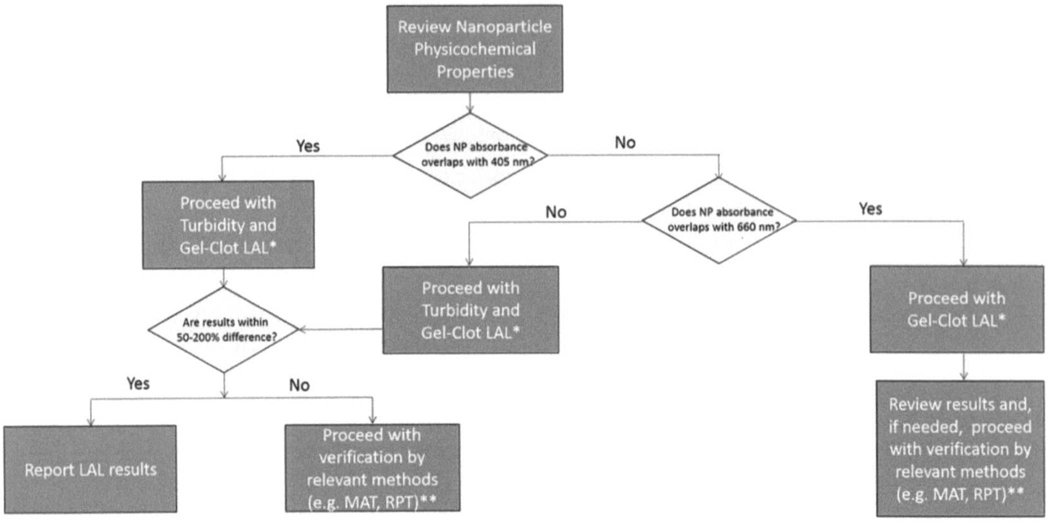

Fig. 1 Decision tree for selecting LAL format. Selection of the LAL format starts with evaluating the nanoparticle physicochemical properties. Two LAL formats are used to analyze the same formulation. If the results are in agreement, the LAL results are reported. If the results disagree, a follow-up assessment is conducted using a biological assay. This decision tree was initially described in reference [21] and is presented here with modifications. *MAT* monocyte activation test, *RPT* rabbit pyrogen test; *According to the USP and FDA guidelines, in case of discrepancy between test results of different LAL methods, the decision is made by the gel-clot assay; **MAT application is limited when a test nanoformulation contains a cytotoxic drug; other methods and controls can be used; apply scientific justification for result interpretation and preparation of appropriate controls

presence of beta-glucans is important to identify potential fungal contamination of pharmaceutical products; it should be performed using specific assays, e.g., Fungitell [22, 23]. When present in the LAL reaction, beta-glucans activate the factor G pathway and result in inaccurate detection and quantification of endotoxin. Filtration through cellulose acetate filters, commonly used during the production of nanotechnology-based formulations, is a common reason for nanoparticle contamination with beta-glucans. Cellulose-based filters are a known source of beta-glucans and increase the rate of false-positive LAL results [24]. The interference from beta-glucans can be overcome by performing sample analysis using the recombinant factor C assay [16–18], by reconstituting lysate in Glucashield buffer, or by diluting the test sample with β-G-blocker reagent (*see* **Note 1**). In the case study presented in Fig. 2, the false-positive interference from beta-glucan in several nanoparticle formulations was solved by using Glucashield buffer for the reconstitution of the lysate before proceeding with the LAL reaction.

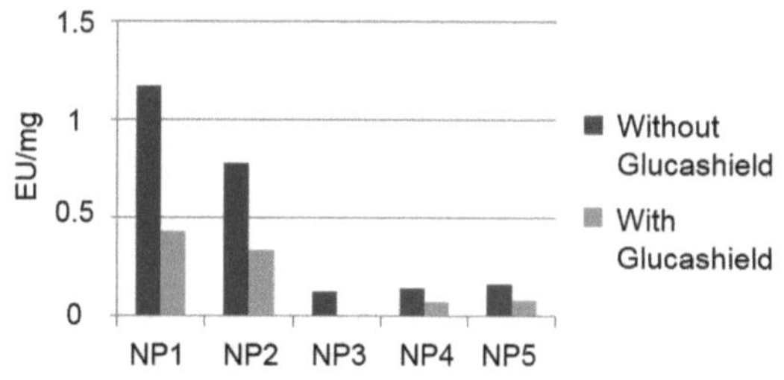

Fig. 2 Overcoming false-positive results caused by beta-glucan contamination. Five nanoparticle formulations (NP1-NP5) were tested by the kinetic turbidity LAL procedure as described earlier [25]. The same nanoparticle sample was analyzed with and without GlucaShield buffer. A decrease in the assay result generated using GlucaShield buffer is indicative of beta-glucan contamination commonly arising from cellulose-based filters used during nanoparticle manufacturing. The mean response of two replicates (%CV < 20) is shown

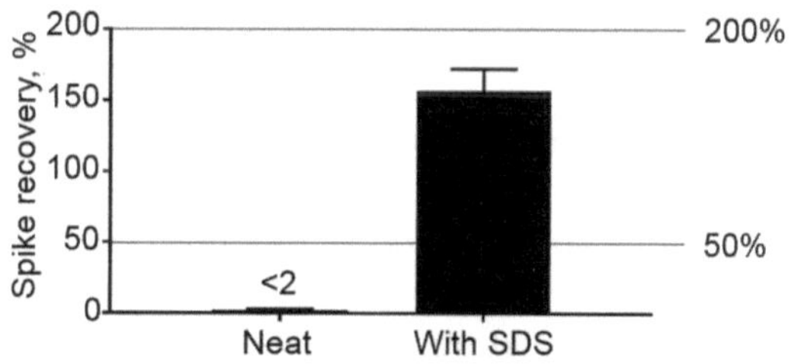

Fig. 3 Overcoming false-negative results due to cationic charge. Cationic DOTAP liposomes were analyzed by the kinetic turbidity LAL procedure as described earlier [25]. The cationic liposome sample (Neat) was analyzed at several dilutions within the assay MVD. The spike recovery of the untreated sample at all dilutions was below 2%. The same sample was also tested at the same dilutions after the addition of 0.5% SDS and showed an acceptable spike recovery. The mean spike recovery ± SD ($N = 2$) is shown

4 Overcoming False-Negative Results Due to the Presence of Cationic Charge

Endotoxin is a heterogeneous and chemically diverse entity. The central part of endotoxin is lipopolysaccharide (LPS), which, in turn, has a complex chemical structure. The biologically active component of LPS is lipid A, composed of a disaccharide backbone, four to six fatty acids, and one or two phosphate groups [1]. Anionic phosphate groups of lipid A can react electrostatically with cationic materials. Such an interaction neutralizes LPS and makes it invisible to the LAL reaction. Therefore, inhibition of endotoxin detection is commonly observed with cationic nanoparticles (e.g., amine-terminated dendrimers, 1,2-dioleoyl-3-trimethylammonium-propane (DOTAP) and other cationic lipid-containing liposomes and lipid nanoparticles). This type of interference can be overcome by dilution not exceeding the MVD. However, dilution alone is often ineffective for many cationic nanomaterials. In some cases, adding sodium dodecyl sulfate (SDS) to such test samples may help to overcome the interference (Fig. 3; *see* **Note 2**).

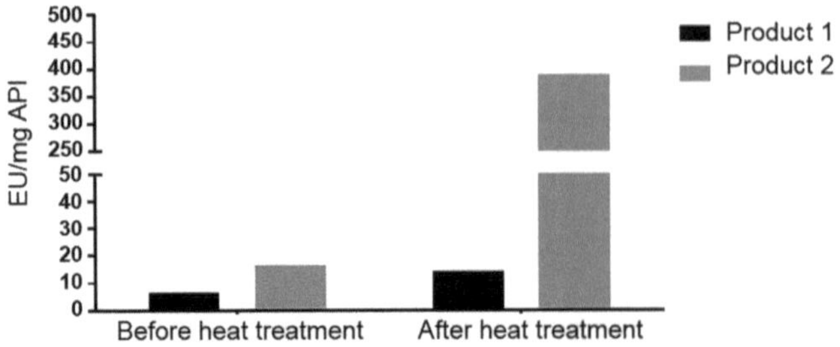

Fig. 4 Example of endotoxin release from liposomes after heat treatment. Two formulations of PEGylated liposomes (product 1 and product 2) were tested using the kinetic turbidity LAL procedure, as described earlier [25]. The composition of these liposomes was identical to that used in Doxil. The active pharmaceutical ingredients (APIs) were different in product 1 and product 2. The same nanoparticle sample was analyzed without heat treatment (before heat treatment) and after 30 min of incubation at 96 °C (after heat treatment). The mean response of two replicates (%CV < 20) is shown

5 Overcoming False-Negative Results Due to Endotoxin Entrapment

Endotoxin is lipophilic and can be entrapped in lipid-based, hollow nanoparticles such as liposomes [1]. The LAL assays detect only endotoxin present in solution and, therefore, can miss endotoxin entrapped in nanoparticles. The entrapment, therefore, may lead to the underestimation of endotoxin in liposome formulations. Processing of a nanoparticle test sample before LAL analysis may be needed to release the entrapped endotoxin (Fig. 4). Here, we describe a procedure which was developed and verified for liposomes with a composition identical to that of the commercial Doxil formulation. The method includes heat treatment. Place the test sample into a borosilicate tube, cover with parafilm, and incubate on a heating block set at 96 °C for 15–30 min (*see* **Note 3**). Allow the sample to cool down to room temperature and centrifuge at 2500 × *g* for 5 min to bring all sample content down to the bottom of the test tube. After that, proceed with preparing dilution and testing according to the standard LAL procedure, for example, like the procedure described in reference [25].

6 Overcoming False-Negative Results Due to Endotoxin Binding to Proteins and Containers

Proteins may bind endotoxin and prevent its interaction with the lysate, causing subsequent inhibition in the LAL assay. When dilution within the MVD is inefficient for overcoming such interference, protein removal using endotoxin-free proteinase can be used to overcome the interference (Fig. 5; *see* **Note 4**). Endotoxin binding to plastic used in containers for sample storage and

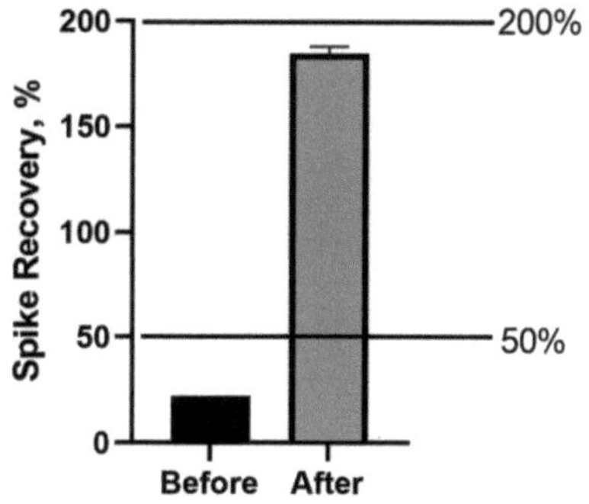

Fig. 5 Example of endotoxin detection in protein-rich nanoparticles before and after digestion with endotoxin-free proteinase. Two protein-based nanoparticle samples were tested using the kinetic turbidity LAL before and after protein digestion with the commercial proteinase kit EndoPrep™. The mean response of two replicates (%CV < 20) is shown. The number above each bar represents the percent spike recovery of control standard endotoxin; note that gel-clot LAL is a semiquantitative assay, and interference with it is assessed by clot formation in respective controls without calculating percent spike recovery

dilutions or pipette tips can also affect the accurate detection of endotoxin [26]. Pyrogen-free borosilicate glass containers and tubes are recommended for sample storage and dilution, respectively. Verified, low endotoxin binding pipette tips are also available from commercial suppliers to assist researchers.

7 The Use of Commercial Dispersion Reagents

Commercial dispersion reagents are available to improve nanoparticle dispersion for endotoxin testing. One such reagent is Pyrosperse™, initially developed for sample solubilization to overcome inhibition or the so-called endotoxin-masking phenomenon (https://bioscience.lonza.com/lonza_bs/US/en/Endotoxin-Detection/p/000000000000186617/PYROSPERSE%E2%84%A2-Dispersing-Agent). Assay validation is recommended to verify the need for such a reagent for endotoxin analysis in nanoparticles. If the LAL assay demonstrates comparable performance between samples tested in the presence or without a dispersing agent, the inclusion of such an agent into the analysis is not necessary and can be skipped to reduce the cost and complexity of the testing. In the example shown in Fig. 6, endotoxin results for a nanoparticle formulation were comparable between various LAL formats regardless of the presence of the dispersing agent. The results for other types of nanomaterials may be different; therefore, the necessity for a dispersing reagent should be determined on a case-by-case basis.

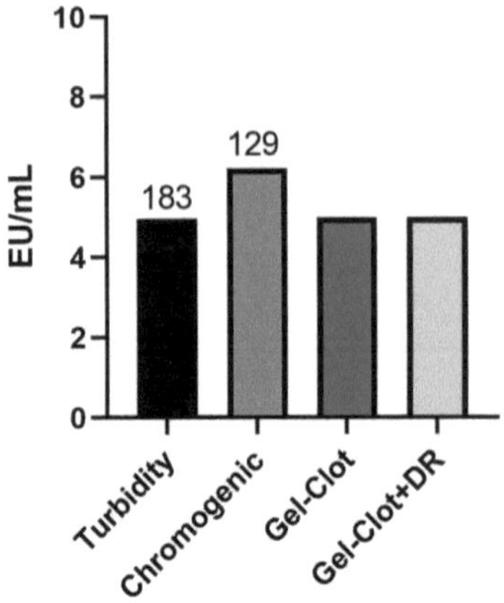

Fig. 6 Example of endotoxin detection in a nanoparticle formulation with and without a dispersing agent. A polymeric micelle nanoparticle sample was tested by the kinetic turbidity, kinetic chromogenic, and gel-clot LAL assays. The gel-clot LAL was conducted for the sample dispersed in either LAL grade water or commercial dispersing reagent (DR) Pyrosperse™. The percent spike recovery is shown above the bar for quantitative assays. Acceptable spike recovery is between 50% and 200%. The acceptance criteria for the gel-clot LAL are detailed in the US Pharmacopoeia [3]. *DR* dispersing reagent

8 Some Experimental Approaches to Remove Endotoxin from Reagents, Equipment, Contaminated Nanoparticle Formulations, and Their Components

Endotoxin removal from equipment and reagents used for nanoparticle synthesis depends on the type of materials and their resistance to depyrogenation procedures. The easiest, most reliable, and widely accessible depyrogenation method is baking at high temperatures (≥200 °C) for at least 30 min. This procedure can be used to depyrogenate glassware, spatulas, and other heat-resistant tools commonly used during the synthesis of nanoparticles. The high temperature is damaging to most nanomaterials. However, some (e.g., carbon nanotubes and certain metallic particles) can tolerate it. The combination of high temperature and an acid or base is used to depyrogenate materials by acid or base hydrolysis, respectively. While different hydrolysis procedures are available, incubation in the presence of 0.05 M HCl for 30 min at 100 °C or in 0.5 M NaOH at 50 °C for 30 min is used most frequently [27]. Gamma-irradiation is another conventional sterilization and depyrogenation process [28]. Some nanoparticles (e.g., citrate-stabilized gold colloids) can tolerate it, while other materials (e.g., silver colloids) do not [29]. Since most cell culture-grade disposable plasticware (e.g., serological pipets, tips, flasks, dishes, and tubes) are sterilized by gamma-irradiation, using these materials during nanoparticle synthesis can help reduce endotoxin in the final product. Both heat and gamma-irradiation procedures are not applicable to nanomaterials containing biological components (recombinant proteins and antibodies), nucleic acids, and peptides. Sterile filtration, autoclaving, ethylene oxide, high hydrostatic pressure, and UV light are additional methods commonly used to sterilize surfaces, medical devices, and some nanomaterials. Some of these procedures, such as ethylene oxide sterilization, are also efficient depyrogenation methods. Similar to gamma-irradiation discussed above, most of these approaches are too harsh for most nanomaterials, especially those containing proteins, peptides, and nucleic acids [28]. Endotoxin can be removed from these nanoparticles and their components using other methods, including, but not limited to, ion exchange chromatography, Triton X-114 extraction, and affinity columns [30–36].

Different sensitivities of nanoparticle carriers, drugs, and targeting ligands to standard sterilization and depyrogenation procedures make terminal sterilization approaches unfeasible for most nanoformulations [28]. However, applying suitable methods to clean individual formulation components followed by pyrogen-free assembly into a final product is often considered an efficient way of producing endotoxin-free nanoformulations. Cavicide, a quaternary ammonium-containing disinfectant, is useful in both sterilizing surfaces and endotoxin removal. Cavicide can be used to clean laboratory surfaces and parts of the synthesis equipment intolerant to high temperatures. If Cavicide is used to clean plastic surfaces or tubing which come in contact with the nanoformulation during a synthetic procedure, rinsing these surfaces with an excess of pyrogen-free water is recommended to remove traces of Cavicide after depyrogenation.

9 Precautions to Avoid Endotoxin Contamination in Nanoformulations

Although various procedures are available for endotoxin removal [31–34], it is easier to prevent contamination than to clean up a contaminated product. The following steps are found useful for avoiding contamination of nanoparticles with endotoxin during synthetic procedures [15, 36]:

(a) Clean and disinfect working surfaces with Cavicide.

(b) Whenever possible, use disposable pyrogen-free materials. Depyrogenate all non-disposable materials (glassware, spatulas, stir-bars, etc.) necessary for the synthetic procedure.

(c) Depyrogenate or use disposable pyrogen-free parts of the equipment that will come in contact with the nanomaterial (columns, tubing, etc.).

(d) Select starting reagents and verify that all chemicals, components, and precursors used during synthesis are pyrogen-free (*see* **Note 5**).

(e) If possible, perform the synthetic procedure in a biological safety cabinet or a clean room facility.

(f) Wear gloves and clean them by spraying and rubbing in Cavicide before touching tubes, reagents, and other materials and parts of equipment used during the synthesis. Change gloves frequently.

(g) Do not cough, sneeze, or breathe directly into reaction tubes. Avoid touching non-sterile surfaces. Change tips and replace pipettes if they accidentally touch the side of the tube, reagent storage bottle, or another non-sterile surface.

10 Notes

1. Recombinant factor C assay kit and β-G-blocker reagents are commercially available products produced by Lonza (catalog numbers 50-658 U and N190, respectively). GlucaShield buffer is a commercially available reagent provided by Associates of Cape Cod (catalog number GB051-5). A test nanomaterial with excitation/emission wavelengths at or around 380/440 nm or light absorbance at or around 440 nm may interfere with the recombinant factor C assay. This interference can be overcome by increasing sample dilution. The dilution of the test sample, however, should not exceed the MVD which is calculated according to the following formula:

   $(\mathrm{EL} \times \text{Sample concentration})/\lambda$, where λ is the sensitivity of the assay.

   When performing the recombinant factor C assay, one needs to consider using inhibition enhancement controls similar to those employed in the traditional LAL assay. β-G-blocker is combined with the test sample at a 1:1 volume ratio. It demonstrates good performance in the LAL kits from the same manufacturer. GlucaShield buffer is used to reconstitute the LAL lysate. Similar to β-G-blocker, this reagent exhibits excellent performance with kits and other LAL reagents produced by the same manufacturer that produces the buffer.

2. Prepare 1% SDS in LAL grade water. Combine equal volumes of the test sample and 1% SDS, then prepare dilutions of this SDS-containing sample in LAL grade water according to the MVD and proceed with the analysis according to the standard LAL protocol. It is important to account for the two-fold dilution during experiment planning and result analysis. For example, if the MVD is 500 and the neat test sample is analyzed at dilutions 5, 50, and 500, you will need to dilute the test sample 2.5-, 25-, and 250-fold. When equal volumes of the nanoparticle at these dilutions are combined with 1% SDS solution and subsequently diluted in water, the final test sample dilutions are 5, 50, and 500. Note that SDS in water, in the

absence of the cationic nanomaterial, will result in spike recovery below 50%; however, spike recovery of the SDS-containing nanoparticle sample SDS falls within 50–200%.

3. Performance verification of this procedure was conducted using control standard endotoxin (CSE), which is known to be less stable than naturally occurring endotoxin. Heat treatment for a period exceeding 30 min results in a decrease in the CSE spike recovery below 50%.
4. Commercial enzymes and kits for endotoxin-free protein digestion are available. One such example is EndoPrep™ (catalog number EDP 4001.01).
5. Most commercially available water for cell culture is pyrogen-free (e.g., Hyclone, catalog number SH30529.02) and is suitable for nanoparticle synthesis. Specialized LAL grade water such as that available from Lonza (catalog number W50-1000) and Associates of Cape Cod (catalog number W100P) can also be used. In general, 18 MΩ-cm per se is not indicative of pyrogen-free water. Bacteria can colonize in water pipes and filters commonly used in laboratory water purification systems. Autoclaving is appropriate for sterilization but does not eliminate endotoxin. Therefore, in-house produced, autoclaved 18 MΩ-cm water can be used instead of commercial pyrogen-free water only after analysis by LAL assay demonstrating endotoxin levels below the assay sensitivity (ideally <0.001 EU/mL).

Acknowledgments

This project has been funded in whole or in part with federal funds from the National Cancer Institute, National Institutes of Health, under contract 75N91010D00024. The content of this publication does not necessarily reflect the views or policies of the Department of Health and Human Services, nor does the mention of trade names, commercial products, or organizations imply endorsement by the US Government.

References

1. Brade H, Opal SM, Vogel SN, Morrison DC (eds) (1999) Endotoxin in health and disease. Marcel Dekker Inc., New York
2. Dobrovolskaia MA, Vogel SN (2002) Toll receptors, CD14, and macrophage activation and deactivation by LPS. Microbes Infect 4(9):903–914. https://doi.org/10.1016/S1286-4579(02)01613-1
3. USP 30 NF 25 (2007) <85> Bacterial endotoxins test. vol 1
4. HHS, US FDA (2012) Guidance for industry. Pyrogen and endotoxins testing: questions and answers. https://www.fda.gov/downloads/drugs/guidances/ucm310098.pdf
5. HHS, US FDA (2015) Guidance for industry and Food and Drug Administration staff. Endotoxin testing recommendations for single-use intraocular ophtalmic devices. https://www.fda.gov/downloads/medicaldevices/deviceregulationandguidance/guidancedocuments/ucm393376.pdf

6. Jones CF, Grainger DW (2009) In vitro assessments of nanomaterial toxicity. Adv Drug Deliv Rev 61(6):438–456. https://doi.org/10.1016/j.addr.2009.03.005
7. Sharma SK (1986) Endotoxin detection and elimination in biotechnology. Biotechnol Appl Biochem 8(1):5–22
8. Crist RM, Grossman JH, Patri AK, Stern ST, Dobrovolskaia MA, Adiseshaiah PP, Clogston JD, McNeil SE (2013) Common pitfalls in nanotechnology: lessons learned from NCI's nanotechnology characterization laboratory. Integr Biol (Camb) 5(1):66–73. https://doi.org/10.1039/c2ib20117h
9. Dobrovolskaia MA, Patri AK, Potter TM, Rodriguez JC, Hall JB, McNeil SE (2012) Dendrimer-induced leukocyte procoagulant activity depends on particle size and surface charge. Nanomedicine (Lond) 7(2):245–256. https://doi.org/10.2217/nnm.11.105
10. Inoue K (2011) Promoting effects of nanoparticles/materials on sensitive lung inflammatory diseases. Environ Health Prev Med 16(3):139–143. https://doi.org/10.1007/s12199-010-0177-7
11. Inoue K, Takano H (2011) Aggravating impact of nanoparticles on immune-mediated pulmonary inflammation. Sci World J 11:382–390. https://doi.org/10.1100/tsw.2011.44
12. Inoue K, Takano H, Yanagisawa R, Hirano S, Kobayashi T, Fujitani Y, Shimada A, Yoshikawa T (2007) Effects of inhaled nanoparticles on acute lung injury induced by lipopolysaccharide in mice. Toxicology 238(2–3):99–110. https://doi.org/10.1016/j.tox.2007.05.022
13. Inoue K, Takano H, Yanagisawa R, Hirano S, Sakurai M, Shimada A, Yoshikawa T (2006) Effects of airway exposure to nanoparticles on lung inflammation induced by bacterial endotoxin in mice. Environ Health Perspect 114(9):1325–1330. https://doi.org/10.1289/ehp.8903
14. Shi Y, Yadav S, Wang F, Wang H (2010) Endotoxin promotes adverse effects of amorphous silica nanoparticles on lung epithelial cells in vitro. J Toxicol Environ Health A 73(11):748–756. https://doi.org/10.1080/15287391003614042
15. Dobrovolskaia MA, McNeil SE (2016) Nanoparticles and endotoxin. In: Dobrovolskaia MA, McNeil SE (eds) Handbook of immunological properties of engineered nanomaterials, vol 1. World Scientific Publishing, Singapore, pp 143–187
16. Alwis KU, Milton DK (2006) Recombinant factor C assay for measuring endotoxin in house dust: comparison with LAL, and (1-->3)-beta-D-glucans. Am J Ind Med 49(4):296–300. https://doi.org/10.1002/ajim.20264
17. Ding JL, Ho B (2010) Endotoxin detection—from limulus amebocyte lysate to recombinant factor C. Subcell Biochem 53:187–208. https://doi.org/10.1007/978-90-481-9078-2_9
18. McKenzie JH, Alwis KU, Sordillo JE, Kalluri KS, Milton DK (2011) Evaluation of lot-to-lot repeatability and effect of assay media choice in the recombinant Factor C assay. J Environ Monit 13(6):1739–1745. https://doi.org/10.1039/c1em10035a
19. Fujita Y, Nabetani T (2014) Iron sulfate inhibits Limulus activity by induction of structural and qualitative changes in lipid A. J Appl Microbiol 116(1):89–99. https://doi.org/10.1111/jam.12349
20. Reich J, Lang P, Grallert H, Motschmann H (2016) Masking of endotoxin in surfactant samples: effects on limulus-based detection systems. Biologicals 44(5):417–422. https://doi.org/10.1016/j.biologicals.2016.04.012
21. Dobrovolskaia MA, Germolec DR, Weaver JL (2009) Evaluation of nanoparticle immunotoxicity. Nat Nanotechnol 4(7):411–414. https://doi.org/10.1038/nnano.2009.175
22. Lyons JL, Roos KL, Marr KA, Neumann H, Trivedi JB, Kimbrough DJ, Steiner L, Thakur KT, Harrison DM, Zhang SX (2013) Cerebrospinal fluid (1,3)-beta-D-glucan detection as an aid for diagnosis of iatrogenic fungal meningitis. J Clin Microbiol 51(4):1285–1287. https://doi.org/10.1128/jcm.00061-13
23. Tran T, Beal SG (2016) Application of the 1,3-beta-D-Glucan (Fungitell) assay in the diagnosis of invasive fungal infections. Arch Pathol Lab Med 140(2):181–185. https://doi.org/10.5858/arpa.2014-0230-RS
24. Henne W, Schulze H, Pelger M, Tretzel J, von Sengbusch G (1984) Hollow-fiber dialyzers and their pyrogenicity testing by limulus amebocyte lysate. Artif Organs 8(3):299–305
25. Neun BW, Dobrovolskaia MA (2011) Detection and quantitative evaluation of endotoxin contamination in nanoparticle formulations by LAL-based assays. Methods Mol Biol 697:121–130. https://doi.org/10.1007/978-1-60327-198-1_12
26. Cooper JF, Resolving LAL (1990) Test interferences. J Parenter Sci Technol 44(1):13–15
27. Sandle T (2011) A practical approach to depyrogenation studies using bacterial endotoxin. J GxP Compliance 15(4):90–96
28. Subbarao N (2016) Nanoparticle sterility and sterilization of nanomaterials. In:

Dobrovolskaia MA, McNeil SE (eds) Handbook of immunological properties of engineered nanomaterials, vol 1, vol 6. World Scientific Publishing Ltd, Singapore, pp 53–75
29. Zheng J, Clogston JD, Patri AK, Dobrovolskaia MA, McNeil SE (2011) Sterilization of silver nanoparticles using standard gamma irradiation procedure affects particle integrity and biocompatibility. J Nanomed Nanotechnol 2011(Suppl 5):001. https://doi.org/10.4172/2157-7439.s5-001
30. Ragab AA, Van De Motter R, Lavish SA, Goldberg VM, Ninomiya JT, Carlin CR, Greenfield EM (1999) Measurement and removal of adherent endotoxin from titanium particles and implant surfaces. J Orthop Res 17(6):803–809. https://doi.org/10.1002/jor.1100170603
31. Dobrovolskaia MA, Neun BW, Clogston JD, Ding H, Ljubimova J, McNeil SE (2010) Ambiguities in applying traditional limulus amebocyte lysate tests to quantify endotoxin in nanoparticle formulations. Nanomedicine (Lond) 5(4):555–562. https://doi.org/10.2217/nnm.10.29
32. London AS, Mackay K, Lihon M, He Y, Alabi BR (2014) Gel filtration chromatography as a method for removing bacterial endotoxin from antibody preparations. Biotechnol Prog 30(6):1497–1501. https://doi.org/10.1002/btpr.1961
33. Ma R, Zhao J, Du HC, Tian S, Li LW (2012) Removing endotoxin from plasmid samples by triton X-114 isothermal extraction. Anal Biochem 424(2):124–126. https://doi.org/10.1016/j.ab.2012.02.015
34. Mack L, Brill B, Delis N, Groner B (2014) Endotoxin depletion of recombinant protein preparations through their preferential binding to histidine tags. Anal Biochem 466:83–88. https://doi.org/10.1016/j.ab.2014.08.020
35. Magalhaes PO, Lopes AM, Mazzola PG, Rangel-Yagui C, Penna TC, Pessoa A Jr (2007) Methods of endotoxin removal from biological preparations: a review. J Pharm Pharm Sci 10(3):388–404
36. Afonin KA, Grabow WW, Walker FM, Bindewald E, Dobrovolskaia MA, Shapiro BA, Jaeger L (2011) Design and self-assembly of siRNA-functionalized RNA nanoparticles for use in automated nanomedicine. Nat Protoc 6(12):2022–2034. https://doi.org/10.1038/nprot.2011.418

Chapter 10

Detection of Beta-Glucan Contamination in Nanoparticle Formulations

Barry W. Neun and Marina A. Dobrovolskaia

Abstract

Beta-glucans with diverse chemical structures are produced by a variety of microorganisms and are commonly found in microbial cell walls. β-(1,3)-D-glucans are present in yeast and fungi, and, for this reason, their traces are commonly used as a sign of yeast or fungal infection or contamination. Despite being less immunologically active than endotoxins, beta-glucans are pro-inflammatory and can activate cytokines and other immunological responses via their cognate pattern recognition receptors. Unlike endotoxins, there is no established threshold pyrogen dose for beta-glucans; as such, their quantity in pharmaceutical products is not regulated. Nevertheless, regulatory agencies recognize the potential contribution of beta-glucans to the immunogenicity of protein-containing drug products and recommend assessing beta-glucans to aid the interpretation of immunotoxicity studies and assess the risk of immunogenicity. The protocol for the detection and quantification of β-(1,3)-D-glucans in nanoparticle formulations is based on a modified limulus amoebocyte lysate assay. The results of this test are used to inform immunotoxicity studies of nanotechnology-based drug products.

Key words Nanoparticles, Beta-glucans, Contamination, Pathogen-associated molecular patterns, Pattern recognition receptors, Inflammation, Immunomodulation

1 Introduction

β-(1,3)-D-glucans are products of the cell walls of some microorganisms, including yeast and fungi. The most common source of β-(1,3)-D-glucans is the fungi *Candida* and *Aspergillus*. β-(1,3)-D-glucans are also found at low levels (<60 pg/mL) in the blood of healthy humans where they are believed to distribute from the gastrointestinal tract following the consumption of β-(1,3)-D-glucan-containing foods (plants, grains, cereal, mushrooms, etc.) [1]. In patients with invasive fungal infections, these levels increase to or above 70 pg/mL [1–6]. β-(1,3)-D-glucans may become undesirable contaminants in pharmaceutical products, where they are introduced during manufacturing through filtration processes utilizing cellulose-based filters, contaminated starting materials, or

Jeffrey D. Clogston et al. (eds.), *Characterization of Nanoparticles Intended for Drug Delivery*, Methods in Molecular Biology, vol. 2789, https://doi.org/10.1007/978-1-0716-3786-9_10,

common excipients such as sucrose. β-(1,3)-D-glucans are not as immunologically potent as bacterial endotoxins. However, they are pro-inflammatory and can also activate the immune system. There is an increasing amount of information suggesting that β-(1,3)-D-glucans may exaggerate endotoxin-mediated toxicities, as well as synergize with other immunologically active impurities introduced into pharmaceutical products during manufacturing and, therefore, lead to adverse immune effects [6–9]. Moreover, the FDA immunogenicity guidance for industry suggests minimizing the levels of β-(1,3)-D-glucans in therapeutic protein formulations to decrease the immunogenicity of these products [10].

Unlike bacterial endotoxins, β-(1,3)-D-glucans are currently not regulated; there is no compendial standard for their detection and no harmonized approach for acceptable levels. Nevertheless, there is a growing trend in the industry and among regulatory authorities worldwide to detect β-(1-3)-D-glucans and understand their safety levels [1, 3, 4, 6–8]. Since many nanotechnology platforms are not immunologically inert, understanding the presence of immunologically reactive contaminants besides endotoxin becomes an essential step in understanding the safety margins for formulations containing such materials [6, 11, 12]. This becomes especially important when such nanomaterials are intended for immunotherapy involving the intentional application of immune checkpoint inhibitors to enhance the immune response.

The assay described in this chapter is based on the commercial kit approved for the detection of β-(1,3)-D-glucans in human serum to diagnose fungal infections [1]. This commercial Glucatell® assay is a derivative of the limulus amebocyte lysate (LAL) assay. The lysate in the LAL assay contains two main proteases named factor C and factor G, which are activated specifically by bacterial endotoxins and β-(1,3)-D-glucans, respectively (Fig. 1). The presence of β-(1,3)-D-glucans in test nanoparticle samples results in a false-positive result in the LAL assay. To avoid such false-positive results, a Glucashield reagent is used to block factor G (*see* Chapter 9). In contrast, to make the LAL assay specific to β-(1,3)-D-glucans, factor C is removed from the lysate. The resultant lysate, also known as Glucatell reagent, contains factor G and, therefore, is specific to β-(1,3)-D-glucans. β-(1,3)-D-glucans in the test nanoparticle sample cause the activation of serine proteases. An activated protease cleaves p-nitroaniline (pNA) from the peptide substrate. The generation of pNA results in a change in sample color (along with the addition of the diazo reagents) and is measured at 540 nm. The intensity of the color is proportional to the amount of β-(1,3)-D-glucans in the test sample. We adapted an R&D version of this assay for the screening of nanoformulations to inform immunotoxicity studies and help create immunologically safer nanomaterials.

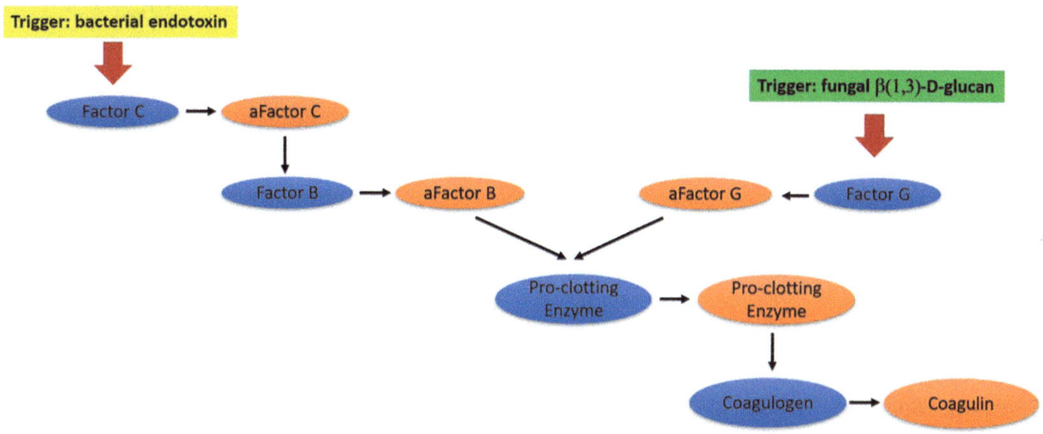

Fig. 1 Key elements of the LAL enzymatic reaction and their specificity to endotoxin and β-(1,3)-D-glucans. LAL is a mixture of serine proteases (factor C, factor B, factor G, pro-clotting enzyme, and coagulogen), which form an enzymatic cascade. There are two arms in the cascade. One arm is triggered by bacterial endotoxins, which specifically activate factor C. The second arm is triggered by β-(1,3)-D-glucans, which specifically activate factor G. The activated factor C and activated factor G, in turn, cleave their respective targets, and the proteolytic cascade continues until it culminates with the production of coagulin which results in turbidity and clotting of the lysate. The presence of β-(1,3)-D-glucans in test samples, therefore, interferes with the accurate quantification of bacterial endotoxins by the LAL assay. To avoid this interference, the factor G pathway is either blocked by glucan masking reagents (e.g., Glucashield) or by changing the assay format to the recombinant factor C. To avoid similar interference from endotoxin during the quantification of β-(1,3)-D-glucans using the LAL assay, factor C is depleted from the LAL lysate. The LAL without factor C is known as Glucatell reagent, which is used in this protocol. Note: a = activated

2 Materials

1. Glucatell kit (Associates of Cape Cod).
2. Borosilicate glass test tubes for sample dilution, 13 × 100 mm.
3. Pyrogen-free pipette tips (1000, 250 μL).
4. Parafilm®.
5. Disposable endotoxin-free glass dilution tubes, 12 × 75 mm.
6. Plate reader capable of reading at 540 nm.
7. Block heater with 96-well plate insert (37 °C).

3 Methods

3.1 Preparation of Reagents

1. Preparation of the standard solution: Add the appropriate volume of reagent-grade water (RGW) to the vial of glucan standard (Pachyman) to make a 100 pg/mL solution (*see* **Note 1**). Vortex for at least 1 min to resuspend homogeneously.

2. Reconstitute one vial of Glucatell reagent with 2.8 mL of Pyrosol. Swirl the vial gently to dissolve completely (*see* **Notes 2** and **3**).
3. Preparation of the diazo coupling reagents: Add the 1 N HCl solution (vial 1A) to the sodium nitrite (*see* **Note 4**).
4. Add 4.0 mL of RGW to the ammonium sulfamate (*see* **Note 5**).
5. Add the contents of the N-methyl-pyrrolidinone to the N-(1-naphthyl) ethylenediamine dihydrochloride (NEDA) (*see* **Notes 6** and **7**).

3.2 Preparation of Standards, Controls, and Test Samples

1. Prepare calibration standards by spiking a known amount of β-(1,3)-D-glucan standard into RGW (Table 1).
2. Prepare the quality control (QC) by spiking a known amount of β-(1,3)-D-glucan standard into RGW (Table 2).
3. Prepare the inhibition enhancement control (IEC) for each dilution of the tested nanomaterial (NP) by spiking a known amount of β-(1,3)-D-glucan standard into the NP test sample (Table 3).
4. Prepare three dilutions (1:5, 1:50, and 1:500) of each NP sample: first, prepare the initial fivefold dilution using RGW; next, perform two serial tenfold dilutions of this initial dilution in RGW (Table 4).

Table 1
Calibration standards

Sample	Nominal concentration (pg/mL)	Preparation procedure
1	40	600 μL RGW + 400 μL stock β-(1,3)-D-glucan standard (100 pg/mL)
2	20	400 μL RGW + 400 μL solution 1
3	10	400 μL RGW + 400 μL solution 2
4	5	400 μL RGW + 400 μL solution 3

Table 2
Quality control

Sample	Nominal concentration (pg/mL)	Preparation procedure
QC	15	567 μL RGW + 100 μL stock β-(1,3)-D-glucan standard (100 pg/mL)

Table 3
Inhibition/enhancement control

Sample	Nominal concentration (pg/mL)	Preparation procedure
IEC	15	283.5 μL NP + 50 μL stock β-(1,3)-D-glucan standard (100 pg/mL)

Table 4
Study samples

Sample	Dilution	Preparation procedure
Nanoparticle A	5	100 μL NP stock + 400 μL RGW
Nanoparticle A	50	50 μL Dil 5 + 450 μL RGW
Nanoparticle A	500	50 μL Dil 50 + 450 μL RGW

3.3 Experimental Procedure

1. Prepare calibration standards, quality controls, inhibition enhancement controls, and test nanomaterials as described above.
2. Plate 50 μL of RGW (blank) or calibration standards in duplicate into the corresponding wells of a 96-well plate (an example plate map is shown in Fig. 2).
3. Plate 50 μL of samples and IEC in duplicate to the corresponding wells of a 96-well plate (see example plate map in Fig. 2).
4. Add 50 μL of Glucatell reagent to each well using a repeater pipette. Cover the plate with a lid and shake by tapping the edge.
5. Place the plate on a heating block or a plate reader preheated to 37 °C ± 1 °C for the recommended incubation time (see Certificate of Analysis) for the curve chosen.
6. Stop the reaction by adding 50 μL of sodium nitrite from vial 1 (*see* **Note 4**) with a repeater pipette. Then add in sequence 50 μL of ammonium sulfamate (*see* **Note 5**), and then 50 μL of N-(1-naphthyl) ethylenediamine dihydrochloride (NEDA) (*see* **Note 6**), using a new pipette tip each time. The color development starts immediately.
7. Place the uncovered plate in the microplate reader and read the optical density at 540–550 nm.

	1	2	3	4	5	6	7	8	9	10	11	12
A												
B		Std 1	Std 1	Sample 1 Dil 5	Sample 1 Dil 5	Sample 1 Dil 500 IEC	Sample 1 Dil 500 IEC	Sample 2 Dil 50 IEC	Sample 2 Dil 50 IEC	Sample 3 Dil 5 IEC	Sample 3 Dil 5 IEC	
C		Std 2	Std 2	Sample 1 Dil 50	Sample 1 Dil 50	Sample 2 Dil 5	Sample 2 Dil 5	Sample 2 Dil 500 IEC	Sample 2 Dil 500 IEC	Sample 3 Dil 50 IEC	Sample 3 Dil 50 IEC	
D		Std 3	Std 3	Sample 1 Dil 500	Sample 1 Dil 500	Sample 2 Dil 50	Sample 2 Dil 50	Sample 3 Dil 5	Sample 3 Dil 5	Sample 3 Dil 500 IEC	Sample 3 Dil 500 IEC	
E		Std 4	Std 4	Sample 1 Dil 5 IEC	Sample 1 Dil 5 IEC	Sample 2 Dil 500	Sample 2 Dil 500	Sample 3 Dil 50	Sample 3 Dil 50	RGW QC IEC	RGW QC IEC	
F		NC	NC	Sample 1 Dil 50 IEC	Sample 1 Dil 50 IEC	Sample 2 Dil 5 IEC	Sample 2 Dil 5 IEC	Sample 3 Dil 500	Sample 3 Dil 500			
G												
H												

Fig. 2 Example plate map. Samples may be positioned on the plate in this order; other plate maps may also be used. Std = standard; NC = negative control; IEC = inhibition/enhancement control; dil = dilution; RGW QC = reagent-grade water quality control

8. Use the plate reader software to plot a standard curve and to calculate the concentration of β-(1,3)-D-glucan in the test samples, quality controls, and IECs.

3.4 Data Report and Interpretation

No compendial procedure or criteria are currently available for the estimation of acceptable levels of β-(1,3)-D-glucans in pharmaceutical products. Below, we discuss the approach used in our laboratory [6, 13]. This approach relies on several assumptions: (a) an average adult weight is 70 kg; (b) the blood volume of such an adult is 5.6 L (or 8% of the body weight); and (c) the entire injected dose stays in circulation. The detected amount of β-(1,3)-D-glucans per 1 mL of the test-nanomaterial is converted into the β-(1,3)-D-glucan dose using the active pharmaceutical ingredient (API) concentration in the stock nanomaterial and the intended API dose. Next, the β-(1,3)-D-glucan dose is converted into the concentration in 1 mL of blood; the latter is compared to the limit (70 pg/mL) used in the clinical diagnostic Fungitell assay where beta-glucan levels are indicative of fungal infection. For example, if this assay determines that the level of β-(1,3)-D-glucans is 1000 pg/mL of a nanoformulation containing 10 mg/mL of API and the API dose is 10 mg/kg, then the beta-glucan dose is 1000 pg/kg. After conversion to the amount per milliliter of blood (1000 pg × 70 kg = 70,000 pg of beta-glucan per 5600 mL of blood), the result is 12.5 pg/mL, which is less than 70 pg/mL. Therefore, we would consider this result normal, as it is within healthy levels of beta-glucans present in human blood from dietary sources. Other approaches are available in the literature [14] and discussed in detail elsewhere [6].

3.5 Acceptance Criteria

1. The linear regression algorithm is used to construct the standard curve. Precision (% CV) and accuracy (percent difference from theoretical) of each calibration standard and quality control should be within 25%.
2. At least three calibration standards should be available for the assay to be considered acceptable.
3. If quality controls fail to meet the acceptance criterion described in criterion 1, the run should be repeated.
4. If the standard curve fails to meet the acceptance criteria described in criteria 1 and 2, the run should be repeated.
5. Precision of the study sample should be within 25%.
6. Precision of the inhibition/enhancement control should be within 25%.
7. Spike recovery indicative of the accuracy of the inhibition/enhancement control should be between 50% and 200%. Spike recovery less than 50% is indicative of inhibition; that above 200% is indicative of either contamination or enhancement.
8. If sample interference is detected, the assay results for this sample are invalid.

4 Notes

1. The β-(1,3)-D-glucan content is stated on the vial label. Refer to the vial label for the volume of water to add.
2. It is critical to avoid vortexing at this step.
3. Use the reconstituted Glucatell reagent within 10 min or place it at 2–8 °C and use within 2 h.
4. This is vial 1 for use in **step 6** of the experimental procedure; it is contained in the Glucatell kit.
5. This is vial 2 for use in **step 6** of the experimental procedure; it is contained in the Glucatell kit.
6. This is vial 3 for use in **step 6** of the experimental procedure; it is contained in the Glucatell kit.
7. All of these solutions should be used on the same day.

Acknowledgment

This project has been funded in whole or in part with federal funds from the National Cancer Institute, National Institutes of Health, under contract 75N91010D00024. The content of this publication does not necessarily reflect the views or policies of the

Department of Health and Human Services, nor does the mention of trade names, commercial products, or organizations imply endorsement by the US Government.

References

1. ACC (2023) FUNGITELL® assay serum test for (1-3)-b-D-glucan. Associates of Cape Cod
2. Ellis M, Al-Ramadi B, Finkelman M, Hedstrom U, Kristensen J, Ali-Zadeh H, Klingspor L (2008) Assessment of the clinical utility of serial beta-D-glucan concentrations in patients with persistent neutropenic fever. J Med Microbiol 57(Pt 3):287–295
3. Odabasi Z, Mattiuzzi G, Estey E, Kantarjian H, Saeki F, Ridge RJ, Ketchum PA, Finkelman MA, Rex JH, Ostrosky-Zeichner L (2004) Beta-D-glucan as a diagnostic adjunct for invasive fungal infections: validation, cutoff development, and performance in patients with acute myelogenous leukemia and myelodysplastic syndrome. Clin Infect Dis 39(2): 199–205
4. Pazos C, Moragues MD, Quindós G, Pontón J, del Palacio A (2006) Diagnostic potential of (1,3)-beta-D-glucan and anti-Candida albicans germ tube antibodies for the diagnosis and therapeutic monitoring of invasive candidiasis in neutropenic adult patients. Rev Iberoam Micol 23(4):209–215
5. Pazos C, Pontón J, Del Palacio A (2005) Contribution of (1->3)-beta-D-glucan chromogenic assay to diagnosis and therapeutic monitoring of invasive aspergillosis in neutropenic adult patients: a comparison with serial screening for circulating galactomannan. J Clin Microbiol 43(1):299–305
6. Neun BW, Cedrone E, Potter TM, Crist RM, Dobrovolskaia MA (2020) Detection of beta-glucan contamination in nanotechnology-based formulations. Molecules 25(15):3367
7. Ławniczek-Wałczyk A, Górny RL (2010) Endotoxins and β-glucans as markers of microbiological contamination--characteristics, detection, and environmental exposure. Ann Agric Environ Med 17(2):193–208
8. Liss B, Cornely OA, Hoffmann D, Dimitriou V, Wisplinghoff H (2016) 1,3-β-D-Glucan contamination of common antimicrobials. J Antimicrob Chemother 71(4):913–915
9. Vigor K, Emerson J, Scott R, Cheek J, Barton C, Bax HJ, Josephs DH, Karagiannis SN, Spicer JF, Lentfer H (2016) Development of downstream processing to minimize beta-glucan impurities in GMP-manufactured therapeutic antibodies. Biotechnol Prog 32(6): 1494–1502
10. USFDA (2014) Guidance for industry immunogenicity assessment for therapeutic protein products. https://www.fda.gov/downloads/drugs/guidances/ucm338856.pdf
11. Dobrovolskaia MA, McNeil SE (2013) Endotoxin and engineered nanomaterials. In: Dobrovolskaia MA, McNeil SE (eds) Handbook of immunological properties of engineered nanomaterials, vol 1. World Scientific Publishing Ltd, Singapoore, pp 143–186
12. Holley CK, Dobrovolskaia MA (2021) Innate immunity modulating impurities and the immunotoxicity of nanobiotechnology-based drug products. Molecules 26(23):7308
13. Neun BW, Dobrovolskaia MA (2019) Understanding endotoxin and beta-glucan contamination in nanotechnology-based drug products. In: Williams KL (ed) Endotoxin detection and control in pharma, limulus, and mammalian systems. Springer, Cham, pp 481–496
14. Schubert C, Moudgal C (2017) Parenteral safety of beta-glucans. In: PDA endotoxins workshop. PDA, Washington, DC

Chapter 11

Detection of Intracellular Complement Activation by Nanoparticles in Human T Lymphocytes

Anna Ilinskaya, Ankit Shah, Andrew Van Dusen, and Marina A. Dobrovolskaia

Abstract

The complement system is complex and includes two main components: the systemic or plasma complement and the so-called intracellular complement or complosome. The complement proteins expressed by the liver and secreted into blood plasma compose the plasma complement system, whereas complement proteins expressed by and functioning inside the cell represent the intracellular complement. The complement system plays an essential role in host defense; however, complement activation may lead to pathologies when uncontrolled. When such undesirable activation of the plasma complement occurs in response to a drug product, it leads to immediate-type hypersensitivity reactions independent of immunoglobulin E. These reactions are often called complement activation-related pseudoallergy (CARPA). In addition to the blood plasma, the complement protein C3 is found in many cells, including lymphocytes, monocytes, endothelial, and even cancer cells. The activation of the intracellular complement generates split products, which are exported from the cell onto the membrane. Since the activation of the intracellular complement in T lymphocytes was found to correlate with autoimmune disorders, and growing evidence is available for the involvement of T lymphocytes in the development of drug-induced hypersensitivity reactions, understanding the ability of nanomaterials to activate intracellular complement may aid in establishing a long-term safety profile for these materials. This chapter describes a flow cytometry–based protocol for detecting intracellular complement activation by engineered nanomaterials.

Key words Nanoparticles, Complement, Complosome, Immunotoxicity, Adjuvants, CARPA, Autoinflammatory response

1 Introduction

The human complement system exists both extracellularly and intracellularly. The extracellular complement system (also known as plasma or systemic complement) is better characterized and consists of more than 50 liver-produced proteins circulating in the bloodstream that help recognize and remove pathogens [1, 2]. Three main pathways—the classical, lectin, and alternative pathways—aid in this host defense process, and such pathways lead

Jeffrey D. Clogston et al. (eds.), *Characterization of Nanoparticles Intended for Drug Delivery*, Methods in Molecular Biology, vol. 2789, https://doi.org/10.1007/978-1-0716-3786-9_11,

to the activation of C3 and C5 convertases [1–3]. These enzymes split C3 and C5 complement components into smaller protein fragments that can trigger an inflammatory response (e.g., C3a, C5a anaphylatoxins), aid in opsonization (e.g., C3b), or form a pore known as the membrane attack complex (sC5b-9) to lyse pathogens. Activation of the plasma complement may occur in response to drug products, which trigger the so-called complement activation-related pseudoallergy (CARPA). CARPA has been reported for many biologics, nucleic acid therapeutics, small molecules, and nanotechnology-based formulations [4, 5]. When left uncontrolled, CARPA may be fatal; therefore, many efforts have been directed at understanding drugs' physicochemical attributes associated with complement activation. It is now well established that many excipients commonly used to prepare pharmaceutical products (e.g., polysorbate 80 and Cremophor EL) can activate the plasma complement system and trigger CARPA [6–8]. In nanomedicine, nanoparticle size, charge, surface area, shape, and the presence of polyethylene glycol (PEG) have been linked to CARPA [9–11]. Strategies for overcoming this toxicity for nanomedicines have been established and discussed elsewhere [12, 13].

The intracellular complement system consists of complement-associated proteins that are active intracellularly. Though the complement system is thought to primarily have immune functions, the intracellular complement system has specific roles in numerous cell types, including immune cells, epithelial, mesothelial, and endothelial cells, fibroblasts, hepatocytes, and pancreatic beta-cells. In CD4+ T cells, the binding of an antigen to a T cell receptor paired with CD28 co-stimulation causes C3b to be exported to the cell surface. Consequently, C3b stimulates CD46, which then leads to the activation of Th1 effector cells. Other immune cells like neutrophils and monocytes contain high levels of C3 and C5, which are also crucial to inflammasome activation in monocytes and macrophages. In nonimmune cells, such as epithelial cells, the intracellular complement protects against oxidative stress. Moreover, in hepatocytes, C3 regulates normal triglyceride levels by promoting low levels of lipophagy (reviewed in [14]). Thus, it can be seen that the intracellular complement system has critical functions across a variety of cell types. One of the key differences between the plasma and intracellular complement systems is in the activation mechanism, in that the convertase is responsible for plasma complement activation, whereas cathepsin L is involved in the activation of intracellular complement [15].

Intracellular complement proteins have various metabolic effects. For instance, activation of the lysosomal C3a receptor activates mTOR, and activated CD46 (a cell receptor and complement regulator) can promote the assembly of mTORC1—a nutrient-sensing complex that regulates cell growth—on lysosomal surfaces. CD46 can also increase the rates of key metabolic pathways like glycolysis and oxidative phosphorylation in effector cells, as well as promote the expression of nutrient channels and metabolic enzymes. Receptors for C3a and C5a can also modulate mitochondrial function. In addition to its metabolic effects, the intracellular complement system can promote autophagy through C3 and C3b and potentially regulate histone-DNA interactions through C3 (reviewed in [14]).

Dysfunction of the intracellular complement system has also been implicated in numerous diseases. For instance, low levels of C3, C5, and CD46 in T cells result in weakened immune responses. Moreover, people with leukocyte adhesion deficiency-1 (LAD-1) do not have high enough C3 levels within their immune cells, which further contributes to their immune deficiency. Overexpression of C3 in T cells may also contribute to autoimmune diseases like rheumatoid arthritis, and overactivity of C5 and its receptor in Th1 cells and cytotoxic T cells may contribute to systemic lupus erythematosus. In addition, overexpression of intracellular factor H and C1 seem to be implicated in clear cell renal cell carcinoma (ccRCC) tumor development and survival, with tumor C1 expression encouraging the recruitment of more suppressive immune cells (reviewed in [14]). As such, dysregulation of intracellular complement proteins has direct impacts on the immune system, which can then lead to the development of disease.

The potential contribution of the intracellular complement to the toxicity of drug products is less understood. Due to the involvement of this complement component in many cellular processes critical for both normal immune function and the development of autoinflammatory responses, the intracellular complement is gaining the attention of toxicologists. While still limited, the existing data suggest that large cationic molecules and nanoparticles possessing cationic moieties on their surface can trigger intracellular complement activation [16]. Interestingly, cellular membrane damage but not cathepsin L is involved in the intracellular complement activation by cationic materials [16]. Thus, it was suggested that the leakage and exposure of the intracellular complement on the cell surface constitute the so-called self-opsonization process through which damaged cells communicate with normal cells about the danger signals in the surrounding environment and the need for clearing the damage [16]. Analysis of the intracellular complement activation by nanomedicines may help understand hypersensitivity and autoinflammatory reactions and contribute to establishing complete safety profiles of these novel materials.

The assay described herein was adapted from the original study by Liszewski et al. [17] and optimized to support nanoparticle characterization [16]. Human peripheral blood mononuclear cells (PBMC) are treated with nanoparticles and controls, and the presence of the complement split products on the cell surface is then monitored by flow cytometry. Although the gating strategy used in this protocol is specific to CD3-positive lymphocytes (T cells), this protocol could be adjusted to detect intracellular complement on the surface of other cells. The assay requires 400 μL of the test nanomaterial at the concentration selected in a cytotoxicity assay before performing this test.

2 Materials

1. Human blood coagulated with Li-heparin, used to prepare PBMC (*see* **Note 1**).
2. Phosphate-buffered saline (PBS).
3. Hanks' balanced salt solution (HBSS).
4. Ficoll-Paque Premium.
5. Pen/Strep solution.
6. Acridine Orange Propidium Iodine (AOPI) staining solution in PBS for automatic cell counter or any reagent allowing to distinguish between live and dead cells using a hemacytometer.
7. Albumin, from bovine serum (BSA).
8. Complete Roswell Park Memorial Institute (RPMI): supplement commercial RPMI with 10% heat-inactivated fetal bovine serum (FBS), 4 mM L-glutamine, and 100 U/mL penicillin.
9. Staining buffer (0.1% BSA in PBS): 50 mg of reagent in 50 mL of ice-cold PBS. Mix well and place on ice. Prepare fresh before each use. Discard any leftovers after the experiment is complete.
10. Purified Mouse Anti-Human CD3 IgG2a (Abcam, cat#ab86883) (use for the positive control).
11. Purified Mouse Anti-Human CD3 IgG2a (eBioscience, cat#16-0037) (use for flow cytometry to gate CD3 positive cells).
12. Purified Mouse Anti-Human C3adesArg, clone 2991, (NEO-Epitope) IgG1 (Hycult Biotech, cat#HM2074).
13. Purified Mouse IgG1 Isotype control (BD Pharmingen, cat# 554121).
14. Purified Mouse IgG2a Isotype control (BD Pharmigen, #554126).
15. Allophycocyanin (APC) Rat Anti-Mouse IgG2a (eBioscience, cat#17-4210).
16. Phycoerythrin (PE) Rat Anti-Mouse IgG1 (eBioscience, cat#12-4015).
17. Polystyrene round bottom tubes, 12 × 75 mm.
18. Pipettes covering the range from 0.05 to 10 mL.

19. 24-well cell culture plate.
20. Polypropylene tubes, 5 mL.
21. Centrifuge with speed 400 × *g* (or ~1440 rpm).
22. Refrigerator, 2–8 °C.
23. Freezer, −20 °C.
24. Cell culture incubator with 5% CO_2 and 95% humidity.
25. Biohazard safety cabinet approved for level II handling of biological material.
26. Inverted microscope.
27. Vortex.
28. Cellometer, source, or hemacytometer to perform cell count.
29. Flow cytometer.

3 Methods

3.1 Preparation of Reagents and Controls

1. Prepare heat-inactivated fetal bovine serum (FBS). Thaw a bottle of FBS at room temperature or overnight at 2–8 °C, and allow it to equilibrate to room temperature. Incubate for 30 min at 56 °C in a water bath, mixing every 5 min to heat-inactivate it. Single-use aliquots may be stored at 2–8 °C for up to 1 month or at a nominal temperature of −20 °C.
2. Prepare complete RPMI as described in Subheading 2. Store the complete RPMI at 2–8 °C protected from light for up to 1 month. Warm the medium in a water bath at 37 °C before use.
3. Prepare staining buffer as described in Subheading 2. Mix well and place on ice. Prepare fresh before each use. Discard any leftovers after the experiment is complete.
4. Prepare the positive control by treating cells for 1 h with 5 μg/test of mouse anti-human CD3 IgG2a (Abcam) (*see* **Note 2**).
5. Prepare the negative control. Use cells without treatment as a negative control.

3.2 Preparation of Test Nanomaterials

1. Prepare test nanomaterials according to the manufacturer's instructions (*see* **Notes 3** and **4**).
2. Titrate nanoparticles in PBMC cultures, starting with the highest reasonable, achievable concentration to select the highest concentration, which, after 1 h of incubation, results in cell viability of no less than 80% (*see* **Note 5**).

3.3 PBMC Isolation

1. Place freshly drawn blood into a 15 or 50 mL conical centrifuge tube, add an equal volume of room temperature PBS, and mix well.
2. Slowly layer the Ficoll-Paque solution underneath the blood/PBS mixture by placing the tip of the pipette containing Ficoll-Paque at the bottom of the blood sample tube. Alternatively, the blood/PBS mixture may be slowly layered over the Ficoll-Paque solution. Use 3 mL of Ficoll-Paque solution per 4 mL of blood/PBS mixture (*see* **Note 6**).
3. Centrifuge for 30 min at 900 × *g*, 18–20 °C, without a brake.
4. Using a sterile pipette, remove the upper layer containing plasma and platelets and discard it.
5. Transfer the mononuclear cell layer into another centrifuge tube using a sterile pipette.
6. Wash cells by adding excess HBSS and centrifuging for 10 min at 400 × *g* at 18–20 °C. The HBSS volume should be approximately three times the volume of the mononuclear layer (*see* **Note 7**).
7. Discard the supernatant and repeat the wash step one more time.
8. Resuspend cells in complete RPMI 1640 medium. Count cells and determine viability using AOPI staining solution. If viability is at least 90%, proceed to the next step.

3.4 Procedure for Cell Treatment with Nanoparticles and Controls

1. Adjust cell count to 1.25×10^6 cells/mL.
2. Aliquot 800 μL of cell suspension prepared in step 1 per each well of a 24-well plate (*see* **Note 8**).
3. Add 200 μL of test nanoparticles or control to appropriate wells.
4. After 1 h of incubation with test materials and controls, harvest the cells and measure the cell viability. If the viability is ≥80%, proceed to the next step. If not, stop the experiment.

3.5 Preparation of Cells for Flow Cytometry

1. Harvest cells into polystyrene round bottom 12 × 75 mm Falcon tubes and centrifuge at 400 × *g* for 5 min.
2. Aspirate the supernatant and wash cells with 2 mL of ice-cold PBS.
3. Resuspend cells in 100 μL of staining buffer.
4. Add 1 μg/test of Purified Mouse Anti-Human C3adesArg (clone 2991) to all tubes, except for tubes with isotype controls and unlabeled cells.
5. Add 1 μg/test of Purified Mouse IgG1 and 0.25 μg/test of Purified Mouse IgG2a in isotype control #1 (*see* **Note 9**).

6. Add 1 μg/test of Purified Mouse IgG1 in isotype control #2 (*see* **Note 9**).
7. Add 0.25 μg/test of CD3 (eBioscience) to all tubes, except for the positive control, isotype control #1, and unlabeled cells.
8. Incubate on ice for 20 min.
9. Wash cells twice with 1 μL of PBS by centrifugation at 400 × *g* at 4 °C for 5 min and resuspend cells in 100 μL of ice-cold staining buffer.
10. Add 0.25 μg/test of PE Rat Anti-Mouse IgG1 and 0.5 μg/test APC Rat Anti-Mouse IgG2a to the tubes.
11. Incubate at 4 °C for 20 min.
12. Wash cells twice with 1 mL of PBS by centrifugation at 400 × *g* for 5 min.
13. Resuspend in 400 μL of PBS.
14. Acquire data using a flow cytometer (*see* **Note 10**).

3.6 Preparation of Instrument for Flow Cytometry

1. Switch on the instrument. Ensure that the sheath tank is full and the waste tank is empty.
2. To get consistent results, run the BD FACSComp software using BD CaliBRITE beads (*see* **Note 11**).
3. Adjust the forward scatter (FSC) vs. side scatter (SSC) dot plot using unlabeled cells to get the population of cells in the plot (Fig. 1a).
4. Set the threshold to remove most of the debris.
5. Gate the lymphocyte population in gate 1 (Fig. 1a).
6. Create an FL2 vs. FL4 dot plot. Limit this dot plot to the cell population from gate 1.
7. Adjust FL2 and FL4 settings to get cells in the left low corner of FL2 vs. FL4 dot plot using unlabeled/untreated cells and isotype control #1.

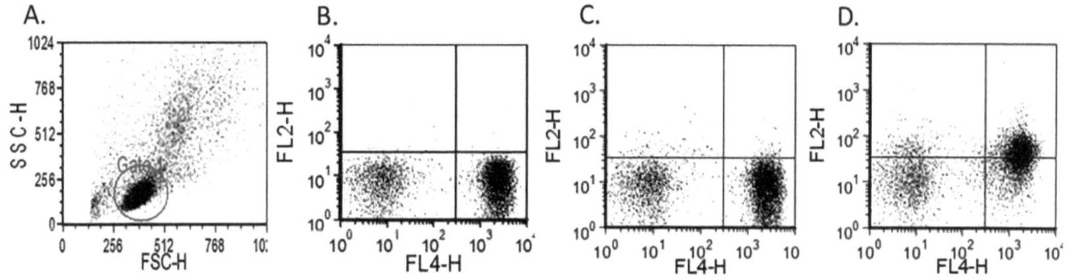

Fig. 1 Example of flow cytometry analysis of C3a expression. (**a**) FSC vs. SSC dot plot and gate to select the cell population of interest, (**b**) isotype control #2 (FL2 vs. FL4), (**c**) negative control (FL2 vs. FL4), and (**d**) positive control (FL2 vs. FL4). C3a-specific fluorescence is monitored in FL2, whereas CD3 is detected in FL4

8. Set up quadrant regions based on isotype control #2 (Fig. 1b).
9. Acquire and save the data. The expected performance of the negative and positive control is shown in Fig. 1c and d, respectively, and that of nanoparticles is in Fig. 2.
10. Follow the instrument closing procedure to shut down the instrument.

3.7 Data Analysis and Report

1. Report the data as the percentage of double-positive cells, which is determined by comparing the number of double-positive cells to that in the negative control (*see* **Note 12**).
2. Compare the results in nanoparticle-treated samples to those in the negative control (*see* **Note 13**).

3.8 Acceptance Criteria

1. The run is acceptable if no nonspecific binding is detected in isotype control #1.
2. The run is acceptable if the positive control is positive, i.e., more than 10% of cells are positive in C3adesArg staining.

4 Notes

1. Obtaining donor blood from healthy volunteers requires institutional approval. Consult with your organization to verify that such approval is available.
2. All antibodies described in this protocol were carefully selected based on the literature and performance qualification experiments. It is the authors' experience that, despite the same target specificity, not all commercially available antibodies perform equally well in this test; moreover, inter-lot variability in antibody performance has been noticed. Our experience agrees with that of the authors of other published studies (verbal communication). The reason for this observation is unknown, and other reagents with good performance may exist. It is recommended to start using the antibodies highlighted in this protocol, then select other reagents with similar specificity based on their performance in the assay.
3. Always wear personal protective equipment and take appropriate precautions when handling your nanomaterial. It is recommended to wear two layers of gloves when handling nanomaterials. Also, follow your facility's recommended disposal procedure for nanomaterials.
4. Some nanomaterials are formulated as a solution stored at 4 °C or frozen. Refrigerated nanomaterials may require mixing prior to use. Frozen nanomaterials are thawed at room temperature and mixed prior to use. Vortexing or vial inversion are commonly used procedures for mixing nanomaterials. Some

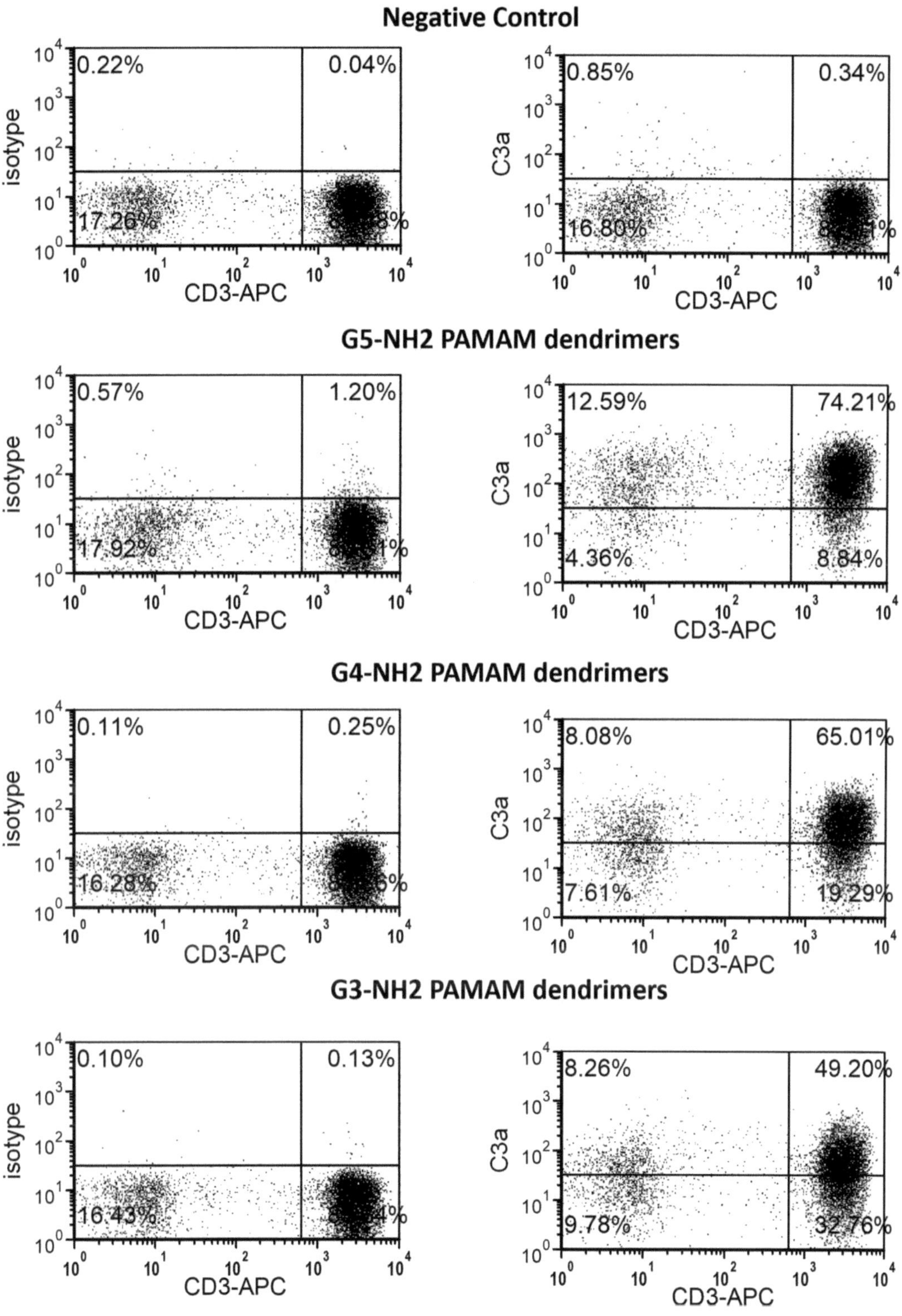

Fig. 2 Example data generated using nanoparticles. The cells were either left untreated (negative control) or treated with amine-terminated polyamidoamine (PAMAM) dendrimers according to the procedure described in this protocol. Each sample was analyzed against its respective isotype control; the percent double-positive cells C3a+CD3+ reflect the exposure of C3a on the surface of T lymphocytes. The generation-dependent effect of cationic PAMAM dendrimers suggests the role of the particle size and the density of terminal surface groups in activating intracellular complement in lymphocytes

nanomaterials are lyophilized or provided as dry powders. These materials require reconstitution prior to use. While saline and PBS are common physiologically relevant solvents, some nanomaterials may require a specialty buffer for reconstitution. Class 2 biological safety cabinets are recommended for all work with nanomaterials.

5. The critical aspect of this test is to assess nanoparticles at noncytotoxic concentrations, i.e., those not affecting cell viability. Therefore, there is no standard concentration for this test. Each nanomaterial must be evaluated for cytotoxicity before testing in this assay. The trypan blue exclusion test, MTT, 7-ADD, Cellometer, or other relevant cell viability assay can be used. Since the dilution factor for this assay is 5, the stock concentration of nanoparticles required for the test should be five times higher than this concentration. For example, if the cell viability test determined that the highest nontoxic to PBMC concentration of nanoparticles is 0.1 mg/mL, then you will need 400 μL of a stock with a concentration of 0.5 mg/mL to perform the test. Two hundred (200) μL of this stock will be used directly to treat the cells, and the rest of the material will be used to prepare three serial 1:5 dilutions of this stock. In this example, the final concentrations tested in the assay will be 0.1, 0.02, 0.004, and 0.0008 mg/mL. If a test nanomaterial is loaded with a cytotoxic drug, the concentrations should reflect the drug and the carrier concentrations. It is also essential to include a carrier control.
6. To maintain the Ficoll-blood interface, holding the tube at a 45° angle is helpful.
7. Four (4) mL of blood/PBS mixture usually results in ~2 mL of the mononuclear layer and requires at least 6 mL of HBSS for the wash step. Using 10 mL of HBSS for each 2 mL of cells is recommended.
8. Plan additional wells for isotype controls. It is recommended that each nanoparticle treatment concentration has its isotype control.
9. It is convenient to prepare an appropriate amount of working solutions of antibodies in a staining buffer and add 100 μL of antibodies containing the staining solution to the cells.
10. All staining procedures should be performed on ice, and the centrifuge temperature must be adjusted to 4 °C.
11. The procedure for instrument preparation described in this chapter is based on the fluorescence activated cell sorter (FACS) Calibur, the most common flow cytometer used in research labs. If another cytometer is used, the procedure specific to that instrument should be followed.

12. Use appropriate software to analyze the acquired data. CellQuest, FCS Express, or other similar software can be used.
13. Statistical analysis helps determine the significance of the difference between the negative control and study samples. Depending on the study design and the number of analyzed samples, Student's t-test and one-way Analysis of Variance (ANOVA) are the most used statistical methods. The Student's t-test compares two groups, whereas ANOVA is more appropriate for comparing three or more groups [18].

Acknowledgment

This project has been funded in whole or in part with federal funds from the National Cancer Institute, National Institutes of Health, under contract 75N91010D00024. The content of this publication does not necessarily reflect the views or policies of the Department of Health and Human Services, nor does the mention of trade names, commercial products, or organizations imply endorsement by the US Government.

References

1. Sahu A, Lambris JD (2001) Structure and biology of complement protein C3, a connecting link between innate and acquired immunity. Immunol Rev 180:35–48
2. Salvador-Morales C, Sim RB (2013) Complement activation. Front Nanobiomed Res 1: 357–384
3. Szebeni J, Alving CR, Rosivall L, Bunger R, Baranyi L, Bedocs P, Toth M, Barenholz Y (2007) Animal models of complement-mediated hypersensitivity reactions to liposomes and other lipid-based nanoparticles. J Liposome Res 17(2):107–117
4. Szebeni J (2014) Complement activation-related pseudoallergy: a stress reaction in blood triggered by nanomedicines and biologicals. Mol Immunol 61(2):163–173
5. Szebeni J (2018) Mechanism of nanoparticle-induced hypersensitivity in pigs: complement or not complement? Drug Discov Today 23(3):487–492
6. Szebeni J (2001) Complement activation-related pseudoallergy caused by liposomes, micellar carriers of intravenous drugs, and radiocontrast agents. Crit Rev Ther Drug Carrier Syst 18(6):567–606
7. Szebeni J (2005) Complement activation-related pseudoallergy: a new class of drug-induced acute immune toxicity. Toxicology 216(2–3):106–121
8. Szebeni J, Muggia F, Gabizon A, Barenholz Y (2011) Activation of complement by therapeutic liposomes and other lipid excipient-based therapeutic products: prediction and prevention. Adv Drug Deliv Rev 63(12):1020–1030
9. Fülöp T, Nemes R, Mészáros T, Urbanics R, Kok RJ, Jackman JA, Cho NJ, Storm G, Szebeni J (2018) Complement activation in vitro and reactogenicity of low-molecular weight dextran-coated SPIONs in the pig CARPA model: correlation with physicochemical features and clinical information. J Control Release 270:268–274
10. Kozma GT, Shimizu T, Ishida T, Szebeni J (2020) Anti-PEG antibodies: properties, formation, testing and role in adverse immune reactions to PEGylated nano-biopharmaceuticals. Adv Drug Deliv Rev 154–155:163–175
11. Unterweger H, Dézsi L, Matuszak J, Janko C, Poettler M, Jordan J, Bäuerle T, Szebeni J, Fey T, Boccaccini AR, Alexiou C, Cicha I (2018) Dextran-coated superparamagnetic iron oxide nanoparticles for magnetic resonance imaging: evaluation of size-dependent imaging properties, storage stability and safety. Int J Nanomedicine 13:1899–1915

12. Szebeni J, Simberg D, González-Fernández Á, Barenholz Y, Dobrovolskaia MA (2018) Roadmap and strategy for overcoming infusion reactions to nanomedicines. Nat Nanotechnol 13(12):1100–1108
13. Szebeni J, Storm G, Ljubimova JY, Castells M, Phillips EJ, Turjeman K, Barenholz Y, Crommelin DJA, Dobrovolskaia MA (2022) Applying lessons learned from nanomedicines to understand rare hypersensitivity reactions to mRNA-based SARS-CoV-2 vaccines. Nat Nanotechnol 17(4):337–346
14. West EE, Kemper C (2023) Complosome - the intracellular complement system. Nat Rev Nephrol 19(7):426–439
15. Dobrovolskaia MA (2022) Lessons learned from immunological characterization of nanomaterials at the Nanotechnology Characterization Laboratory. Front Immunol 13:984252
16. Ilinskaya AN, Shah A, Enciso AE, Chan KC, Kaczmarczyk JA, Blonder J, Simanek EE, Dobrovolskaia MA (2019) Nanoparticle physicochemical properties determine the activation of intracellular complement. Nanomedicine 17:266–275
17. Liszewski MK, Kolev M, Le Friec G, Leung M, Bertram PG, Fara AF, Subias M, Pickering MC, Drouet C, Meri S, Arstila TP, Pekkarinen PT, Ma M, Cope A, Reinheckel T, Rodriguez de Cordoba S, Afzali B, Atkinson JP, Kemper C (2013) Intracellular complement activation sustains T cell homeostasis and mediates effector differentiation. Immunity 39(6): 1143–1157
18. Mishra P, Singh U, Pandey CM, Mishra P, Pandey G (2019) Application of student's t-test, analysis of variance, and covariance. Ann Card Anaesth 22(4):407–411

Chapter 12

Analysis of Nanoparticles' Potential to Induce Autoimmunity

Barry W. Neun, Timothy M. Potter, Christina Robinson, Simone Difilippantonio, Elijah Edmondson, and Marina A. Dobrovolskaia

Abstract

Autoimmune responses are characterized by the presence of antibodies and lymphocytes specific to self or so-called autoantigens. Among such autoantigens is DNA; therefore, screening for antibodies recognizing single- and/or double-stranded DNA is commonly used to detect and classify autoimmune diseases. While autoimmunity affects both sexes, females are generally more affected than males, which is recapitulated in some animal models. A variety of factors, including genetic predisposition and the environment, contribute to the development of autoimmune disorders. Since certain drug products may also contribute to the development of autoimmunity, understanding a drug's potential to trigger an autoimmune response is of interest to immunotoxicology. However, models to study autoimmunity are limited, and it is generally agreed that no model can accurately predict autoimmunity in humans. Herein, we present an in vivo protocol utilizing the SJL/J mouse model to study nanoparticles' effects on the development of autoimmune responses. The protocol is adapted from the literature describing the use of this model to study chemically induced lupus.

Key words Nanoparticles, Autoimmunity, SJL/J mice, ds-DNA antibody

1 Introduction

Immunological tolerance or unresponsiveness to self-antigens is a fundamental property of the healthy immune system. When such control over responses against self-antigens fails, immune reactions against autologous antigens occur. Autoimmunity is defined as the host's immune response against self-antigens due to alterations in self-tolerance or regulation of lymphocytes. It is characterized by the presence of self-reactive lymphocytes and antibodies. Autoimmune disorders are chronic, progressive, and self-perpetuating (i.e., characterized by epitope spreading); they may be systemic or restricted to certain organs, depending on the expression of self-antigens recognized by self-reactive antibodies or lymphocytes.

Jeffrey D. Clogston et al. (eds.), *Characterization of Nanoparticles Intended for Drug Delivery*, Methods in Molecular Biology, vol. 2789, https://doi.org/10.1007/978-1-0716-3786-9_12,

Host genetic background and environment, including but not limited to infections and exposure to xenobiotics, contribute to the development of autoimmunity.

Autoimmune disorders are complex polygenic diseases in which multiple genes contribute, including genes encoding major histocompatibility complex (MHC). For example, individuals with human leukocyte antigen (HLA) DQB1*3, DRB1*15, DRB1*11 and DRB1*10 alleles are more frequently reported to have systemic lupus erythematosus (SLE) with skin, nephrological, hematological, and neurological manifestations, respectively, than individuals without these alleles [1]. The lupus-like disease can also develop in response to drug products; these diseases are commonly referred to as drug-induced lupus (DIL) or drug-induced autoimmunity (DIA) [2, 3]. Over 100 drugs have been linked to DIA, and this number is expected to increase in the future [2]; however, no standard approaches are currently available for testing DIA [4, 5]. Certain HLA types, e.g., HLA-DR4 and HLA-DR0301, and Complement C4 null allele were reported among factors that increase the risk of DIA [2].

SJL/J, also known as Swiss Jim Lambert, is an inbred strain of immunocompetent mice whose genetic background predisposes these animals to several diseases, including but not limited to experimental autoimmune encephalomyelitis (EAE) and systemic lupus erythematosus (SLE), which manifests consistently as lupus nephritis (https://www.jax.org/strain/000686). The SJL/J mouse model is commonly used to study spontaneous and chemically induced SLE. SJL/J mice develop kidney disease with immune complexes and histopathologic features similar to human lupus nephritis, including crescentic glomerulonephritis [6] (Fig. 1). Female SJL/L mice are more affected than males, with increased renal pathology, increased anti-dsDNA antibodies, and increased mortality [7–9], similar to SLE in humans in which the male-to-female SLE ratio is 1–9, and the disease severity is also greater in females [9].

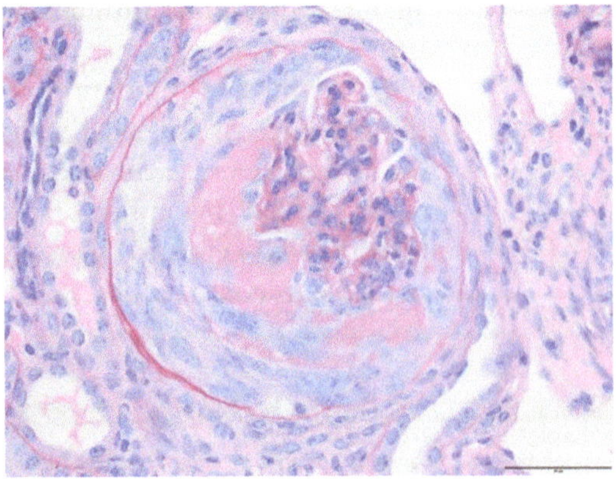

Fig. 1 Example H&E stained kidney section. This section demonstrates crescentic glomerulonephritis commonly observed in drug-induced lupus

Inducible SLE in the SJL/J mouse model develops by week 15, and about a third of the animals remain alive by week 35 [7–9]. The protocol presented herein was adapted from Smith et al. [9]. In this protocol, control or nanoparticle samples are administered intraperitoneally (i.p.) to SJL/J mice; the blood is collected before the injection and at five-week intervals up to weeks 21 or 35. At the end of the study, the animals are humanely euthanized, and histopathology analysis is performed on kidney sections stained with hematoxylin & eosin (H&E) or Periodic acid–Schiff (PAS) stain; the collected blood sera are analyzed for the presence of anti-dsDNA antibodies.

The data generated in this model are not used to predict autoimmunity in human patients. Rather, they serve as a model for evaluating environmental triggers of SLE and lupus nephritis to estimate the propensity of a test nanomaterial to cause autoimmune responses in a genetically susceptible host.

2 Materials

1. An 8-week-old female mouse of the SJL/J strain.
2. Phosphate-buffered saline (PBS) (negative control).
3. Pristane (positive control).
4. Vehicle (the same buffer used for nanoparticles).
5. Nanomaterials.
6. Syringe.
7. Blood collection tubes.
8. Mouse anti-dsDNA IgG detection ELISA kits.
9. Mouse anti-nuclear antibody detection ELISA kits.

3 Methods

3.1 Preparation of Study Samples and Controls

1. Use Pristane as a positive control (*see* **Notes 1** and **2**).
2. Prepare test nanomaterials by reconstituting the stock or dry powder in the vehicle at a concentration sufficient to deliver the desired dose (*see* **Notes 3–6**).

3.2 Experimental Procedure

1. On day 1 of the study, determine animals' body weight and randomize them into treatment groups with 10 females each (*see* **Notes** 7 and **8**).
2. Collect 100 μL of blood from each animal via the submandibular vein.
3. Allow the blood to clot, then spin in a centrifuge for 5 min at 10,000 × *g* to separate clots from serum.

4. Collect the serum and transfer it to fresh tubes; label the tubes as baseline (time 0) and store the collected sera at −20 °C.
5. Administer 0.5 mL of each control and test nanomaterials via i.p. injection (*see* **Note 8**).
6. Collect 100 μL of blood from each animal via the submandibular vein on weeks 8, 16, 21, 26, 31, and 35 (*see* **Note 9**); process the blood and store the sera as described in **steps 3** and **4** above.
7. All sera are analyzed by commercial ELISA kits for the presence of anti-dsDNA antibodies.
8. Collect and fix kidney tissues in 10% neutral buffered formalin for subsequent H&E and PAS staining and analysis [10] (*see* **Note 10**).

3.3 Data Analysis and Interpretation

1. For ELISA analysis, the percent coefficient of variation (%CV) should be less than 25%; it is calculated for each control or test sample analyzed by ELISA according to the following formula:

$$\%CV = \frac{\text{Standard deviation}}{\text{Mean}} \times 100\%$$

2. Renal histopathology is quantified using H&E and PAS stained 4 μm thick sections by scoring glomerular changes, renal tubule changes, and immune cell infiltrates according to the following grading criteria:

Glomerular changes: 0—normal; 1—minimally increased PAS-positive mesangial matrix; 2—mild mesangial expansion (mesangial matrix <2 nuclei diameter); 3—moderate mesangial expansion, mesangial matrix wide <4 nucleus diameter, crescentic glomeruli but infrequent; 4—severe mesangial expansion with frequent crescentic glomeruli.

Tubular changes: 0—within normal limits; 1—minimal tubular degeneration often focal; 2—mild tubular degeneration present with multifocal regions containing tubular degeneration, regeneration, and necrosis; 3—moderate tubular degeneration and necrosis with frequent proteinaceous casts; 4—marked tubular changes with frequent tubular necrosis.

Tubulointerstitial inflammation: 0—within normal limits; 1—minimally increased inflammatory infiltrates composed predominantly of lymphocytes and plasma cells often focally forming 3–5 cell thick perivascular cuffs; 2—mildly increased inflammatory infiltrates that are multifocal; 3—moderately increased inflammatory infiltrates that form thick perivascular cuffs multifocally that are prominent with the 4× objective; 4—markedly increased inflammatory

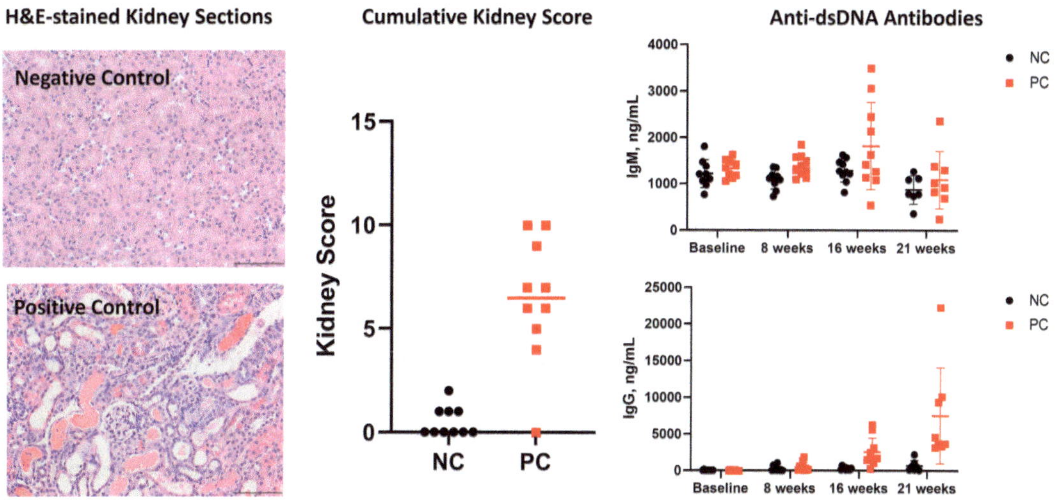

Fig. 2 Expected results. The study was conducted according to the protocol described in this chapter. Examples of H&E stained kidney sections, cumulative kidney score, and anti-dsDNA antibodies are shown. Each dot represents a mean response obtained from one animal

infiltrates that widely separate vessels from adjacent renal parenchyma.

3. The response is considered positive when the test result is twofold or more higher than the baseline (negative control).
4. The test is considered valid when the negative control is negative, and the positive control is positive in all tests, including kidney score and anti-dsDNA antibody titers (Fig. 2).

4 Notes

1. Work with the positive control and engineered nanomaterials requires personal protective equipment (gloves, safety goggles, lab coat) and should be performed in a biological safety cabinet.
2. Pristane is provided by the manufacturer as a stock solution that is sterile filtered and intended for animal use. You will need 0.5 mL of the stock per animal.
3. Weighing dry nanomaterials should be performed in a biological safety cabinet.
4. A common vehicle used is saline, but other buffers, e.g., PBS, can be used if appropriate for the given nanoparticle.
5. Some nanomaterials require specific formulation buffers which may contain sucrose, dextrose, or other excipients. In this case, a vehicle control should be prepared and tested in addition to the negative control.

6. The dose of nanomaterial should match the intended clinical dose. For dose conversion between species and estimation of the human equivalent dose, consult the FDA Guidance for Industry [5]. Additional doses, including the maximum tolerated dose, when known, and lower doses should be considered.
7. The route and frequency of nanomaterial administration should match the clinically intended route and frequency. This protocol uses a single injection via the i.p. route. Should other routes (e.g., intravenous [i.v.]) or frequencies be used, other, more relevant positive controls should be considered because pristane is the optimal control for a single injection via the i.p. route.
8. Conducting the in vivo experiment described in this protocol requires approval from the institutional Animal Care and Use Committee (ACUC). Changes in dose level, frequency and/or route of administration, or controls require ACUC approval before study initiation. Each nanomaterial requires ACUC review and approval prior to use in animals.
9. By week 21, the animals in the positive control group will develop lupus symptoms and may need to be removed from the study. The study for other test groups may be continued up to week 35. Daily observation of animals by a veterinarian is recommended to ensure proper care and termination of animals experiencing discomfort.
10. Pathology analysis is normally performed at the end of the study (week 21 or 35) to validate and quantify renal glomerular lesions; however, some animals may be removed from the study earlier per the veterinarian's discretion.

Acknowledgments

This project has been funded in whole or in part by federal funds from the National Cancer Institute, National Institutes of Health, under contract 75N91019D00024. The content of this publication does not necessarily reflect the views or policies of the Department of Health and Human Services, nor does the mention of trade names, commercial products, or organizations imply endorsement by the US Government. All animal procedures have been approved by the NCI-Frederick ACUC. NCI-Frederick is accredited by Association for Assessment and Accreditation of Laboratory Animal Care (AAALAC) International and follows the Public Health Service Policy for the Care and Use of Laboratory Animals. Animal care was provided in accordance with the procedures outlined in the "Guide for Care and Use of Laboratory Animals" (National Research Council; 1996; National Academy Press; Washington, DC).

References

1. Wadi W, Elhefny NE, Mahgoub EH, Almogren A, Hamam KD, Al-Hamed HA, Gasim GI (2014) Relation between HLA typing and clinical presentations in systemic lupus erythematosus patients in Al-Qassim region. Saudi Arabia Int J Health Sci (Qassim) 8(2): 159–165
2. Solhjoo M, Goyal A, Chauhan K (2023) Drug-induced lupus erythematosus. In: StatPearls. StatPearls Publishing LLC., Treasure Island. StatPearls Publishing Copyright © 2023
3. Dedeoglu F (2009) Drug-induced autoimmunity. Curr Opin Rheumatol 21(5):547–551
4. USFDA (2006) Guidance for industry: S8 immunotoxicity studies for human pharmaceuticals. USFDA, Silver Spring
5. USFDA (2023) Guidance for industry: non-clinical evaluation of the immunotoxic potential of pharmaceuticals. USFDA, Silver Spring
6. Yeung CK, Wong KL, Wong WS, Ng MT, Chan KW, Ng WL (1984) Crescentic lupus glomerulonephritis. Clin Nephrol 21(5): 251–258
7. Perry D, Sang A, Yin Y, Zheng YY, Morel L (2011) Murine models of systemic lupus erythematosus. J Biomed Biotechnol 2011: 271694
8. Satoh M, Richards HB, Shaheen VM, Yoshida H, Shaw M, Naim JO, Wooley PH, Reeves WH (2000) Widespread susceptibility among inbred mouse strains to the induction of lupus autoantibodies by pristane. Clin Exp Immunol 121(2):399–405
9. Smith DL, Dong X, Du S, Oh M, Singh RR, Voskuhl RR (2007) A female preponderance for chemically induced lupus in SJL/J mice. Clin Immunol 122(1):101–107
10. Gregorios JHB, Faldas MM (2016) Histopathologic techniques. Katha Publishing Company, Incorporated, Makati

Chapter 13

Analysis of Nanoparticles' Effects on Drug-Induced Psoriasis

Ankit Shah, Chelsea Sanders, Simone Difilippantonio, Elijah Edmondson, and Marina A. Dobrovolskaia

Abstract

Psoriasis, an auto-inflammatory disorder, has major manifestations in the skin but can affect other organs. Currently, this condition has no cure, and the treatments include anti-inflammatory medications. Nanoparticles are widely used for drug delivery and have found successful applications in therapy for cancer and infectious diseases. Nanoparticles can also be used to deliver anti-inflammatory drugs to sites of inflammation. Moreover, some nanotechnology platforms possess intrinsic anti-inflammatory properties and may benefit the therapy of inflammation-driven disorders. Herein, we present a protocol to study nanotechnology concepts' anti-inflammatory properties in a chemically-induced psoriasis model.

Key words Nanoparticles, Auto-inflammatory reactions, Skin, Hypersensitivity, Inflammation, Cytokines, Psoriasis

1 Introduction

Psoriasis is a chronic auto-inflammatory disease. The most common manifestation of psoriasis is in the skin, where rapid proliferation of keratinocytes in the epidermis leads to the thickening of the skin and formation of red, scaled patches; however, nails and joints may also be affected. Genetic background, environmental triggers (e.g., infections, xenobiotics, and trauma), and social factors (e.g., emotional stress) contribute to the development of psoriasis.

Certain medications (e.g., imiquimod, interferons, hydroxychloroquine, beta-blockers, and monoclonal antibody targeting PD-1 protein) may trigger the development of this disease or exaggerate the pre-existing condition [1]. The underlying mechanisms of drug-induced or drug-exaggerated psoriasis are incompletely understood. Both direct and indirect mechanisms have been described. For example, imiquimod, a small molecule antiviral compound, stimulates the innate immunity pattern

Jeffrey D. Clogston et al. (eds.), *Characterization of Nanoparticles Intended for Drug Delivery*, Methods in Molecular Biology, vol. 2789, https://doi.org/10.1007/978-1-0716-3786-9_13,

recognition receptor toll-like receptor 7 (TLR7), activating the NF-kB pathway and producing pro-inflammatory cytokines (e.g., TNF, IL-8, IL-1β, and IL-6) [2]. While this response is beneficial in treating certain viral diseases, it may lead to undesirable skin inflammation when it occurs in the skin. An indirect mechanism involving the so-called Koebner phenomenon was described for metoprolol and doxycycline [3, 4]. In this mechanism, the drug-mediated hypersensitivity or photoallergic reactions lead to skin cell damage with the subsequent mounting of the inflammatory response.

Currently, there is no cure for psoriasis, and the symptoms are managed by anti-inflammatory medications to reduce inflammation. Drugs with broad anti-inflammatory and immunosuppressive properties, e.g., corticosteroids, methotrexate, and cyclosporin, are commonly used to treat psoriasis. However, medications specifically targeting T helper 17 (Th17) cells have also been investigated. For example, the FDA recently approved a monoclonal antibody neutralizing cytokine IL-17 produced by Th17 cells; this antibody, secukinumab, has shown efficacy in patients with severe psoriasis [5].

Nanotechnology is considered for drug delivery, vaccines, and diagnostic agents [6–9]. Among other indications, nanoparticles are investigated for topical drug delivery to alleviate psoriatic skin lesions with success in both preclinical and clinical studies [10]. Some nanoparticles possess intrinsic anti-inflammatory properties, which may benefit psoriasis therapy. For example, an iron oxide nanoformulation, Feraheme, used in the clinic to treat iron deficiency anemia in patients with chronic kidney disease was shown to directly inhibit T cell activation and proliferation and suppress IL-17 induction in response to pro-inflammatory stimuli [11]. When applied topically to the imiquimod-inflamed skin, Feraheme decreased the psoriatic score, similar to the hydrocortisone cream, suggesting the potential benefits of these particles for psoriasis therapy and warranting optimization of the formulation for topical applications [12].

The protocol described herein is intended to study the anti-inflammatory properties of nanomaterials in a mouse model of imiquimod-induced psoriasis. Two test modes—prophylactic and therapeutic—are available (Fig. 1). The prophylactic mode is intended to study nanomaterials' ability to prevent the development of psoriasis, whereas the therapeutic mode assesses nanomaterials' effects on established psoriatic skin lesions. The data generated in this model are used to estimate the propensity of a test nanomaterial to prevent the development or reduce the severity of psoriatic skin lesions.

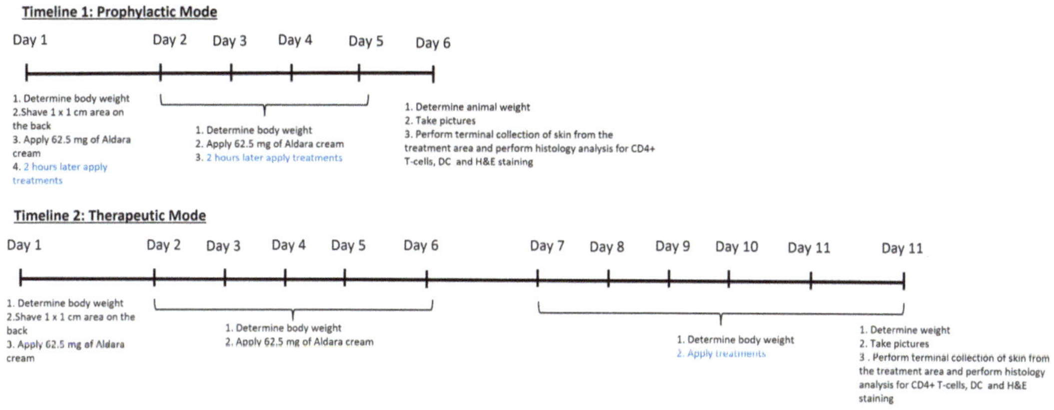

Fig. 1 Study timelines. Two timelines are possible for the protocol described in this chapter. Timeline 1—prophylactic mode—involves the administration of test materials concurrently with the Aldara cream and is used to assess the sample's potential to reduce the development of inflammatory skin lesions. Timeline 2—therapeutic mode—involves the administration of Aldara cream for 1 week to establish inflammation, then applying test materials to assess their ability to reduce the established inflammation

2 Materials

1. An 8–10-week-old male and female mouse of the A/J strain.
2. Aldara cream (*see* **Note 1**).
3. Hydrocortisone cream (positive control).
4. Vaseline or baby oil (negative control).
5. Vehicle (the same buffer used for nanoparticles).
6. Nanomaterials.

3 Methods

3.1 Preparation of Study Samples and Controls

1. Prepare test nanomaterials by reconstituting the stock or dry powder in a vehicle at a concentration sufficient to deliver the desired dose (*see* **Notes 2–6**).
2. Prepare controls (*see* **Note 7**).

3.2 Experimental Procedure

3.2.1 Prophylactic Mode

1. On day 1 of the study, determine animals' body weight and randomize them into treatment groups with ten animals each (five males and five females) (Fig. 1; *see* **Notes 8** and **9**).
2. Shave a 1 × 1 cm area on the back and remove remaining fur using Nair.
3. House animals individually with E-collars.
4. Apply 62.5 mg of Aldara cream and spread evenly across the shaved area using a spatula.

5. Two hours later, apply positive control, vehicle control, or nanoparticle treatments.
6. Repeat **steps 4** and **5** once a day on days 2–6 (*see* **Note 10**).
7. Collect skin from the treatment area and perform histology analysis for CD4+ T cells, dendritic cells (DC), and hematoxylin and eosin (H&E) staining [13].

3.2.2 Therapeutic Mode

1. On day 1 of the study, determine animals' body weight and randomize them into groups with ten animals each (five males and five females) (Fig. 1; *see* **Notes 8** and **9**).
2. Shave a 1 × 1 cm area on the back and remove remaining fur using Nair.
3. House animals individually with E-collars.
4. Apply 62.5 mg of Aldara cream and spread evenly across the shaved area using a spatula.
5. Repeat **step 4** once a day on days 2–6 (*see* **Note 10**).
6. Apply positive control, vehicle control or test nanomaterials on day 7 (*see* **Note 10**).
7. Repeat **step 6** once a day on days 8–12 (*see* **Note 10**).
8. Collect skin from the treatment area and perform histology analysis for CD4+ T cells, DC, and H&E staining [13].

3.3 Data Analysis and Interpretation

1. Dermatitis, hyperkeratosis, acanthosis, and erythema are scored independently on a scale from 0 to 4, where 0 = none, 1 = minimal, 2 = mild, 3 = moderate, and 4 = marked.
2. The data is reported as cumulative psoriatic score calculated as the sum of scores described above. The cumulative score serves as a measure of the severity of inflammation (scale 0–12).
3. The response in the treatment group is compared to that in the negative control (Fig. 2; *see* **Notes 11** and **12**).

4 Notes

1. Aldara cream contains imiquimod, a prescription medication used to treat HPV-mediated genital warts. Hydrocortisone cream is available over the counter and does not require a prescription.
2. Work with Aldara and hydrocortisone creams and engineered nanomaterials requires personal protective equipment (gloves, safety goggles, lab coat) and should be performed in a biological safety cabinet.
3. Weighing dry nanomaterials should be performed in a biological safety cabinet.

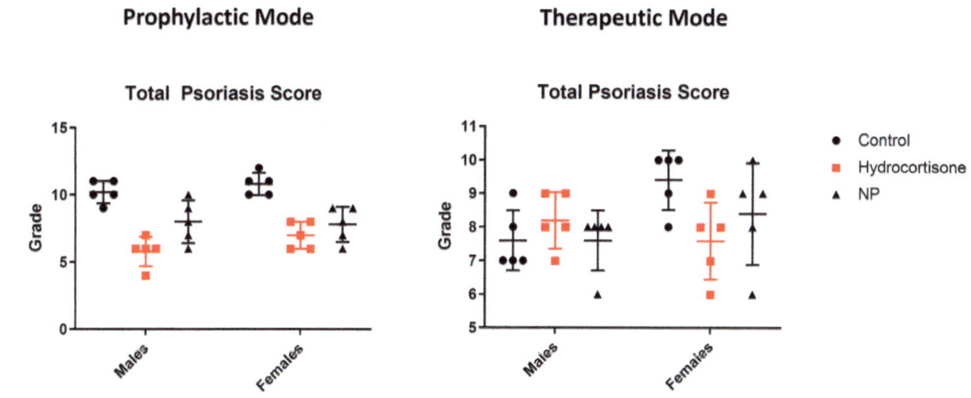

Fig. 2 Potential outcomes. The results of a study conducted according to the protocol described in this chapter are shown. Two timelines were used. The results shown in the prophylactic mode have been adapted from an earlier study by our laboratory [12]

4. The typical vehicle is saline, but other buffers can be used if appropriate for the given nanoparticle. Phosphate-buffered saline (PBS) can be used for topical application; however, baby oil or Vaseline allows for better retention of the treatment on the skin.
5. Some nanomaterials require specific formulation buffers, which may contain various excipients. In this case, a vehicle control should be prepared and tested in addition to the negative control.
6. The dose of nanomaterial should match the intended clinical dose. For dose conversion between species and estimation of the human equivalent dose, consult the FDA Guidance for Industry [14]. Additional doses, including the maximum tolerated dose, when known, and lower doses should be considered.
7. If using Aldara or hydrocortisone, no special preparation is needed. If using different controls, follow the manufacturer's instructions.
8. The route and frequency of nanomaterial administration should match the clinically intended route and frequency. This protocol uses a topical route. Should other routes (e.g., i.v.) or frequencies be used, other, more relevant positive controls should be considered.
9. Conducting the in vivo experiment described in this protocol requires approval from the Institutional Animal Care and Use Committee (ACUC). Changes in dose level, frequency and/or route of administration, or controls require ACUC approval before the study initiation. Each nanomaterial requires ACUC review and approval prior to use in animals.

10. It is recommended to determine animals' body weight every time before applying controls and test nanomaterials.
11. The statistical significance of the observed effect is determined using an appropriate statistical method. Student's t-test for comparison between two groups and one-way analysis of variance (ANOVA) for multiple comparisons are commonly used methods [15].
12. A physiologically significant difference is one showing two or more fold differences in responses between the control and a test group.

Acknowledgment

This project has been funded in whole or in part by federal funds from the National Cancer Institute, National Institutes of Health, under contract 75N91019D00024. The content of this publication does not necessarily reflect the views or policies of the Department of Health and Human Services, nor does mention of trade names, commercial products, or organizations imply endorsement by the US Government. All animal procedures have been approved by the NCI-Frederick ACUC. NCI-Frederick is accredited by Association for Assessment and Accreditation of Laboratory Animal Care (AAALAC) International and follows the Public Health Service Policy for the Care and Use of Laboratory Animals. Animal care was provided in accordance with the procedures outlined in the "Guide for Care and Use of Laboratory Animals" (National Research Council; 1996; National Academy Press; Washington, D.C.).

References

1. Balak DM, Hajdarbegovic E (2017) Drug-induced psoriasis: clinical perspectives. Psoriasis (Auckl) 7:87–94
2. Li ZJ, Sohn KC, Choi DK, Shi G, Hong D, Lee HE, Whang KU, Lee YH, Im M, Lee Y, Seo YJ, Kim CD, Lee JH (2013) Roles of TLR7 in activation of NF-κB signaling of keratinocytes by imiquimod. PLoS One 8(10):e77159
3. Doyon JB, Liu KJ, Berman RA (2017) Metoprolol-induced total body erythroderma. J Gen Intern Med 32(2):221–222
4. Kuznetsov AV, Weisenseel P, Flaig MJ, Ruzicka T, Prinz JC (2011) Photoallergic erythroderma due to doxycycline therapy of erythema chronicum migrans. Acta Derm Venereol 91(6):734–736
5. Krueger JG, Wharton KA Jr, Schlitt T, Suprun M, Torene RI, Jiang X, Wang CQ, Fuentes-Duculan J, Hartmann N, Peters T, Koroleva I, Hillenbrand R, Letzkus M, Yu X, Li Y, Glueck A, Hasselberg A, Flannery B, Suárez-Fariñas M, Hueber W (2019) IL-17A inhibition by secukinumab induces early clinical, histopathologic, and molecular resolution of psoriasis. J Allergy Clin Immunol 144(3):750–763
6. Alexander S, Moghadam MG, Rothenbroker M, Chou LYT (2023) Addressing the in vivo delivery of nucleic-acid nanostructure therapeutics. Adv Drug Deliv Rev 199:114898
7. Chopra H, Mohanta YK, Rauta PR, Ahmed R, Mahanta S, Mishra PK, Panda P, Rabaan AA, Alshehri AA, Othman B, Alshahrani MA, Alqahtani AS, Al Basha BA, Dhama K (2023) An insight into advances in developing

nanotechnology based therapeutics, drug delivery, diagnostics and vaccines: multidimensional applications in tuberculosis disease management. Pharmaceuticals (Basel) 16(4):581

8. Kim SJ, Puranik N, Yadav D, Jin JO, Lee PCW (2023) Lipid nanocarrier-based drug delivery systems: therapeutic advances in the treatment of lung cancer. Int J Nanomedicine 18:2659–2676
9. Puri S, Mazza M, Roy G, England RM, Zhou L, Nourian S, Anand Subramony J (2023) Evolution of nanomedicine formulations for targeted delivery and controlled release. Adv Drug Deliv Rev 200:114962
10. Dadwal N, Kurmi BD, Singh D, Singh A (2023) Recent advancements in novel formulations of anti-psoriatic agents for effective delivery: clinical importance and patent survey. Recent Pat Nanotechnol. https://doi.org/10.2174/1872210517666230601124620. Epub ahead of print. PMID: 37264655
11. Shah A, Mankus CI, Vermilya AM, Soheilian F, Clogston JD, Dobrovolskaia MA (2018) Feraheme® suppresses immune function of human T lymphocytes through mitochondrial damage and mitoROS production. Toxicol Appl Pharmacol 350:52–63
12. Shah A, Cedrone E, Sanders CL, Butcher D, DeFrancesco AM, DeGrange CL, Difilippantonio S, Edmondson EF, Dobrovolskaia MA (2019) The potential utility of iron oxide nanoparticles for the treatment of skin inflammation in a mouse model of psoriasis. Precis Nanomed 2(1):249–255
13. Chen W, Bamford RN, Edmondson EF, Waldmann TA (2022) IL15 and anti-PD-1 augment the efficacy of agonistic intratumoral anti-CD40 in a mouse model with multiple TRAMP-C2 tumors. Clin Cancer Res 28(10):2082–2093
14. USFDA (2023) Guidance for industry: non-clinical evaluation of the immunotoxic potential of pharmaceuticals
15. Mishra P, Singh U, Pandey CM, Mishra P, Pandey G (2019) Application of student's t-test, analysis of variance, and covariance. Ann Card Anaesth 22(4):407–411

Chapter 14

Detection of Nanoparticle-Mediated Total Oxidative Stress in T Cells Using CM-H_2DCFDA Dye

Ankit Shah and Marina A. Dobrovolskaia

Abstract

Oxidative stress is commonly observed in cells following exposure to nanoparticles. Both negative (e.g., cytotoxicity and inflammation) and beneficial (e.g., anti-inflammatory and tumor growth inhibiting) responses have been linked in the literature to oxidative stress, emphasizing the importance of developing methodologies to study this phenomenon in cells following their exposure to nanoparticles. In the protocol described herein, primary human T cells isolated from the peripheral blood of healthy donor volunteers are treated with nanoparticles and controls, and the generation of reactive oxygen species is detected by flow cytometry using CM-H_2DCFDA reagent.

Key words Nanoparticles, T cells, Oxidative stress, Flow cytometry

1 Introduction

The generation of reactive oxygen species (ROS) and its control by antioxidant systems are a part of normal cellular metabolism [1]. However, disturbance of redox homeostasis and accumulation of ROS damage cellular organelles and DNA, eventually leading to cell death [2]. Another consequence of excessive ROS generation is the activation of stress-related pathways and the release of pro-inflammatory cytokines [1]. Induction of oxidative stress is a common toxicity associated with nanoparticles. For example, endocrine-disrupting effects of zinc oxide nanoparticles in mice were attributed to oxidative stress [3]; suppression of T cell activation by iron oxide nanoparticles coincided with mitochondrial ROS generation [4]; and lung injury and inflammation in response to quartz and single-wall carbon nanotubes were associated with ROS generation [5, 6]. Other studies linking nanoparticle toxicity to oxidative stress have also been conducted and reviewed in detail elsewhere [7]. Some researchers investigated the oxidative stress–mediated cytotoxicity of nanoparticles as a tool to eliminate cancer

Jeffrey D. Clogston et al. (eds.), *Characterization of Nanoparticles Intended for Drug Delivery*, Methods in Molecular Biology, vol. 2789, https://doi.org/10.1007/978-1-0716-3786-9_14,

cells (reviewed in [8–10]). Other beneficial properties of nanoparticle-mediated oxidative stress have been described in the context of psoriasis therapy [11]. The commonality of oxidative stress as the mechanism underlying both toxic and beneficial effects of nanomaterials emphasizes the importance of methodologies used to study nanoparticles.

The purpose of this protocol is to understand the propensity of nanomaterials to induce oxidative stress in human T lymphocytes. T cells are isolated from human blood anti-coagulated with Li-heparin using RosetteSep™ reagent, followed by gradient separation using Ficoll-Paque Plus solution. The blood is obtained from three healthy donor volunteers. Specimens from each donor are processed separately. Isolated T cells are incubated with test nanoparticles, and the generation of ROS is detected using CM-H_2DCFDA reagent. CM-H_2DCFDA reagent is a cell-permeable fluorogenic probe that is non-fluorescent when in a resting state but emits a strong fluorescent signal in fluorescein isothiocyanate (FITC) fluorescence upon oxidation. Per the researcher's discretion, cell lines or peripheral blood mononuclear cells (PBMC) can be used instead of primary T cells.

2 Materials

1. Phosphate-buffered saline (PBS).
2. Ficoll-Paque Premium.
3. Fetal bovine serum (FBS).
4. Roswell park memorial institute (RPMI) 1640.
5. Pen/Strep solution.
6. L-glutamine.
7. Live and dead cell stain.
8. Bovine serum albumin (BSA).
9. CM-H_2DCFDA reagent.
10. Hank's balanced salt solution (HBSS).
11. Dimethyl sulfoxide (DMSO).
12. RosetteSep™.
13. Tert-butyl hydroperoxide (TBHP).
14. N-Acetyl-L-(+)-cysteine (NAC).
15. Biohazard safety cabinet approved for level II handling of biological materials.
16. Flow cytometer.
17. Centrifuge with speed 1200 × *g*.

3 Methods

3.1 Preparation of Study Samples and Controls

1. Use PBS as a negative control.
2. Prepare a vehicle control (*see* **Note 1**).
3. Prepare a 50 mM intermediate dilution of TBHP by adding 3.22 μL of the 70% stock (7.78 M) to 496.8 μL of complete RPMI. Add 4 μL of the intermediate TBHP dilution to appropriate wells and adjust the sample volume to 1 mL; use this solution as a positive control.
4. Prepare a 250 mM interim stock of the inhibition control by dissolving 40.8 mg NAC in 1 mL DMSO. Use 16 μL of the prepared stock solution of NAC in the appropriate wells containing 800 μL cells/media (1:50 dilution) (*see* **Note 2**).
5. Prepare a 100 μM intermediate solution of CM-H_2DCFDA by adding 100 μL of DMSO to CM-H_2DCFDA reagent. Next, add 5 μL of the intermediate CM-H_2DCFDA dilution to 1 mL of cell suspension. Alternatively, use 300 μL of medium with 5 μM CM-H_2DCFDA per sample.
6. Prepare test nanomaterials (*see* **Note 3**).

3.2 Preparation of T Cells

1. Collect whole blood from three healthy donor volunteers into Li-heparin blood collection tubes; discard the first 10 cc (*see* **Note 4**).
2. Place freshly drawn blood into 15 or 50 mL conical centrifuge tubes, add 50 μL RosetteSep per mL of blood, mix well, and incubate the tubes at room temperature for 20 min.
3. Add an equal amount of room temperature PBS with 2% FBS to the blood and mix well.
4. Slowly layer the Ficoll-Paque solution underneath the blood/PBS mixture by placing the tip of the pipet containing Ficoll-Paque at the bottom of the blood sample tube. Alternatively, the blood/PBS mixture may be slowly layered over the Ficoll-Paque solution (*see* **Notes 5** and **6**).
5. Centrifuge for 20 min at 1200 × *g*, 18–20 °C, without brake (*see* **Note 7**).
6. Using a sterile pipet, remove the upper layer containing plasma and platelets and discard it.
7. Using a fresh sterile pipet, transfer the T lymphocyte cell layer into another centrifuge tube.
8. Wash cells by adding an excess of HBSS and centrifuging for 10 min at 400 × *g*, 18–20 °C. The HBSS volume should be approximately three times the volume of the mononuclear layer (*see* **Note 8**).

9. Discard the supernatant and repeat the wash step one more time.
10. Resuspend cells in complete RPMI 1640 medium and determine cell viability using live–dead cell staining (*see* **Notes 9** and **10**).

3.3 Experimental Procedure

1. Adjust cell count to 1×10^6 cells/mL.
2. Aliquot 500 μL of cell suspension per well of 48-well plates (Fig. 1).
3. Pretreat "Inhibition control" wells with NAC for 1 h at 37 °C/5% CO_2 before treatment with nanoparticles. Do not wash cells after incubation.
4. Add 100 μL of test nanoparticles and control to appropriate wells and incubate for the appropriate length of the assay at 37 °C/5% CO_2.
5. Add 100 μL of complete media to "unlabeled cells" wells (*see* **Note 11**).
6. Wash cells once with 3 mL of PBS.
7. Add the CM-H_2DCFDA reagent to all samples except "unlabeled cells" and incubate for 30 min at 37 °C/5% CO_2.
8. Wash cells once with 3 mL of PBS.
9. Reconstitute cells at 1×10^6 cells/mL with PBS and analyze samples by flow cytometry (Fig. 2).

3.4 Data Analysis and Acceptance Criteria

1. Report the data as the geometric mean (GeoMean) of treated samples/GeoMean (NC). Alternatively, percent positive cells can be reported if GeoMean is not statistically different (*see* **Note 12**).
2. The run is acceptable if GeoMean (PC)/GeoMean (NC) $\geq$ 1.5.
3. The run is acceptable if GeoMean (PC + NAC)/GeoMean (PC) $\geq$ 1.

4 Notes

1. Vehicle control is the buffer or media used to formulate test nanomaterials. Common excipients used are trehalose, sucrose, and albumin. However, other reagents and materials are also used alone or in combination. The vehicle control should match the formulation buffer of the test nanomaterial by both composition and concentration. This control can be skipped if nanoparticles are stored in PBS.
2. The final working concentration of NAC in test samples is 5 mM.

	1	2	3	4	5	6
A	NC	VC	PC	PC + NAC	TS (0.008 mg/mL)	TS + NAC (0.008 mg/mL)
B	TS (0.04 mg/mL)	TS + NAC (0.04 mg/mL)	TS (0.2 mg/mL)	TS + NAC (0.2 mg/mL)	TS (2 mg/mL)	TS + NAC (2 mg/mL)
C	Unlabeled Cells					
D						

Fig. 1 Example of a plate map. Sample positioning on the plate is at the researcher's discretion. One example is shown in this fig. NC = negative control; VC = vehicle control; PC = positive control; NAC = N-acetyl-L-cysteine; TS = test sample; shaded cells = inhibition control

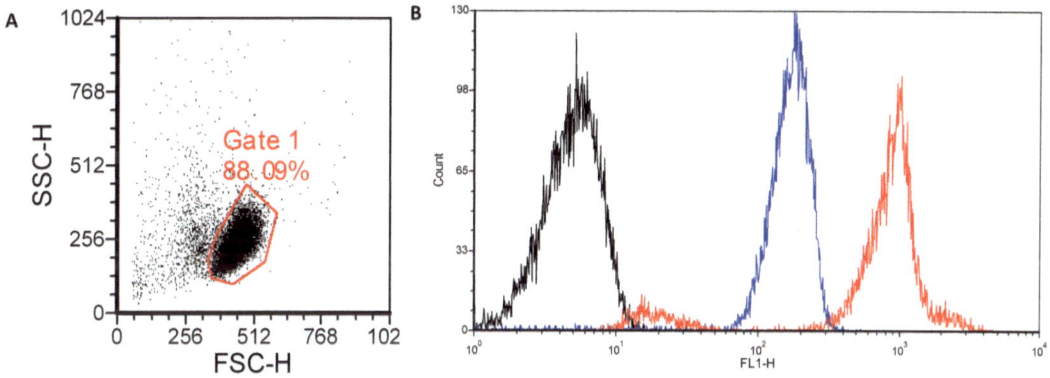

Fig. 2 Detection of total oxidative stress in lymphocytes. Gating strategy (**a**) and an example histogram showing cell count in the green fluorescent channel (**b**). Black line = untreated cells; blue line = negative control; red line = positive control

3. When the experiment is done in 24-well plates, the assay requires 1.5 mL of nanoparticles dissolved/resuspended in a complete culture medium at a concentration five times higher than the highest final test concentration. The concentration is selected based on the plasma concentration of the nanoparticle at the intended therapeutic dose. Considerations for estimating theoretical plasma concentration were reviewed elsewhere [12].
4. The collection and handling of human blood require approval from the institutional biosafety committee. For the best results, whole blood should be used within 1 h after collection.

Prolonged storage (>2 h) of whole blood will lead to a decrease in cell function.

5. Use 1 mL of Ficoll-Paque solution per 2 mL of blood/PBS mixture. For example, 15 mL Ficoll-Paque is required for 15 mL of the blood diluted with 15 mL PBS containing 2% FBS.
6. To maintain the Ficoll–blood interface, it is helpful to hold the tube at a 45° angle.
7. Depending on the type of centrifuge, one also may need to set the acceleration speed to a minimum.
8. Usually, 20 mL of blood/PBS mixture results in ~1 mL of the T cell layer and requires at least 3 mL of HBSS for the wash step. We use 5 mL of HBSS per each 1 mL of cells.
9. Viability of 90% or above is required prior to proceeding to the next experimental step.
10. A similar assay can be applied to the PBMC.
11. To inhibit the effects of the positive control, it is essential to perform PC treatment immediately after NC treatment for an additional 30 min. More prolonged incubation after NC treatment is ineffective when tested for inhibition on PC.
12. Use appropriate software to analyze the data acquired using a flow cytometer. CellQuest, FCS Express, or other flow cytometry software can be used.

Acknowledgment

This project has been funded in whole or in part by federal funds from the National Cancer Institute, National Institutes of Health, under contract 75N91019D00024. The content of this publication does not necessarily reflect the views or policies of the Department of Health and Human Services, nor does the mention of trade names, commercial products, or organizations imply endorsement by the US Government.

References

1. Finkel T, Holbrook NJ (2000) Oxidants, oxidative stress and the biology of ageing. Nature 408(6809):239–247
2. Kohen R, Nyska A (2002) Oxidation of biological systems: oxidative stress phenomena, antioxidants, redox reactions, and methods for their quantification. Toxicol Pathol 30(6): 620–650
3. Santacruz-Márquez R, Flaws JA, Sánchez-Peña LDC, Hernández-Ochoa I (2023) Exposure to zinc oxide nanoparticles increases estradiol levels and induces an antioxidant response in antral ovarian follicles in vitro. Toxics 11(7): 602
4. Shah A, Mankus CI, Vermilya AM, Soheilian F, Clogston JD, Dobrovolskaia MA (2018) Feraheme® suppresses immune function of human T lymphocytes through mitochondrial damage and mitoROS production. Toxicol Appl Pharmacol 350:52–63

5. Shvedova AA, Kisin E, Murray AR, Johnson VJ, Gorelik O, Arepalli S, Hubbs AF, Mercer RR, Keohavong P, Sussman N, Jin J, Yin J, Stone S, Chen BT, Deye G, Maynard A, Castranova V, Baron PA, Kagan VE (2008) Inhalation vs. aspiration of single-walled carbon nanotubes in C57BL/6 mice: inflammation, fibrosis, oxidative stress, and mutagenesis. Am J Physiol Lung Cell Mol Physiol 295(4): L552–L565
6. Vallyathan V, Castranova V, Pack D, Leonard S, Shumaker J, Hubbs AF, Shoemaker DA, Ramsey DM, Pretty JR, McLaurin JL et al (1995) Freshly fractured quartz inhalation leads to enhanced lung injury and inflammation. Potential role of free radicals. Am J Respir Crit Care Med 152(3):1003–1009
7. Horie M, Tabei Y (2021) Role of oxidative stress in nanoparticles toxicity. Free Radic Res 55(4):331–342
8. Li Y, Yang J, Sun X (2021) Reactive oxygen species-based nanomaterials for cancer therapy. Front Chem 9:650587
9. Zhao N, Xin H, Zhang L (2021) Advanced biomedical applications of reactive oxygen species-based nanomaterials in lung cancer. Front Chem 9:649772
10. Zheng X, Jin Y, Liu X, Liu T, Wang W, Yu H (2021) Photoactivatable nanogenerators of reactive species for cancer therapy. Bioact Mater 6(12):4301–4318
11. Shah A, Cedrone E, Sanders CL, Butcher D, DeFrancesco AM, DeGrange CL, Difilippantonio S, Edmondson EF, Dobrovolskaia MA (2019) The potential utility of iron oxide nanoparticles for the treatment of skin inflammation in a mouse model of psoriasis. Precis Nanomed 2(1):249–255
12. Dobrovolskaia MA, McNeil SE (2013) Understanding the correlation between in vitro and in vivo immunotoxicity tests for nanomedicines. J Control Release 172(2):456–466

Chapter 15

Detection of Induction of Mitochondrial Oxidative Stress by Nanoparticles in T Cells Using MitoSOX Red Dye

Ankit Shah and Marina A. Dobrovolskaia

Abstract

The induction of oxidative stress by engineered nanomaterials has been associated with cytotoxic and inflammatory responses, damaging healthy cells and tissues. In contrast, when directed against cancer and autoinflammatory diseases, some nanomaterials inducing oxidative stress have also been reported as potential therapies for these disorders. Therefore, studying oxidative stress has become a popular tool not only in toxicology and immunotoxicology but in other areas of biology as well, including those related to developing novel therapies. Total oxidative stress may result from multiple cellular organelles. The protocol described herein allows for the analysis of oxidative stress in mitochondria.

Key words Nanoparticles, Oxidative stress, Mitochondria, T cells

1 Introduction

Accumulation of reactive oxygen species (ROS) in cells resulting from an altered balance between ROS production and detoxification leads to oxidative stress. It is a powerful mechanism underlying stress-related cell death and inflammation [1, 2]. Various nanoparticles have been shown to induce total oxidative stress with subsequent inflammatory and cytotoxic effects in healthy cells and tissues [3–5]. In contrast, the induction of oxidative stress by nanomaterials was investigated as a tool to eliminate cancer cells (reviewed in [6]) or reduce inflammatory responses in psoriasis [7]. Total oxidative stress involves multiple cellular organelles, including mitochondria, lysosomes, endoplasmic reticulum, and peroxisomes. Various methods and best practices for analyzing oxidative stress have been described [8–10]. One such protocol for studying total oxidative stress is available in Chapter 14 of this book.

The protocol described herein allows for determining the involvement of mitochondria in oxidative stress. T lymphocytes

Jeffrey D. Clogston et al. (eds.), *Characterization of Nanoparticles Intended for Drug Delivery*, Methods in Molecular Biology, vol. 2789, https://doi.org/10.1007/978-1-0716-3786-9_15,

isolated from Li-heparin anticoagulated human blood are used as a model. Isolated T cells are incubated with test nanoparticles, and the generation of mitochondrial ROS is detected using the MitoSOX reagent. MitoSOX reagent is a cell-permeable fluorogenic probe that is non-fluorescent when in resting cells but emits a strong fluorescent signal detectable in the phycoerythrin (PE) fluorescent channel of a flow cytometer upon oxidation. Moreover, the dye localizes explicitly in the mitochondria and is, therefore, suitable for mitochondria-specific oxidative stress detection. A recent study of iron oxide nanoparticles demonstrated the induction of mitochondrial oxidative stress in the absence of detectable total oxidative stress [11], thereby emphasizing the importance of using multiple approaches. If desired, this protocol can be adapted for the analysis of mitochondrial oxidative stress in other primary cells and cell lines.

2 Materials

1. Phosphate-buffered saline (PBS).
2. Ficoll-Paque Premium.
3. Fetal bovine serum (FBS).
4. Roswell park memorial institute (RPMI) 1640.
5. Pen/Strep solution.
6. L-glutamine.
7. Live and dead cell stain.
8. Bovine serum albumin (BSA).
9. MitoSOX Red reagent.
10. Hank's balanced salt solution (HBSS).
11. Dimethyl sulfoxide (DMSO).
12. RosetteSep™.
13. Tert-butyl hydroperoxide (TBHP).
14. N-Acetyl-L-(+)-cysteine (NAC).
15. Cell culture incubator with 5% CO_2 and 95% humidity.
16. Biohazard safety cabinet approved for level II handling of biological materials.
17. Flow cytometer.
18. Centrifuge with a speed of 1200 × *g*.

3 Methods

3.1 Preparation of Study Samples and Controls

1. Use PBS as a negative control.
2. Prepare the vehicle control (*see* **Note 1**).
3. Prepare a 50 mM intermediate dilution of TBHP by adding 3.22 μL of the 70% stock (7.78 M) to 496.8 μL of complete RPMI. Add 4 μL of the intermediate TBHP dilution to appropriate wells and adjust the sample volume to 1 mL; use this solution as a positive control.
4. Prepare a 250 mM interim stock of the inhibition control by dissolving 40.8 mg NAC in 1 mL DMSO. Use 16 μL of the prepared stock solution of NAC in the appropriate wells containing 800 μL cells/media (1:50 dilution) (*see* **Note 2**).
5. Prepare an intermediate 5 mM MitoSOX Red reagent solution by adding 13 μL of DMSO to the MitoSOX reagent. Add 1 μL of the intermediate MitoSOX dilution to 1 mL of cell suspension. Alternatively, use 300 μL of medium with 5 μM MitoSOX per sample.
6. Prepare test nanomaterials (*see* **Note 3**).

3.2 Preparation of T Cells

1. Collect whole blood from three healthy donor volunteers into Li-heparin blood collection tubes; discard the first 10 cc (*see* **Note 4**).
2. Place freshly drawn blood into 15 or 50 mL conical centrifuge tubes, add 50 μL of RosetteSep per mL of blood, mix well, and incubate the tubes at room temperature for 20 min.
3. Add an equal amount of room temperature PBS with 2% FBS to the blood and mix well.
4. Slowly layer the Ficoll-Paque solution underneath the blood/PBS mixture by placing the tip of the pipet containing Ficoll-Paque at the bottom of the blood sample tube. Alternatively, the blood/PBS mixture may be slowly layered over the Ficoll-Paque solution (*see* **Notes 5** and **6**).
5. Centrifuge for 20 min at 1200 × *g*, 18–20 °C, without a brake (*see* **Note 7**).
6. Using a sterile pipet, remove the upper layer containing plasma and platelets and discard it.
7. Using a fresh sterile pipet, transfer the T lymphocyte cell layer into another centrifuge tube.
8. Wash cells by adding an excess of HBSS and centrifuging for 10 min at 400 × *g*, 18–20 °C. The HBSS volume should be approximately three times the volume of the mononuclear layer (*see* **Note 8**).

9. Discard the supernatant and repeat the wash step once more.
10. Resuspend cells in complete RPMI 1640 medium and determine cell viability using live–dead cell staining (*see* **Notes 9** and **10**).

3.3 Experimental Procedure

1. Adjust the cell count to 1×10^6 cells/mL.
2. Aliquot 500 μL of cell suspension per well of 48-well plates (Fig. 1).
3. Pretreat "Inhibition control" wells with NAC for 1 h at 37 °C/5% CO_2 before treatment with nanoparticles. Do not wash cells after incubation.
4. Add 100 μL of test nanoparticles and control to appropriate wells and incubate for the appropriate length of the assay at 37 °C/5% CO_2.
5. Add 100 μL of complete media to "unlabeled cells" wells (*see* **Note 11**).
6. Wash cells once with 3 mL of PBS.
7. Add the MitoSOX Red reagent to all samples except "unlabeled cells" and incubate for 30 min at 37 °C/5% CO_2.
8. Wash cells once with 3 mL of PBS.
9. Reconstitute cells at 1×10^6 cells/mL with PBS and analyze samples by flow cytometry (Fig. 2).

3.4 Data Interpretation

1. Report the data as the geometric mean (GeoMean) of treated samples/GeoMean (NC). Alternatively, percent positive cells can be reported if GeoMean is not statistically different (*see* **Note 12**).
2. The run is acceptable if GeoMean (PC)/GeoMean (NC) is ≥ 1.5.
3. The run is acceptable if GeoMean (PC + NAC)/GeoMean (PC) is <1.

4 Notes

1. Vehicle control is the buffer or media used to formulate test nanomaterials. Common excipients used are trehalose, sucrose, and albumin. However, other reagents and materials are also used alone or in combination. The vehicle control should match the formulation buffer of the test nanomaterial in both composition and concentration. This control can be skipped if nanoparticles are stored in PBS.
2. The final working concentration of NAC in test samples is 5 mM.

	1	2	3	4	5	6
A	NC	VC	PC	PC + NAC	TS (0.008 mg/mL)	TS + NAC (0.008 mg/mL)
B	TS (0.04 mg/mL)	TS + NAC (0.04 mg/mL)	TS (0.2 mg/mL)	TS + NAC (0.2 mg/mL)	TS (2 mg/mL)	TS + NAC (2 mg/mL)
C	Unlabeled Cells					
D						

Fig. 1 Example of a plate map. Sample positioning on the plate is at the researcher's discretion. One example is shown in this fig. NC = negative control; VC = vehicle control; PC = positive control; NAC = N-acetyl-L-cysteine; TS = test sample; shaded cells = inhibition control

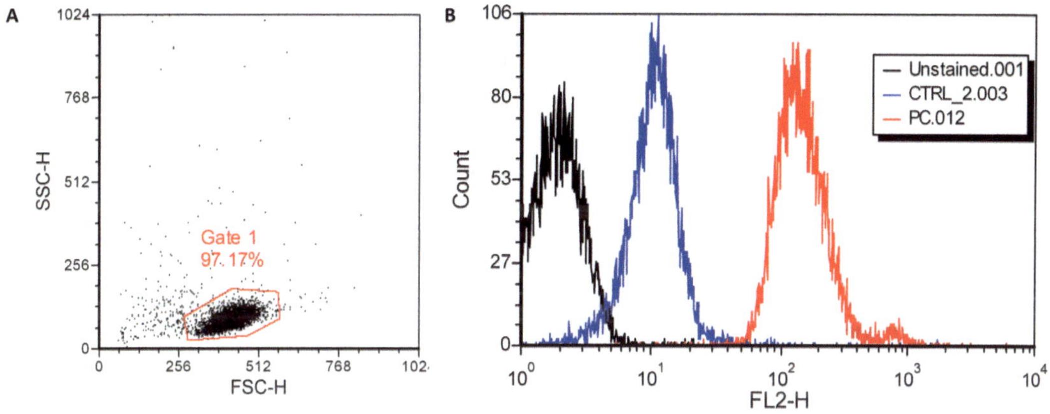

Fig. 2 Detection of mitochondrial oxidative stress in lymphocytes. Gating strategy (**a**) and an example histogram showing cell count in the red fluorescent channel (**b**). Black line = untreated cells; blue line = negative control; red line = positive control

3. When the experiment is done in 24-well plates, the assay requires 1.5 mL of nanoparticles dissolved/resuspended in a complete culture medium at a concentration five times higher than the highest final test concentration. The concentration is selected based on the plasma concentration of the nanoparticle at the intended therapeutic dose. Considerations for estimating theoretical plasma concentration were reviewed elsewhere [12].
4. The collection and handling of human blood require approval from the institutional biosafety committee. For the best results, whole blood should be used within 1 h after collection.

Prolonged storage (>2 h) of whole blood will lead to a decrease in cell function.

5. Use 1 mL of Ficoll-Paque solution per 2 mL of blood/PBS mixture. For example, 15 mL Ficoll-Paque is required for 15 mL of the blood diluted with 15 mL PBS containing 2% FBS.
6. To maintain the Ficoll–blood interface, it is helpful to hold the tube at a 45° angle.
7. Depending on the type of centrifuge, one may also need to set the acceleration speed to a minimum.
8. Usually, 20 mL of blood/PBS mixture results in ~1 mL of the T cell layer and requires at least 3 mL of HBSS for the wash step. We use 5 mL of HBSS per each 1 mL of cells.
9. Viability of 90% or above is required prior to proceeding to the next experimental step.
10. A similar assay can be applied to peripheral blood mononuclear cells (PBMC).
11. To inhibit the effects of the positive control, it is essential to perform PC treatment immediately after NC treatment for an additional 30 min. More prolonged incubation after NC treatment is ineffective when tested for inhibition on PC.
12. Use appropriate software to analyze the data acquired using a flow cytometer. CellQuest, FCS Express, or other flow cytometry software can be used.

Acknowledgment

This project has been funded in whole or in part by federal funds from the National Cancer Institute, National Institutes of Health, under contract 75N91019D00024. The content of this publication does not necessarily reflect the views or policies of the Department of Health and Human Services, nor does the mention of trade names, commercial products, or organizations imply endorsement by the US Government.

References

1. Kohen R, Nyska A (2002) Oxidation of biological systems: oxidative stress phenomena, antioxidants, redox reactions, and methods for their quantification. Toxicol Pathol 30(6): 620–650
2. Finkel T, Holbrook NJ (2000) Oxidants, oxidative stress and the biology of ageing. Nature 408(6809):239–247
3. Horie M, Tabei Y (2021) Role of oxidative stress in nanoparticles toxicity. Free Radic Res 55(4):331–342
4. Shvedova AA, Kisin E, Murray AR, Johnson VJ, Gorelik O, Arepalli S, Hubbs AF, Mercer RR, Keohavong P, Sussman N, Jin J, Yin J, Stone S, Chen BT, Deye G, Maynard A, Castranova V, Baron PA, Kagan VE (2008) Inhalation vs. aspiration of single-walled

carbon nanotubes in C57BL/6 mice: inflammation, fibrosis, oxidative stress, and mutagenesis. Am J Physiol Lung Cell Mol Physiol 295(4):L552–L565

5. Vallyathan V, Castranova V, Pack D, Leonard S, Shumaker J, Hubbs AF, Shoemaker DA, Ramsey DM, Pretty JR, McLaurin JL et al (1995) Freshly fractured quartz inhalation leads to enhanced lung injury and inflammation. Potential role of free radicals. Am J Respir Crit Care Med 152(3):1003–1009
6. Li Y, Yang J, Sun X (2021) Reactive oxygen species-based nanomaterials for cancer therapy. Front Chem 9:650587
7. Shah A, Cedrone E, Sanders CL, Butcher D, DeFrancesco AM, DeGrange CL, Difilippantonio S, Edmondson EF, Dobrovolskaia MA (2019) The potential utility of iron oxide nanoparticles for the treatment of skin inflammation in a mouse model of psoriasis. Precis Nanomed 2(1):249–255
8. Gaggini M, Sabatino L, Vassalle C (2020) Conventional and innovative methods to assess oxidative stress biomarkers in the clinical cardiovascular setting. Biotechniques 68(4): 223–231
9. Katerji M, Filippova M, Duerksen-Hughes P (2019) Approaches and methods to measure oxidative stress in clinical samples: research applications in the cancer field. Oxidative Med Cell Longev 2019:1279250
10. Murphy MP, Bayir H, Belousov V, Chang CJ, Davies KJA, Davies MJ, Dick TP, Finkel T, Forman HJ, Janssen-Heininger Y, Gems D, Kagan VE, Kalyanaraman B, Larsson NG, Milne GL, Nyström T, Poulsen HE, Radi R, Van Remmen H, Schumacker PT, Thornalley PJ, Toyokuni S, Winterbourn CC, Yin H, Halliwell B (2022) Guidelines for measuring reactive oxygen species and oxidative damage in cells and in vivo. Nat Metab 4(6):651–662
11. Shah A, Mankus CI, Vermilya AM, Soheilian F, Clogston JD, Dobrovolskaia MA (2018) Feraheme® suppresses immune function of human T lymphocytes through mitochondrial damage and mitoROS production. Toxicol Appl Pharmacol 350:52–63
12. Dobrovolskaia MA, McNeil SE (2013) Understanding the correlation between in vitro and in vivo immunotoxicity tests for nanomedicines. J Control Release 172(2):456–466

Chapter 16

Detection of Nanoparticle-Mediated Change in Mitochondrial Membrane Potential in T Cells Using JC-1 Dye

Ankit Shah and Marina A. Dobrovolskaia

Abstract

Alterations in mitochondrial membrane potential are associated with the generation of reactive oxygen species and cell death. While eliminating cancer cells is beneficial for cancer therapy, cytotoxicity to healthy cells may limit the therapeutic applications of mitochondria-damaging nanoparticles. Due to the critical role mitochondria play in cell viability and function, it is important to detect such alterations when studying nanomaterials for therapeutic applications. The protocol described herein utilizes JC-1 dye to detect nanoparticle-mediated changes in mitochondrial membrane potential and is intended to support mechanistic immunotoxicology studies.

Key words Nanoparticles, T cells, Oxidative stress, Mitochondria, Membrane potential

1 Introduction

Alteration in mitochondrial physiology is commonly linked to apoptotic cell death. Under stressed conditions, the mitochondrial permeability transition pores open, leading to a decrease in mitochondrial membrane potential (ΔΦm) [1]. The change in ΔΦm further activates various signaling cascades, leading to cytochrome C release, ATP depletion, and the accumulation of reactive oxygen species (ROS) [1, 2]. The combination of mitochondrial membrane potential loss and mitochondrial ROS leads to organelle-specific toxicity that ultimately causes cell death. This mechanism can be used in combating cancer, but it can also lead to undesirable cytotoxicity. For example, HA-MoO_2-Dox nanoconjugates and gadolinium oxide nanoparticles were effective in reducing the number of breast cancer cells via destabilization of mitochondria and changes in the mitochondrial membrane potential [3, 4]; likewise, silver nanoparticles and cadmium telluride quantum dots were cytotoxic to cervical and hepatocellular carcinoma cells,

Jeffrey D. Clogston et al. (eds.), *Characterization of Nanoparticles Intended for Drug Delivery*, Methods in Molecular Biology, vol. 2789, https://doi.org/10.1007/978-1-0716-3786-9_16,

respectively, via the mechanism involving mitochondrial ROS and membrane destabilization [5, 6]. In contrast, iron oxide nanoparticles were toxic and suppressed the activation of healthy human T lymphocytes by inducing mitochondrial ROS and changes in mitochondrial membrane potential [7]; similarly, the mitochondrial membrane alteration mechanism was reported to underline the toxicity of graphene oxide nanoparticles to normal rat embryonic fibroblasts and human epithelial breast cells [8]. These examples demonstrate the need for methodologies to study nanoparticles' effects on cells in health and disease to understand the therapeutically beneficial and toxic effects of exposure to these novel materials.

This protocol aims at providing experimental tools to help understand nanomaterials' propensity to induce mitochondrial damage as assessed by changes in the mitochondrial membrane potential. T lymphocytes are isolated from healthy human blood anti-coagulated with Li-heparin using RosetteSep™ reagent from STEMCELL Technologies, followed by gradient separation using Ficoll-Paque Plus solution. Blood is obtained from three healthy donor volunteers. Specimens from each donor are processed separately. Isolated T cells are incubated with test nanoparticles, and the changes in mitochondrial membrane potential ($\Delta\Phi m$) are detected by flow cytometry using a cell-permeable fluorogenic probe JC-1, which fluoresces in the red channel (e.g., FL-2) when present in aggregates, but is detected in the green channel (e.g., FL-1) when present in monomer (Fig. 1).

2 Materials

1. Phosphate-buffered saline (PBS).
2. Ficoll-Paque Premium.
3. Fetal bovine serum (FBS).
4. Roswell Park Memorial Institute (RPMI) 1640.
5. Pen/Strep solution.
6. L-glutamine.
7. Reagent allowing to distinguish between live and dead cells.
8. Bovine serum albumin (BSA).
9. MitoProbe™ JC-1 Assay kit (Molecular Probes, M34152).
10. Hank's balanced salt solution (HBSS).
11. Dimethyl sulfoxide (DMSO).
12. RosetteSep™ (STEMCELL Technologies, 15061).
13. Tert-butyl hydroperoxide (TBHP).
14. N-Acetyl-L-(+)-cysteine (NAC).

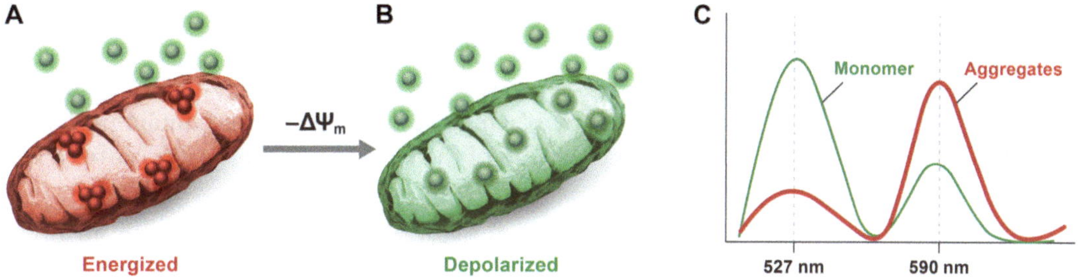

Fig. 1 Principle of JC-1-based assay. (**a**) Healthy mitochondria accumulate JC-1 inside mitochondrial cristae. (**b**) Damaged mitochondria accumulate JC-1 monomers outside mitochondria. (**c**) The transition of JC-1 dye from aggregates to monomer shows different fluorescence in the red and green fluorescent channels (*see* **Note 16**)

15. Cell culture incubator with 5% CO_2 and 95% humidity.
16. Biohazard safety cabinet approved for level II handling of biological material.
17. Instrument or hemocytometer to perform cell count.
18. Flow cytometer with red and green fluorescent channels.
19. Centrifuge with a speed of 1200 × *g*.

3 Methods

3.1 Preparation of Study Samples and Controls

1. Prepare the positive control (carbonyl cyanide m-chlorophenyl hydrazone (CCCP) [provided in the JC-1 Assay kit]) by adding 1 μL of 50 mM CCCP stock to appropriate wells during the staining protocol and incubating the sample at 37 °C for 5 min.
2. Use PBS as a negative control.
3. Prepare a 200 μM intermediate stock of JC-1 reagent by adding 230 μL of DMSO to the vial provided with the kit; next, add 10 μL of the intermediate JC-1 dilution to 1 mL of cell suspension. Alternatively, use 300 μL of medium with 2 μM JC-1 per sample and incubate for 15–30 min at 37 °C.
4. Prepare test nanomaterials (*see* **Notes 1** and **2**).
5. Prepare the vehicle control (*see* **Note 3**).

3.2 Collection and Handling of Whole Blood for Culture

1. Collect whole blood from three healthy donor volunteers who have not been on any medication and are clear from infection for at least 2 weeks before blood donation. Use Li-heparin tubes and discard the first 10 cc (*see* **Note 4**).

3.3 Isolation of T Cells

1. Place freshly drawn blood into 15 or 50 mL conical centrifuge tubes, add 50 μL of RosetteSep per each mL of blood, mix well, and incubate the tubes at room temperature for 20 min.

2. Add an equal amount of room temperature PBS with 2% FBS to the blood and mix well.
3. Slowly layer the Ficoll-Paque solution underneath the blood/PBS mixture by placing the tip of the pipet containing Ficoll-Paque at the bottom of the blood sample tube (*see* **Notes 5** and **6**).
4. Centrifuge for 20 min at 1200 × *g*, 18–20 °C, without brake (*see* **Note 7**).
5. Using a sterile pipet, remove the upper layer containing plasma and platelets and discard it.
6. Using a fresh sterile pipet, transfer the T lymphocyte cell layer into another centrifuge tube.
7. Wash cells by adding an excess of HBSS and centrifuging for 10 min at 400 × *g*, 18–20 °C. The HBSS volume should be approximately three times the volume of the mononuclear layer (*see* **Note 8**).
8. Discard the supernatant and repeat the wash step once more.
9. Resuspend cells in complete RPMI 1640 medium and determine cell viability (*see* **Notes 9** and **10**).

3.4 Experimental Procedure

1. Adjust the cell count to 1×10^6 cells/mL.
2. Aliquot 400 μL of cell suspension prepared in the previous step per well of a 48-well plate.
3. Add 100 μL of test nanoparticles and controls to appropriate wells and incubate for an appropriate length of the assay at 37 °C/5% CO_2. Add 100 μL of complete media to "unlabeled cells" wells (*see* **Note 11**).
4. Wash cells once with 3 mL of PBS.
5. Add the JC-1 reagent to all samples except "unlabeled cells" and incubate for 30 min at 37 °C/5% CO_2.
6. Wash cells once with 3 mL of PBS.
7. Reconstitute cells in 300 μL PBS and analyze samples by flow cytometry (Fig. 2 and *see* **Note 12**).

3.5 Data Interpretation

1. Use appropriate software to analyze the data acquired by a flow cytometer (*see* **Note 13**).
2. Report the data as the percentage of positive cells in the green fluorescent channel, indicating the cells with damaged mitochondria (*see* **Notes 14** and **15**).
3. The run is acceptable if PC shows more than 80% of cells in the green channel and NC shows more than 70% in the red channel (*see* **Note 14**).

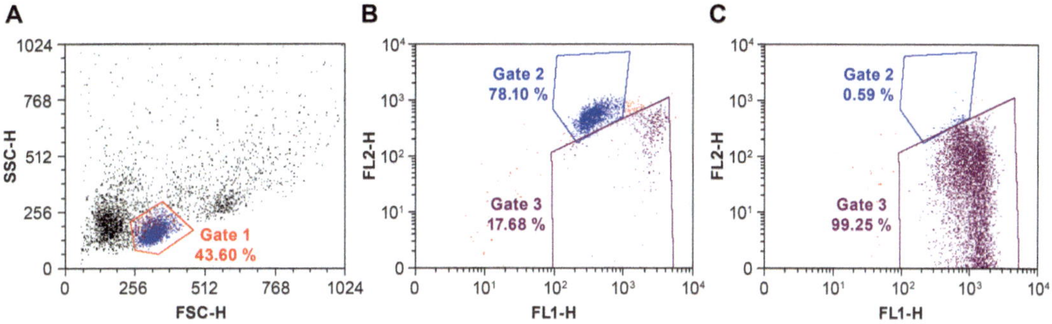

Fig. 2 Example data generated in this assay. (**a**) Gating of lymphocytes containing JC-1 aggregates (gate 2) and JC-1 monomer (gate 3) using negative control (NC) and positive control (PC) samples, respectively, in the red (FL2) vs. green (FL1) channels; the data are shown in a dot plot format. (**b**) Untreated control sample stained with JC-1 reagent. (**c**) CCCP-treated sample stained with JC-1 reagent; the blue color highlights the NC, and the purple color highlights the PC. SSC-side scatter; FSC-forward scatter; FL-fluorescent channel

4 Notes

1. All work involving nanomaterials requires personal protective equipment (lab coat, safety goggles, and gloves) and should be performed in a biological safety cabinet.
2. When the experiment is done in 24-well plates, the assay requires 1.5 mL of nanoparticles dissolved/resuspended in complete culture medium at a concentration five times higher than the highest final test concentration. The concentration is selected based on the plasma concentration of the nanoparticle at the intended therapeutic dose. For the purpose of this protocol, this concentration is called the "theoretical plasma concentration." Considerations for estimating theoretical plasma concentration were reviewed elsewhere [9].
3. Vehicle control is the buffer or media used to formulate test nanomaterials. Common excipients used are trehalose, sucrose, and albumin. However, other reagents and materials are also used alone or in combination. The vehicle control should match the formulation buffer of the test nanomaterial by both composition and concentration. This control can be skipped if nanoparticles are stored in PBS.
4. For best results, whole blood should be used within 1 h after collection. Prolonged storage (>2 h) of whole blood will lead to a decrease in cell function.
5. Alternatively, the blood/PBS mixture may be slowly layered over the Ficoll-Paque solution. Use 1 mL of Ficoll-Paque solution per 2 mL of blood/PBS mixture. For example, 15 mL Ficoll-Paque is required for 15 mL of blood diluted with 15 mL PBS with 2% FBS in a 50 mL tube.

6. To maintain the Ficoll–blood interface, it is helpful to hold the tube at a 45° angle.
7. Using a brake would result in vibration and disturb the separation of cells. Depending on the type of centrifuge, one may also need to set acceleration speed to minimum.
8. Usually, 20 mL of blood/PBS mixture results in ~1 mL of the T cell layer and requires at least 3 mL of HBSS for the wash step; using 5 mL of HBSS per each 1 mL of cells is preferable.
9. If the viability is at least 90%, proceed to the experiment.
10. The similar assay can be applied to PBMC.
11. For the positive control inhibition, it is essential to perform PC treatment immediately after NC treatment for an additional 30 min. Longer incubation after NC treatment is ineffective when tested for inhibition of PC.
12. It is critical to acquire samples on a flow cytometer within 15 min after staining, as longer incubation may result in the release of JC-1 aggregates from healthy mitochondria and may fail acceptance criteria for NC staining.
13. Cell Quest and FCS Express are commonly used software. Other software can also be used to analyze the data acquired on a flow cytometer.
14. FL-1 channel is commonly used for the detection of green fluorescence, and FL-2 is used for red fluorescence; however, depending on the model and make of a flow cytometer, the channel number may be different. It is, therefore, recommended to consult the operator's manual of a cytometer available in a given research laboratory.
15. Alternatively, the ratio of the mean fluorescent intensity (MFI) in red/green channels can be used as an assessment of healthy/damaged mitochondria in each treatment group.
16. This figure is available in a purple and orange pseudo-color scheme via request to the corresponding author.

Acknowledgment

This project has been funded in whole or in part by federal funds from the National Cancer Institute, National Institutes of Health, under contract 75N91019D00024. The content of this publication does not necessarily reflect the views or policies of the Department of Health and Human Services, nor does the mention of trade names, commercial products, or organizations imply endorsement by the US Government.

References

1. Hirsch T, Marzo I, Kroemer G (1997) Role of the mitochondrial permeability transition pore in apoptosis. Biosci Rep 17(1):67–76
2. Rao VK, Carlson EA, Yan SS (2014) Mitochondrial permeability transition pore is a potential drug target for neurodegeneration. Biochim Biophys Acta 1842(8):1267–1272
3. Xia HY, Li BY, Kankala RK, Chen AZ, Wang SB (2023) Hybrid nanoarchitectonics of molybdenum dioxide (MoO(2)) and doxorubicin (DOX) for synergistic chemo-photothermal-based breast carcinoma therapy. Colloids Surf B Biointerfaces 227:113387
4. Akhtar MJ, Ahamed M, Alhadlaq H, Alrokayan S (2019) Toxicity mechanism of gadolinium oxide nanoparticles and gadolinium ions in human breast cancer cells. Curr Drug Metab 20(11):907–917
5. Katubi KM, Alzahrani FM, Ali D, Alarifi S (2019) Dose- and duration-dependent cytotoxicity and genotoxicity in human hepato carcinoma cells due to CdTe QDs exposure. Hum Exp Toxicol 38(8):914–926
6. Nadhe SB, Tawre MS, Agrawal S, Chopade BA, Sarkar D, Pardesi K (2020) Anticancer potential of AgNPs synthesized using Acinetobacter sp. and Curcuma aromatica against HeLa cell lines: a comparative study. J Trace Elem Med Biol 62:126630
7. Shah A, Mankus CI, Vermilya AM, Soheilian F, Clogston JD, Dobrovolskaia MA (2018) Feraheme® suppresses immune function of human T lymphocytes through mitochondrial damage and mitoROS production. Toxicol Appl Pharmacol 350:52–63
8. Imarah AA, Jabir MS, Abood AH, Sulaiman GM, Albukhaty S, Mohammed HA, Khan RA, Al-Kuraishy HM, Al-Gareeb AI, Al-Azzawi WK, Najm MAA, Jawad SF (2023) Graphene oxide-induced, reactive oxygen species-mediated mitochondrial dysfunctions and apoptosis: high-dose toxicity in normal cells. Nanomedicine (Lond) 18(11):875–887
9. Dobrovolskaia MA, McNeil SE (2013) Understanding the correlation between in vitro and in vivo immunotoxicity tests for nanomedicines. J Control Release 172(2):456–466

Chapter 17

Detection of Antigen Presentation by Murine Bone Marrow–Derived Dendritic Cells After Treatment with Nanoparticles

Enping Hong and Marina A. Dobrovolskaia

Abstract

Nanoparticles are frequently considered in vaccine applications due to their ability to co-deliver multiple antigens and adjuvants to antigen-presenting cells. Some nanoparticles also have intrinsic adjuvant properties that further enhance their ability to stimulate immune cells. The delivery of tumor-specific antigens to antigen-presenting cells (APCs) with subsequent antigenic peptide presentation in the context of class I major histocompatibility complex (MHC-I) molecules represents an essential effort in developing nanotechnology-based cancer vaccines. Experimental models are, therefore, needed to gauge the efficiency of nanotechnology carriers in achieving peptide antigen delivery to APCs and presentation in the context of MHC-I. The assay described herein utilizes a model antigen ovalbumin and model APCs, murine bone marrow–derived dendritic cells. The 25-D1.16 antibody, specific to the ovalbumin (OVA) MHC-I peptide SIINFEKL, recognizes this peptide presented in the context of the murine H2-K^b class I MHC molecule, allowing the presentation of this antigen on APCs to be detected by flow cytometry after nanoparticle delivery.

Key words Nanoparticles, Vaccines, Dendritic cells, Antigen presentation

1 Introduction

Nanoparticles have been extensively tested as vehicles for the delivery of vaccines and immunotherapies and as adjuvants for the treatment of cancer [1–5]. An essential effort in the development of therapeutic cancer vaccines is the delivery of antigens to antigen-presenting cells (APCs) such that tumor-specific peptides are presented in the context of class I major histocompatibility complex (MHC) molecules. Such delivery and presentation promote the formation of cytotoxic T lymphocytes to attack and eliminate cancer cells; the strategies and challenges in this area have been discussed extensively in the literature [6–8].

The purpose of this protocol is to determine the ability of nanoparticles to deliver antigens to APCs, such that an antigenic class I peptide sequence is adequately presented in the context of

Jeffrey D. Clogston et al. (eds.), *Characterization of Nanoparticles Intended for Drug Delivery*, Methods in Molecular Biology, vol. 2789, https://doi.org/10.1007/978-1-0716-3786-9_17,

MHC. For the purposes of this assay, the model antigen ovalbumin (OVA) is used in the context of murine cells. The protocol requires the synthesis of nanoparticles that deliver SIINFEKL ($OVA_{257\text{-}264}$), the immunodominant class I peptide derived from the OVA protein, either in its peptide form or as part of a larger molecule (such as the whole OVA protein). The 25-D1.16 antibody, which binds to SIINFEKL presented in the context of the mouse H2-K^b MHC molecule [9], is used to detect antigen presentation on APCs after nanoparticle delivery. The results of this protocol may be used to infer the ability of the nanoparticle system to deliver other similar antigens.

Bone marrow–derived dendritic cells (BMDCs) are used for the antigen presentation assay. Bone marrow cells are collected from 8–12-week-old C57BL6 mice and cultured in a medium supplemented with murine granulocyte–macrophage-colony-stimulating factor (GM-CSF) for differentiation into BMDCs. After 7 days of culture, differentiated BMDCs are treated with nanoparticles and relevant controls for 24 h. After incubation, cells are then stained with 25-D1.16 or isotype control and analyzed for fluorescence by flow cytometry. Although cell lines can be used instead of primary BMDCs, care must be taken to select cell lines that express the H2-K^b MHC molecule since SIINFEKL is an H2-K^b-restricted antigen. The DC2.4 cell line, which is derived from C57BL6 mice and is capable of presenting OVA-derived antigens [10], is one example of an appropriate cell line.

2 Materials

1. Phosphate-buffered saline (PBS), Ca^{2+}/Mg^{2+} free.
2. Cell culture–grade water, endotoxin-free.
3. Fetal bovine serum (FBS).
4. Roswell park memorial institute (RPMI) 1640.
5. Pen/Strep solution.
6. L-glutamine.
7. Any reagent allowing to distinguish between live and dead cells.
8. Bovine serum albumin (BSA).
9. Formaldehyde, 20%.
10. Recombinant murine granulocyte-macrophage colony stimulating factor (GM-CSF).
11. SIINFEKL positive control: $OVA_{257\text{-}264}$ class I peptide.

12. 25-D1.16 phycoerythrin (PE) antibody: Anti-Mouse OVA257-264 (SIINFEKL) peptide bound to H-2Kb PE (eBioscience, 12-5743-82).
13. Isotype PE antibody: Mouse IgG1 k Isotype PE (eBioscience, 12-4724-41).
14. Culture dishes and plates.
15. Polystyrene round-bottom tubes, 12 × 75 mm.
16. Single-edge razor blades.
17. 5 mL sterile, individually wrapped syringes.
18. 25G needles.
19. Cell strainer, 40 μm.
20. Red blood cell ammonium-chloride-potassium (ACK) lysis buffer.
21. 70% ethanol.
22. Biohazard safety cabinet approved for level II handling of biological material.
23. Hemocytometer or an automated instrument to perform cell count.
24. Flow cytometer.
25. Centrifuge with speed 400 × *g*.
26. C57BL6 mice (*see* **Notes 1–3**).

3 Methods

3.1 Preparation of Study Samples and Controls

1. Prepare heat-inactivated fetal bovine serum (hiFBS). Thaw a bottle of FBS at room temperature or overnight at 2–8 °C, and allow it to equilibrate to room temperature. Incubate the thawed FBS for 30 min at 56 °C in a water bath, mixing every 5 min to heat-inactivate it (*see* **Note 4**).
2. Prepare RPMI complete medium by supplementing the stock medium with the following reagents: 10% hiFBS; 2 mM L-glutamine; 100 U/mL penicillin; 100 μg/mL streptomycin (*see* **Note 5**).
3. Prepare GM-CSF stock solution by reconstituting commercial lyophilized recombinant murine GM-CSF in endotoxin-free cell culture–grade water to a concentration of 0.5 mg/mL (*see* **Note 6**).
4. Prepare fluorescense activated cell sorting (FACS) staining buffer by diluting neat FBS in 1× sterile Ca^{2+}/Mg^{2+} free PBS to a final concentration of 1% FBS, e.g., by adding 0.5 mL FBS to 49.5 mL 1× PBS (*see* **Note 7**).

5. Prepare the fixative solution on the day of the experiment by diluting 20% formaldehyde in 1× sterile Ca^{2+}/Mg^{2+} free PBS to a final concentration of 2% formaldehyde, e.g., by adding 1 mL 20% formaldehyde to 9 mL PBS (*see* **Note 7**).
6. Prepare the stock positive control by adding 1 mL of endotoxin-free water to 1 mg of the SIINFEKL peptide (OVA257-264) (*see* **Note 8**).
7. Use PBS as a negative control.
8. Prepare the vehicle control (*see* **Note 9**).
9. Prepare test nanomaterials (*see* **Note 10**).

3.2 Isolation and Counting Bone Marrow Cells

1. Position the euthanized mouse on its back and rinse the fur thoroughly with 70% alcohol (*see* **Notes 1** and **2**).
2. Cut a slit in the fur just below the rib cage without cutting the peritoneal membrane.
3. Firmly grasp the skin and peel it back to expose the hind limbs. Trim away the skin after the limbs are fully exposed.
4. Using sterile sharp dissecting scissors, cut around the hip joint of each limb to free it from the pelvis. Cut through ligaments and excess tissue, being careful not to break the limb bones in the process.
5. After freeing the limbs, cut away as much muscle tissue from the bone as possible. Leave the footpads on the limbs at this stage. Use Kimwipes to aid in this process by gently massaging the bone to clean away residual tissue.
6. Sterilize the cleaned bones and attached footpads by complete immersion in 70% ethanol for 1 min.
7. Transfer the cleaned and sterilized mouse legs to a cell culture dish with 20 mL of media (*see* **Note 11**).
8. Using a razor blade, cut away the footpad at the ankle joint. Discard the footpad.
9. Using the same razor blade, cut the femurs near the hip joint and knee joint to expose the interior marrow shaft (*see* **Note 12**).
10. Using a 5 cc syringe with a 25G needle, draw up 5 mL of complete media.
11. Gently flush out the marrow into the culture dish and continue flushing until bones are white instead of red (*see* **Notes 13** and **14**).
12. Insert a 40 μm cell strainer into the opening of a 50 mL conical tube. Using a 10 mL pipette, collect all the media in the culture dish and pipette it through the cell strainer to collect the filtered cells into the conical tube.

13. Centrifuge collected cells at 400 × *g* for 5 min.
14. Resuspend the cells in 5 mL of ACK lysis buffer for the hypotonic lysis of red blood cells.
15. Immediately centrifuge the cells at 400 × *g* for another 5 min, then discard the supernatant and resuspend cells in 10 mL of RPMI complete medium.
16. Perform a cell count and viability assay (*see* **Note 15**).

3.3 Generation of Immature Dendritic Cells

1. Resuspend BMDCs at 1 × 10^6 cells/mL in RPMI complete medium supplemented with 20 ng/mL GM-CSF (*see* **Note 16**).
2. Distribute all media containing bone marrow cells into 6-well plates, dispensing 5–6 mL per well.
3. Culture bone marrow cells in an incubator maintained at 37 ° C, 5% CO_2, and 95% humidity for 5 days.
4. On day 5, media in the 6-well plates should be yellow or orange–yellow in color, indicating cell proliferation, activation, and nutrient consumption. Dislodge non-adherent and semi-adherent cells by pipetting the media in each well up and down using a 1 mL pipette. Collect all non-adherent and semi-adherent cells into 50 mL conical tubes, and discard the plates with adherent cells.
5. Centrifuge collected cells at 400 × *g* for 10 min.
6. Aspirate the supernatant and resuspend cells in 100 mL of RPMI complete medium, freshly supplemented with 20 ng/mL GM-CSF.
7. Distribute cells into new 6-well plates, dispensing 5–6 mL per well.
8. Continue culturing cells in the incubator for two more days (*see* **Note 17**).
9. Collect and wash all non-adherent and semi-adherent cells as in **steps 4** and **5** described in this section.
10. Resuspend collected BMDCs in 20 mL of RPMI complete medium and determine cell count and viability.

3.4 Experimental Procedure

1. Adjust BMDC cell count to 1.25 × 10^6 cells/mL.
2. Aliquot 800 μL of BMDC suspension per designated well of a 24-well plate.
3. Add 200 μL PBS to the negative control well. Add 200 μL of the appropriate vehicle control to wells marked as "VC." Add 200 μL of complete media to wells marked as "unlabeled cells."
4. Add 200 μL of test nanoparticles and control to designated wells and incubate for 24 h at 37 °C, 5% CO_2, and 95% humidity.

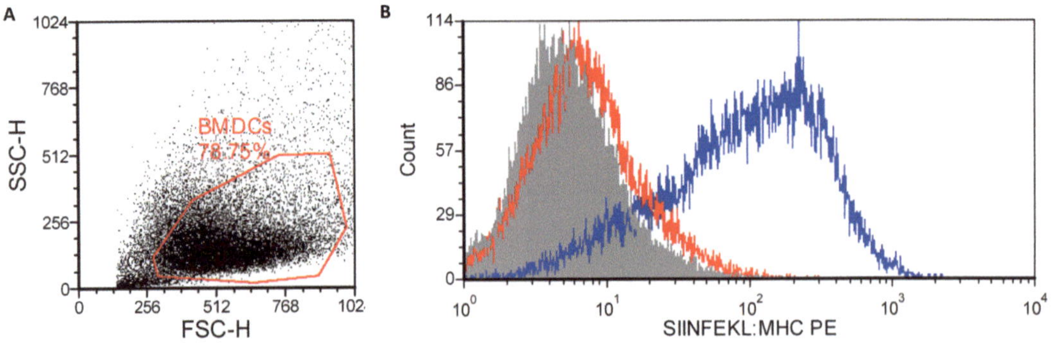

Fig. 1 Expected results. (**a**) Gating example for BMDC cell populations. (**b**) Detection of SIINFEKL presentation on APCs after pulsing with 1 μg/mL SIINFEKL peptide. The gray histogram shows the appearance of the negative control sample with 25-D1.16 staining, the blue line is the positive control sample stained with 25-D1.16 antibody, and the red line is the positive control sample with isotype control staining. FSC-forward scatter; SSC-side scatter

5. Two hours before the end of **step 4** (test nanoparticle incubation), add 200 μL of media with 5 μg/mL of SIINFEKL to each well marked as "PC."
6. At the end of the incubation period, collect all cells from the plate and transfer samples into pre-labeled polystyrene round-bottomed 12 × 75 mm Falcon tubes.
7. Prepare 25-D1.16 PE staining solution by adding 2 μL of the antibody stock solution to 798 μL of FACS staining buffer (for a working concentration of 0.5 μg/mL).
8. Prepare the isotype PE staining solution by adding 2 μL of the antibody stock solution to 798 μL of FACS staining buffer (for a working concentration of 0.5 μg/mL).
9. Wash cells with 2 mL PBS and aspirate/decant supernatant.
10. Resuspend cells in 100 μL of the appropriate staining solution.
11. Incubate cells at room temperature, protected from light, for 30 min.
12. Wash cells with 1 mL PBS and aspirate/decant supernatant.
13. Fix cells with fixative (2% formaldehyde) for 15 min at room temperature and analyze by flow cytometry (*see* **Note 18**, Fig. 1).

3.5 Data Analysis and Report

1. Use appropriate software to analyze the data acquired using a flow cytometer (*see* **Notes 19** and **20**).
2. Report the data as GeoMean, percentage of 25-D1.16 positive cells, and include flow cytometry plots where appropriate.

4 Notes

1. This protocol utilizes 8–12-week-old C57BL/6 males or females. Each set of mouse femurs and tibiae generates a large number of BMDCs (>50 × 10^6 cells); therefore, there is no need to pool cells.
2. An animal facility accredited by association for assessment and accreditation of laboratory animal care (AAALAC) International and following the Public Health Service Policy for the Care and Use of Laboratory Animals (Health Research Extension Act of 1985, Public Law 99–158, 1986) is recommended to house animals needed for this study. Animal care must be provided in accordance with the procedures outlined in the Guide for Care and Use of Laboratory Animals (National Research Council, 1996; National Academy Press, Washington, DC). All animal protocols should be approved by the Institutional Animal Care and Use Committee.
3. All work involving cells and engineered nanomaterials should be performed in a biological safety cabinet.
4. Single-use aliquots may be stored at 2–8 °C for up to 1 month or at a nominal temperature of −20 °C indefinitely. If heat-inactivated FBS is used, the heat inactivation step may be omitted.
5. Store at 2–8 °C protected from light for no longer than 1 month; before use, warm the medium in a water bath to 37 °C.
6. It is recommended to prepare small aliquots at 5 μL each and store them at −80 °C for up to 12 months. On the day of the experiment, add freshly thawed GM-CSF to complete RPMI to a final concentration of 20 ng/mL.
7. Formaldehyde solutions must be prepared in a fume hood. It is recommended to store these solutions at 2–8 °C.
8. It is recommended to aliquot this solution and store it at −20 °C for up to 6 months. On the day of the experiment, add freshly thawed SIINFEKL peptide to RPMI complete medium to a final concentration of 5 μg/mL.
9. Vehicle control is the buffer or media used to formulate test nanomaterials. Common excipients used are trehalose, sucrose, and albumin. However, other reagents and materials are also used alone or in combination. Vehicle control should match the formulation buffer of the test nanomaterial by both composition and concentration. This control can be skipped if nanoparticles are stored in PBS.

10. When the experiment is done in 24-well plates, the assay requires 1.5 mL of nanoparticles dissolved/resuspended in a complete culture medium at a concentration five times higher than the highest final test concentration. The concentration is selected based on the plasma concentration of the nanoparticle at the intended therapeutic dose. For the purpose of this protocol, this concentration is called "theoretical plasma concentration." Considerations for estimating theoretical plasma concentration are reviewed elsewhere [11].
11. From this point onward, only manipulate the legs with sterile forceps and a razor. Thoroughly sterilize gloves, and do not touch the media in the dish. To sterilize tools, immerse them in 70% ethanol for at least 5 min.
12. Cut the tibiae near the other end of the knee joint to expose the marrow of the tibiae. Keep all parts of the limbs (hip joint, knee joint, femurs, and tibiae) submerged in the media.
13. For each limb section, use forceps to keep the bone steady and insert the bevel of the needle into the joint opening/marrow shaft.
14. A brief media flush in the joint openings will suffice to dislodge available cells from the joint sections.
15. The anticipated yield is approximately 50×10^6 bone marrow cells per hindlimb.
16. For example, cells isolated from two hindlimbs of a single animal are resuspended in 100 mL of complete medium, supplemented with 4 μL of recombinant mouse GM-CSF at the stock concentration of 0.5 mg/mL.
17. After the final 2-day culture, cell media should once again be yellow in color.
18. Settings and operation procedures may vary between different flow cytometers. It is, therefore, recommended to follow the operator's manual for the cytometer available in the individual researcher's laboratory. The example of gating and expected results is shown in Fig. 1.
19. FlowJo, CellQuest, and FCS Express software are commonly used, but any other flow cytometry software can be used.
20. Set up a gate excluding NC (Negative Control) cells but including most PC cells. Cells in this gate may be designated 25-D1.16 positive, as shown in the example in Fig. 2.

Acknowledgment

This project has been funded in whole or in part by federal funds from the National Cancer Institute, National Institutes of Health,

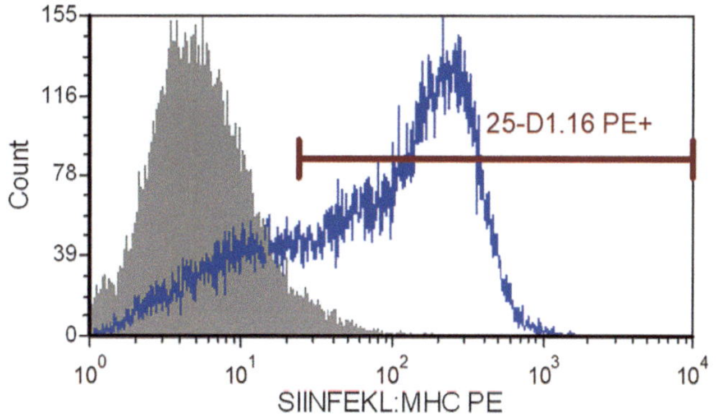

Fig. 2 Gating example for 25-D1.16 positive cells. The gray histogram shows negative control sample with 25-D1.16 staining, and the blue line shows positive control sample with 25-D1.16 staining

under contract 75N91019D00024. The content of this publication does not necessarily reflect the views or policies of the Department of Health and Human Services, nor does the mention of trade names, commercial products, or organizations imply endorsement by the US Government.

References

1. Hashemi Goradel N, Nemati M, Bakhshandeh A, Arashkia A, Negahdari B (2023) Nanovaccines for cancer immunotherapy: focusing on complex formation between adjuvant and antigen. Int Immunopharmacol 117:109887
2. Hong E, Dobrovolskaia MA (2019) Addressing barriers to effective cancer immunotherapy with nanotechnology: achievements, challenges, and roadmap to the next generation of nanoimmunotherapeutics. Adv Drug Deliv Rev 141:3–22
3. Vishweshwaraiah YL, Dokholyan NV (2022) mRNA vaccines for cancer immunotherapy. Front Immunol 13:1029069
4. Yi Y, Yu M, Li W, Zhu D, Mei L, Ou M (2023) Vaccine-like nanomedicine for cancer immunotherapy. J Control Release 355:760–778
5. Zhang X, Yang B, Ni Q, Chen X (2023) Materials engineering strategies for cancer vaccine adjuvant development. Chem Soc Rev 52(9): 2886–2910
6. Go S, Jung M, Lee S, Moon S, Hong J, Kim C, Chung Y, Kim BS (2023) A personalized cancer nanovaccine that enhances T-cell responses and efficacy through dual interactions with dendritic cells and T cells. Adv Mater 35: e2303979
7. Slezak SL, Worschech A, Wang E, Stroncek DF, Marincola FM (2010) Analysis of vaccine-induced T cells in humans with cancer. Adv Exp Med Biol 684:178–188
8. Smith PL, Piadel K, Dalgleish AG (2021) Directing T-cell immune responses for cancer vaccination and immunotherapy. Vaccines (Basel) 9(12):1392
9. Porgador A, Yewdell JW, Deng Y, Bennink JR, Germain RN (1997) Localization, quantitation, and in situ detection of specific peptide–MHC Class I complexes using a monoclonal antibody. Immunity 6:715–726
10. Shen Z, Reznikoff G, Dranoff G, Rock KL (1997) Cloned dendritic cells can present exogenous antigens on both MHC class I and class II molecules. J Immunol 158:2723–2730
11. Dobrovolskaia MA, McNeil SE (2013) Understanding the correlation between in vitro and in vivo immunotoxicity tests for nanomedicines. J Control Release 172(2):456–466

Chapter 18

Antigen-Specific Stimulation of CD8$^+$ T Cells by Murine Bone Marrow–Derived Dendritic Cells After Treatment with Nanoparticles

Enping Hong and Marina A. Dobrovolskaia

Abstract

The assessment of antigen presentation by dendritic cells and subsequent antigen-dependent activation of T lymphocytes is a critical step underlying the efficacy of nanoparticle-based therapeutic vaccines. Since nanoparticle physicochemical properties determine their interactions with the immune system, the early stages of nanotechnology-based vaccine development commonly involve optimizing the particles' properties to create a formulation with desired stability, antigen release, targeting of desired cell populations, and efficacy. To accelerate this process, in vitro models suitable for the rapid assessment of a novel vaccine candidate's efficacy are highly desirable. One such model is described in this protocol. Herein, nanoparticles are formulated to deliver a model antigen, SIINFEKL ($OVA_{257-264}$), the immunodominant class I peptide derived from ovalbumin. These nanoparticles are added to the culture of murine bone marrow–derived dendritic cells, which are subsequently co-incubated with CD8$^+$ T cells from OT-I transgenic mice. The efficient antigen presentation by dendritic cells results in the antigen-dependent proliferation of CD8$^+$ T cells, which is detected by flow cytometry.

Key words Nanoparticles, Vaccines, T lymphocytes, Antigens, Dendritic cells

1 Introduction

Extensive efforts have been made to develop nanoparticles as carriers for therapeutic cancer vaccines to the immune system's antigen-presenting cells (APCs) [1–5]. The stimulation of antigen-specific T cell responses is critical for the efficacy of therapeutic cancer vaccines.

The purpose of this protocol is to use nanoparticles for delivering vaccines to APCs and measuring the antigen-specific T cell responses that are instigated. For the purposes of this assay, the model antigen ovalbumin (OVA) is used in the context of murine cells. The protocol requires the synthesis of nanoparticles that deliver a model antigen, SIINFEKL ($OVA_{257-264}$), the

Jeffrey D. Clogston et al. (eds.), *Characterization of Nanoparticles Intended for Drug Delivery*, Methods in Molecular Biology, vol. 2789, https://doi.org/10.1007/978-1-0716-3786-9_18,

immunodominant class I peptide derived from OVA, either in its peptide form or as part of a larger molecule (such as the whole OVA protein). $CD8^+$ T cells recognizing SIINFEKL are purified from the transgenic OT-I mouse model [6] and activated using nanoparticle-treated APCs.

The experiment is performed by co-culturing nanoparticle-treated APCs with antigen-specific $CD8^+$ T cells. Bone marrow–derived dendritic cells (BMDCs) are used to process and present antigens delivered by test nanoparticles. Bone marrow cells are collected from 8–12-week-old C57BL/6 mice and cultured in a medium supplemented with murine granulocyte–macrophage-colony-stimulating factor (GM-CSF) for differentiation into BMDCs. After 7 days of culture, differentiated BMDCs are treated with nanoparticles and relevant controls for 24 h. To determine the impact of nanoparticle treatment in the presence of an adjuvant, the immune adjuvant lipopolysaccharide (LPS) may be added during the treatment step. Treated BMDCs are then washed and co-incubated with purified SIINFEKL-specific $CD8^+$ T cells obtained from OT-I transgenic mice; these cells are pre-loaded in vitro with carboxyfluorescein succinimidyl ester (CFSE) to track cell proliferation. After a further 3-day culture, supernatants are collected, and cells are fixed and analyzed by flow cytometry. The readouts for this experiment include measures of cell proliferation as determined by cell count and CFSE dilution and the determination of IFN-γ and IL-2 concentrations in cell supernatants by enzyme-linked immunosorbent assay (ELISA). The results of this protocol may be used to infer the ability of the nanoparticle system to deliver other similar antigens.

2 Materials

1. Phosphate-buffered saline (PBS), Ca^{2+}/Mg^{2+} free.
2. Cell culture-grade water, endotoxin-free.
3. Fetal bovine serum (FBS).
4. Roswell park memorial institute (RPMI) 1640.
5. Pen/Strep solution.
6. L-glutamine.
7. MEM nonessential amino acids solution, 100×.
8. Sodium pyruvate, 100 mM.
9. MEM vitamin solution, 100×.
10. 2-mercaptoethanol, 55 mM.
11. Any reagent allowing to distinguish between live and dead cells.

12. Bovine serum albumin (BSA).
13. Formaldehyde, 20%.
14. Recombinant murine GM-CSF.
15. SIINFEKL, OVA 257-264 class I peptide.
16. Lipopolysaccharide (LPS) from E. coli K12.
17. CellTrace CFSE Cell Proliferation kit (Thermo Fisher Scientific, C34570).
18. CD8a^{+} T Cell Isolation kit, mouse (Miltenyi Biotec, 130-104-075).
19. Anti-mouse CD8a FITC (eBioscience, 11-0081-81).
20. Anti-mouse V alpha 2 TCR PE (eBioscience, 12-5812-80).
21. Rat IgG2A k isotype control FITC (eBioscience, 11-4321-80).
22. Rat IgG2A k isotype control PE (eBioscience, 12-4321-41).
23. 6 μm polystyrene beads (Polysciences Inc., 07312-5).
24. Cell culture dishes and plates.
25. Polystyrene round-bottom tubes, 12 × 75 mm.
26. Single-edge razor blades.
27. 1 mL and 5 mL sterile, individually wrapped syringes.
28. 25G needles.
29. Cell strainer, 40 μm.
30. Red blood cell ammonium-chloride-potassium (ACK) lysis buffer.
31. 70% ethanol.
32. Polypropylene tubes.
33. LS columns for separation (Miltenyi Biotec, 130-042-401).
34. Cell culture incubator with 5% CO_2 and 95% humidity.
35. Biohazard safety cabinet approved for level II handling of biological material.
36. Instrument to perform cell count.
37. Flow cytometer.
38. Centrifuge with a speed of 400 × *g*.
39. Separation magnet capable of handling LS columns.
40. C57BL/6 and OT-I TCR transgenic mice (*see* **Notes 1–5**).

3 Methods

3.1 Preparation of Study Samples and Controls

1. Prepare heat-inactivated fetal bovine serum (hiFBS). Thaw a bottle of FBS at room temperature or overnight at 2–8 °C and allow it to equilibrate to room temperature. Incubate the

thawed FBS for 30 min at 56 °C in a water bath, mixing every 5 min to heat-inactivate it (*see* **Note 6**).

2. Prepare RPMI complete medium by supplementing the stock medium with the following reagents: 10% hiFBS; 2 mM L-glutamine; 100 U/mL penicillin; 100 μg/mL streptomycin (*see* **Note 7**).
3. Prepare T cell medium by adding the following reagents to RPMI 1640: 10% hiFBS (heat-inactivated); 2 mM L-glutamine; 100 U/mL penicillin; 100 μg/mL streptomycin; 1× MEM nonessential amino acids; 1 mM sodium pyruvate; 0.4× MEM vitamin solution; 50 μM 2-mercaptoethanol (*see* **Note 8**).
4. Prepare T cell isolation buffer by combining the following reagents: 1× PBS; 0.5% bovine serum albumin; 2 mM EDTA (*see* **Note 9**).
5. Prepare stock GM-CSF solution by reconstituting commercial lyophilized recombinant murine GM-CSF in endotoxin-free cell culture-grade water to a concentration of 0.5 mg/mL (*see* **Note 10**).
6. Prepare fluorescence activated cell sorting (FACS) staining buffer by diluting neat FBS in 1× sterile Ca^{2+}/Mg^{2+} free PBS to a final concentration of 1% FBS, e.g., by adding 0.5 mL FBS to 49.5 mL 1× PBS (*see* **Note 11**).
7. Prepare the flow cytometry fixative solution in a fume hood by diluting 20% formaldehyde in 1× sterile Ca^{2+}/Mg^{2+} free PBS to a final concentration of 2% formaldehyde, e.g., by adding 1 mL 20% formaldehyde to 9 mL PBS (*see* **Note 11**).
8. Prepare LPS stock solution by reconstituting 1 mg of the lyophilized LPS in 1 mL of endotoxin-free water (*see* **Note 12**).
9. Prepare 5 mM working solution of CFSE by adding 18 μL of dimethyl sulfoxide (DMSO) to a vial of CFSE (*see* **Note 13**).
10. Prepare a positive control by reconstituting 1 mg of the SIIN-FEKL peptide ($OVA_{257\text{-}264}$) in 1 mL of endotoxin-free water (*see* **Note 14**).
11. Use PBS as a negative control.
12. Prepare the vehicle control (*see* **Note 15**).
13. Prepare test nanomaterials (*see* **Note 16**).

3.2 Isolation and Counting Bone Marrow Cells

1. Position the euthanized mouse on its back and rinse the fur thoroughly with 70% alcohol (*see* **Notes 1–5**).
2. Cut a slit in the fur just below the rib cage without cutting the peritoneal membrane.
3. Firmly grasp the skin and peel it back to expose the hind limbs. Trim away the skin after the limbs are fully exposed.

4. Using sterile sharp dissecting scissors, cut around the hip joint of each limb to free it from the pelvis. Cut through ligaments and excess tissue, being careful not to break the limb bones in the process.
5. After freeing the limbs, cut away as much muscle tissue from the bone as possible. Leave the footpads on the limb at this stage. Use Kimwipes to aid in this process by gently massaging the bone to clean away residual tissue.
6. Sterilize the cleaned bones and attached footpads by complete immersion in 70% ethanol for 1 min.
7. Transfer cleaned and sterilized mouse legs to a cell culture dish with 20 mL of media (*see* **Note 17**).
8. Using a razor blade, cut away the footpad at the ankle joint. Discard the footpad.
9. Using the same razor blade, cut the femur near the hip joint and knee joint to expose the interior marrow shaft (*see* **Note 18**).
10. Using a 5 cc syringe with a 25G needle, draw up 5 mL of complete media.
11. Gently flush out the marrow into the culture dish and continue flushing until the bones are white instead of red (*see* **Notes 19** and **20**).
12. Insert a 40 μm cell strainer into the opening of a 50 mL conical tube. Using a 10 mL pipette, collect all the media in the culture dish and pipette it through the cell strainer to collect the filtered cells into the conical tube.
13. Centrifuge the collected cells at 400 × *g* for 5 min.
14. Resuspend the cells in 5 mL of ACK lysis buffer for the hypotonic lysis of red blood cells.
15. Immediately centrifuge the cells at 400 × *g* for another 5 min, then discard the supernatant and resuspend the cells in 10 mL of RPMI complete medium.
16. Perform a cell count and viability assay (*see* **Note 21**).

3.3 Generation of Immature Dendritic Cells

1. Resuspend BMDCs at 1×10^6 cells/mL in RPMI complete medium supplemented with 20 ng/mL GM-CSF (*see* **Note 22**).
2. Distribute all media containing bone marrow cells into 6-well plates, dispensing 5–6 mL per well.
3. Culture bone marrow cells in an incubator maintained at 37 ° C, 5% CO_2, and 95% humidity for 5 days.
4. On day 5, media in the 6-well plates should be yellow or orange–yellow in color, indicating cell proliferation, activation,

and nutrient consumption. Dislodge non-adherent and semi-adherent cells by pipetting the media in each well up and down using a 1 mL pipette. Collect all non-adherent and semi-adherent cells into 50 mL conical tubes, and discard the plates with adherent cells.

5. Centrifuge collected cells at 400 × *g* for 10 min.
6. Aspirate the supernatant and resuspend cells in 100 mL of RPMI complete medium, freshly supplemented with 20 ng/mL GM-CSF.
7. Distribute cells into new 6-well plates, dispensing 5–6 mL per well.
8. Continue culturing cells in the incubator for two more days (*see* **Note 23**).
9. Collect and wash all non-adherent and semi-adherent cells as in **steps 4** and **5** described in this section.
10. Resuspend collected BMDCs in 20 mL of RPMI complete medium and determine cell count and viability.

3.4 Treatment with Nanoparticles and Controls

1. Adjust BMDC cell count to 250,000 cells/mL in RPMI complete medium.
2. Seed 100 μL BMDCs (25,000 cells) in each designated well of a 96-well U-bottomed plate. Set up all experimental groups in triplicate (Fig. 1).
3. Add 50 μL PBS to two sets of negative control (NC) wells.

	1	2	3	4	5	6	7
A	NC	PC	VC	TS (0.008 mg/mL)	TS (0.04 mg/mL)	TS (0.2 mg/mL)	TS (2 mg/mL)
B	NC	PC	VC	TS (0.008 mg/mL)	TS (0.04 mg/mL)	TS (0.2 mg/mL)	TS (2 mg/mL)
C	NC	PC	VC	TS (0.008 mg/mL)	TS (0.04 mg/mL)	TS (0.2 mg/mL)	TS (2 mg/mL)
D	NC	PC	VC	TS (0.008 mg/mL)	TS (0.04 mg/mL)	TS (0.2 mg/mL)	TS (2 mg/mL)
E	NC	PC	VC	TS (0.008 mg/mL)	TS (0.04 mg/mL)	TS (0.2 mg/mL)	TS (2 mg/mL)
F	NC	PC	VC	TS (0.008 mg/mL)	TS (0.04 mg/mL)	TS (0.2 mg/mL)	TS (2 mg/mL)

Fig. 1 Example plate map. NC is the negative control, VC is the vehicle control, PC is the positive control, and TS is the test sample. Shaded cells highlight samples with 50 ng/mL LPS added

4. Add 50 μL of 4 μg/mL SIINFEKL to two sets of positive control (PC) wells.
5. For one set of negative control (NC + LPS) wells, add 50 μL of LPS (200 ng/mL in complete RPMI medium).
6. For one set of positive control wells (PC + LPS), also add 50 μL of LPS.
7. Add 50 μL of test nanoparticles at 4× the desired final concentration to designated wells (*see* **Notes 24** and **25**).
8. Incubate BMDCs for 24 h at 37 °C, 5% CO_2, and 95% humidity.
9. After 24 h incubation, centrifuge the plate at 400 × *g* for 5 min.
10. Using a multichannel pipette, without disturbing the cell pellet, very carefully remove 150 μL of supernatant and discard. Add 200 μL of fresh RPMI complete medium (*see* **Note 26**).
11. Centrifuge the plate at 400 × *g* for 5 min.
12. Using a multichannel pipette, without disturbing the cell pellet, carefully remove 100 μL of supernatant and discard (*see* **Note 27**).

3.5 Extraction and Homogenization of OT-I Spleen

1. Position the euthanized OT-I mouse on its right side (with the left side facing up) and rinse the fur thoroughly with 70% alcohol (*see* **Notes 1–5**).
2. Using a pair of sterilized forceps, lift up the skin on the left flank. Cut away the fur just below the rib cage without cutting the peritoneal membrane.
3. Cut a slit in the peritoneal membrane without damaging the spleen (*see* **Note 28**).
4. Gently lift out the spleen using the forceps, cutting away supporting tissue.
5. Transfer the spleen to a 15 mL tube containing cold T cell medium.
6. In a 6-well plate, fill one well with 5 mL of T cell medium.
7. Place a 40 μm cell strainer into the well such that the bottom of the cell strainer is submerged in the medium.
8. Place the spleen into the cell strainer.
9. Remove the plunger of a 1 mL syringe (*see* **Note 29**).
10. Using the flat end of the plunger, homogenize the spleen by crushing it against the cell strainer (*see* **Note 30**).
11. Remove and discard the cell strainer and plunger.
12. Transfer the splenic homogenate to a 15 mL tube.

3.6 CD8$^+$ T Cell Labeling and Purification

1. Centrifuge the homogenate from a single spleen at 400 × *g* for 5 min. Resuspend cells in 500 μL PBS (*see* **Note 31**).
2. Add 1 μL of 5 mM CFSE to a separate volume of 500 μL PBS and vortex to mix well.
3. Add the 500 μL of diluted CFSE to the 500 μL volume of splenic homogenate and mix by pipetting up and down five times.
4. Incubate at 37 °C for 10 min. Protect cells from light at all stages after this step.
5. Quench the CFSE labeling reaction by adding 5 mL of ice-cold T cell media.
6. Centrifuge cells at 400 × *g* for 5 min. Wash cells with 5 mL of PBS and centrifuge again with the same settings.
7. Resuspend labeled cells in 5 mL of T cell media and determine the nucleated cell count (*see* **Note 32**).
8. Resuspend 50 × 10^6 cells in 200 μL T cell isolation buffer in a 1.7 mL tube (*see* **Note 33**).
9. Add 50 μL of Biotin-Antibody Cocktail, mix well, and incubate for 5 min at 2–8 °C.
10. Add 150 μL T cell isolation buffer to the cells.
11. Vortex the vial of Anti-Biotin MicroBeads for 10 s.
12. Add 100 μL of Anti-Biotin MicroBeads to the cells, mix well, and incubate for 10 min at 2–8 °C.
13. Place the LS column in the magnetic field of the QuadroMACS magnet.
14. Prepare the column by adding 3 mL of T cell isolation buffer and discarding the flow-through.
15. Apply the cell suspension onto the column.
16. Collect the flow-through containing unlabeled cells, representing the enriched CD8$^+$ T cells.
17. Wash the column with another 3 mL of T cell isolation buffer, collecting this fraction as well.
18. Resuspend purified CD8$^+$ T cells in 5 mL of T cell medium.
19. Determine cell count and viability.

3.7 Setting Up Cell Stimulation Experiment

1. Adjust the purified CD8$^+$ T cell count to 5 × 10^5 cells/mL T cell medium (50,000 cells per 100 μL).
2. Add 100 μL of purified CD8$^+$ T cells to the plate of treated BMDCs.
3. Centrifuge the plate for 15 s at 400 × *g* to pellet cells.
4. Incubate T cell and BMDC co-culture for 66–72 h at 37 °C, 5% CO_2, and 95% humidity (*see* **Note 34**).

3.8 Assessing T Cell Purity After Isolation

1. Prepare the CD8 staining mix by adding 1 μL each of anti-mouse CD8a labeled with either FITC or APC and anti-mouse V alpha 2 TCR PE antibodies to 400 μL of FACS staining buffer (*see* **Note 36**).
2. Prepare the isotype staining mix by adding 1 μL each of isotype control FITC and isotype control PE antibodies to 400 μL of FACS staining buffer.
3. Add 0.5×10^6 purified T cells to each of two 5 mL round-bottom FACS tubes.
4. Add 1 mL PBS to each tube and centrifuge at 400 × *g* for 5 min.
5. Aspirate or decant the supernatant.
6. Resuspend cells in 100 μL of either CD8 or isotype staining mix from **step 2** above.
7. Stain cells for 30 min at room temperature protected from light.
8. Add 1 mL PBS to each tube and centrifuge at 400 × *g* for 5 min.
9. Aspirate or decant the supernatant.
10. Resuspend cells in 200 μL of fixative (2% formaldehyde) and analyze by flow cytometry (*see* **Notes 37** and **38**).

3.9 Ending Experiment

1. At the end of co-culture, centrifuge the plate at 400 × *g* for 5 min.
2. Carefully remove 150 μL of supernatant per well and transfer it to another 96-well plate for storage.
3. Freeze supernatants at −80 °C for storage (*see* **Note 39**).
4. Add 100 μL of 3% formaldehyde to cells and mix well to fix the cells (*see* **Note 40**).
5. Incubate at room temperature for 15 min to fix, protected from light.
6. Transfer all samples to racked cluster tubes (*see* **Note 41**).
7. Vortex the vial of 6 μm beads to mix well (*see* **Note 42**).
8. To a tube of 10 mL PBS, add three drops of beads and vortex to mix (*see* **Note 42**).
9. Determine bead concentration using a cell counter (*see* **Note 42**).
10. Add 50 μL of pre counted beads to each tube (*see* **Note 42**).
11. Analyze cells and beads by flow cytometry (*see* **Notes 38**, **43**, and **44**).

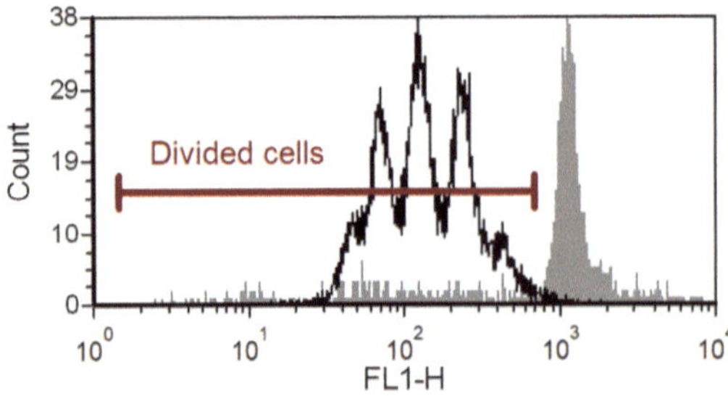

Fig. 2 Detection of CFSE dilution in dividing CD8$^+$ T cells. Gray histogram represents the negative control incubated with untreated BMDCs, with the high CFSE fluorescence intensity indicating a lack of cell division. Black line shows the positive control incubated with SIINFEKL-treated BMDCs. CFSE dilution in divided cells is represented by the detected peaks in the FL1 channel, with distinct and decreasing CFSE fluorescence intensities compared to the negative control

3.10 Data Analysis and Report

1. Use appropriate software to analyze the data acquired by a flow cytometer (*see* **Note 45**).
2. Draw a gate on all cells with lower fluorescence intensity than untreated cells for the "divided cells" gate (Fig. 2).
3. Collect the following data from all samples: Bead gate event count, T cell gate event count, FL1 gMFI of T cells, and % of divided T cells/total T cells.
4. Determine the true cell count for each well using the following formula: True cell count = True bead concentration per well / Bead event count × T cell event count.
5. Analyze changes in treated samples and controls compared to the NC.
6. Report and analyze the following three metrics: True T cell count, CFSE gMFI, and % divided T cells.

4 Notes

1. This protocol utilizes two types of mice: 8–12-week-old C57BL/6 male or female mice for BMDC culture, and 8–12-week-old OT-I TCR transgenic female mice (C57BL/6-Tg (TcraTcrb)1100Mjb/J, The Jackson Laboratory).
2. To perform animal work, a laboratory should be accredited by association for assessment and accreditation of laboratory animal care (AAALAC) International and follow the Public Health Service Policy for the Care and Use of Laboratory Animals (Health Research Extension Act of 1985, Public Law 99-158, 1986).
3. Animal care should be provided in accordance with the procedures outlined in the Guide for Care and Use of Laboratory

Animals (National Research Council, 1996; National Academy Press, Washington, D.C.).

4. All animal protocols must be approved by the Institutional Animal Care and Use Committee (ACUC).
5. Working with animals, cells, and engineered nanomaterials requires personal protective equipment (lab coat, gloves, and safety goggles).
6. Single-use aliquots may be stored at 2–8 °C for up to 1 month or at a nominal temperature of −20 °C indefinitely. If heat-inactivated FBS is used, the heat inactivation step may be omitted.
7. Store at 2–8 °C protected from light for no longer than 1 month; before use, warm the medium in a water bath to 37 °C.
8. Store at 2–8 °C protected from light for no longer than 1 month; before use, warm the medium in a water bath to 37 °C.
9. Filter the T cell isolation buffer before use and store at 2–8 °C.
10. It is recommended to prepare small aliquots at 5 μL each and store them at −80 °C for up to 12 months. On the day of the experiment, a freshly thawed GM-CSF aliquot is added to the complete RPMI medium to a final concentration of 20 ng/mL.
11. Store these solutions at 2–8 °C. Formaldehyde solutions must be prepared in a fume hood.
12. It is recommended to aliquot and store this stock at −20 °C for up to 6 months. On the day of the experiment, a freshly thawed aliquot of the LPS stock should be diluted in RPMI complete medium to a final concentration of 200 ng/mL.
13. It is recommended to store this vial at −20 °C, protected from light.
14. It is recommended to aliquot this solution and store at −20 °C for up to 6 months. On the day of the experiment, an aliquot of the freshly thawed SIINFEKL peptide should be diluted in PBS to a final concentration of 4 μg/mL.
15. Vehicle control is the buffer or media used to formulate test nanomaterials. Common excipients used are trehalose, sucrose, and albumin. However, other reagents and materials are also used alone or in combination. Vehicle control should match the formulation buffer of the test nanomaterial by both composition and concentration. This control can be skipped if nanoparticles are stored in PBS.
16. When the experiment is done in 24-well plates, the assay requires 1.5 mL of nanoparticles dissolved/resuspended in a

complete culture medium at a concentration five times higher than the highest final test concentration. The concentration is selected based on the plasma concentration of the nanoparticle at the intended therapeutic dose. For the purpose of this protocol, this concentration is called "theoretical plasma concentration." Considerations for estimating theoretical plasma concentration were reviewed elsewhere [7].

17. From this point onward, only manipulate the legs with sterile forceps and a razor. Thoroughly sterilize gloves, and do not touch the media in the dish. To sterilize tools, immerse them in 70% ethanol for at least 5 min.
18. Cut the tibiae near the other end of the knee joint to expose the marrow of the tibiae. Keep all parts of the limbs (hip joint, knee joint, femurs, and tibiae) submerged in the media.
19. For each limb section, use forceps to keep the bone steady and insert the bevel of the needle into the joint opening/marrow shaft.
20. A brief media flush in the joint openings will suffice to dislodge available cells from the joint sections.
21. The anticipated yield is approximately 50×10^6 bone marrow cells per hindlimb.
22. For example, cells isolated from two hindlimbs of a single animal are resuspended in 100 mL of complete medium, supplemented with 4 μL of recombinant mouse GM-CSF at the stock concentration of 0.5 mg/mL.
23. After the final 2-day culture, cell media should once again be yellow in color.
24. In treatment conditions where the addition of LPS is desired, add 50 μL of 200 ng/mL LPS.
25. In all samples where LPS is not added, add 50 μL of complete RPMI medium to make up the final volume to 200 μL.
26. Do not use a vacuum-driven aspirator or manifold to wash the cells. BMDCs are loosely adherent and will be removed under vacuum.
27. The final volume in the BMDC wells should be 100 μL of complete RPMI medium.
28. The spleen is visible just under the peritoneal membrane.
29. Do so by extending the syringe within the wrapper, carefully peeling back the wrapper, and removing the plunger by its stem without touching the flat end of the plunger.
30. Move the plunger against the strainer in a circular motion until no red chunks remain.
31. Process spleens separately if working with several samples.

32. Acridine orange propidium iodide (AOPI) staining and a cellometer, hemocytometer, or other stain and instrument can be used as long as they are appropriate for the staining and enumeration of live and dead cells.
33. $CD8^+$ T cell separation is performed according to the manufacturer's protocol. From this step onward, an example of 50×10^6 nucleated spleen cells is used. The exact cell counts used during the experiment are determined based on cell yield and experiment requirements.
34. Monitor the cell culture periodically. As cells are activated, the media will change color from pink to yellow. Excessive yellow color may indicate overstimulation and may warrant early termination of the experiment.
35. The purity of isolated OT-I $CD8^+$ T cells may be verified by flow cytometry immediately after purification. This procedure generally produces $CD8^+$ populations with over 95% purity. Nevertheless, it is important for investigators to verify that the isolation procedure works when setting it up for the first time.
36. Anti-CD8-FITC can be used for T cells not previously loaded with CFSE. If the analysis of CFSE-loaded T cells is wanted, anti-CD8-APC should be used. Other fluorophores not overlapping with CFSE and PE fluorescence can be used.
37. After fixation, the cells can be stored at 2–8 °C.
38. It is recommended to follow the operator's manual for the cytometer available in the individual researcher's lab because settings and operation requirements may vary between different instruments.
39. Supernatants can be further analyzed for cytokines such as IFNγ and IL-2 by ELISA, which is not detailed in this protocol.
40. This treatment gives a final formaldehyde concentration of 2% in a volume of 150 μL.
41. Samples are now ready to be analyzed by flow cytometry; run them within 0–4 days after the experiment.
42. Perform this step just before flow cytometry analysis when cell count using an internal bead control is desired.
43. When the gates on the $CD8^+$ T cell population and on the population of 6 μm beads are set up correctly, each population is distinct and easily distinguished (Fig. 3).
44. T cells are low in forward scatter (FSC) and side scatter (SSC), while beads are low in FSC but high in SSC. It is important to recognize that the T cell population will change in FSC and SSC as T cells are activated and expand (compare Fig. 3a and b).

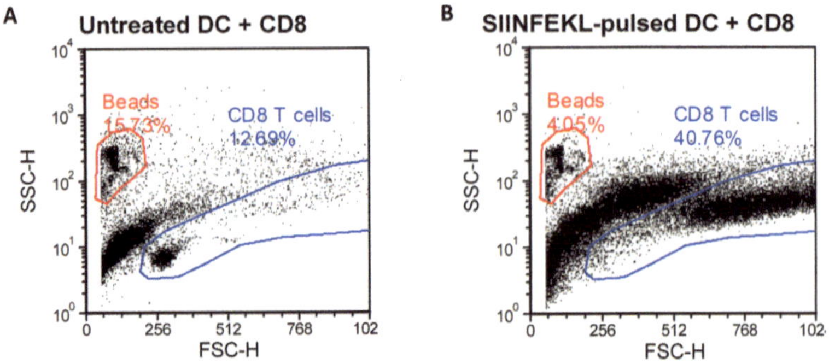

Fig. 3 Gating example for T cell and bead populations. (**a**) CD8+ T cells co-cultured with untreated BMDCs. (**b**) CD8+ T cells co-cultured with SIINFEKL-treated BMDCs. The T cell population increases in both size (FSC) and granularity (SSC). The bead population remains in the same position. The beads have high SSC and are separate from the rest of the cells/debris. The SSC is set here to a logarithmic scale. Specific settings may vary between users and instruments

Therefore, it is important to adjust FSC and SSC settings to account for this change.

45. FlowJo, CellQuest, and FCSExpress are commonly used, but other software can also be used.

Acknowledgment

This project has been funded in whole or in part by federal funds from the National Cancer Institute, National Institutes of Health, under contract 75N91019D00024. The content of this publication does not necessarily reflect the views or policies of the Department of Health and Human Services, nor does the mention of trade names, commercial products, or organizations imply endorsement by the US Government.

References

1. Slezak SL, Worschech A, Wang E, Stroncek DF, Marincola FM (2010) Analysis of vaccine-induced T cells in humans with cancer. Adv Exp Med Biol 684:178–188
2. Smith PL, Piadel K, Dalgleish AG (2021) Directing T-cell immune responses for cancer vaccination and immunotherapy. Vaccines (Basel) 9(12):1392
3. Vishweshwaraiah YL, Dokholyan NV (2022) mRNA vaccines for cancer immunotherapy. Front Immunol 13:1029069
4. Yi Y, Yu M, Li W, Zhu D, Mei L, Ou M (2023) Vaccine-like nanomedicine for cancer immunotherapy. J Control Release 355:760–778
5. Zhang X, Yang B, Ni Q, Chen X (2023) Materials engineering strategies for cancer vaccine adjuvant development. Chem Soc Rev 52(9): 2886–2910
6. Clarke SR, Barnden M, Kurts C, Carbone FR, Miller JF, Heath WR (2000) Characterization of the ovalbumin-specific TCR transgenic line OT-I: MHC elements for positive and negative selection. Immunol Cell Biol 78:110–117
7. Dobrovolskaia MA, McNeil SE (2013) Understanding the correlation between in vitro and in vivo immunotoxicity tests for nanomedicines. J Control Release 172(2):456–466

Chapter 19

Detection of Pre-Existing Antibodies to Polyethylene Glycol and PEGylated Liposomes in Human Serum

Barry W. Neun and Marina A. Dobrovolskaia

Abstract

Polyethylene glycol, or PEG, is common in consumer products, over-the-counter medications, food, and pharmaceutical products. Concerns about PEG immunogenicity and the subsequent negative impact of pre-existing and product-induced antibodies often shadow the benefits of using PEG in nanotechnology-based products. Such anti-PEG antibodies contribute to the accelerated blood clearance of PEGylated nanomedicines and result in premature drug release and antibody-mediated toxicities. Recent data demonstrated that using PEG in COVID-19 lipid nanoparticle-mRNA vaccines is associated with an induction of anti-PEG antibodies in healthy individuals, further contributing to the development or boosting of pre-existing antibodies and increasing the risks of antibody-mediated toxicities to other products containing PEG. Therefore, monitoring the levels of pre-existing and product-induced anti-PEG antibodies provides mechanistic insights for pharmacology, toxicology, and immunological studies of PEGylated drug products.

Key words Nanoparticles, Polyethylene glycol, PEG, PEGylated liposomes

1 Introduction

Poly(ethylene glycol) (PEG) is commonly used in the pharmaceutical industry to modify recombinant proteins and nanoparticle surfaces to improve hydrophilicity and decrease their recognition by the immune system [1–4]. PEGylated therapeutics and nanoparticles are generally considered as more stealth than their un-PEGylated counterparts. Despite improved protection from immune recognition, the immune system is still able to identify these products and mount an antibody response against them. Such immune response may result in the development of anti-drug antibodies (ADA) and, among antibodies specific to biological drugs or nanocarriers, include the formation of antibodies to the PEG itself [5–7]. Moreover, several reports suggested the existence of naturally occurring antibodies in the blood of healthy donor volunteers [8]. The physiological significance of the anti-PEG antibodies is

Jeffrey D. Clogston et al. (eds.), *Characterization of Nanoparticles Intended for Drug Delivery*, Methods in Molecular Biology, vol. 2789, https://doi.org/10.1007/978-1-0716-3786-9_19,

incompletely understood. However, several studies suggested that they may affect the clearance of PEGylated products (e.g., Accelerated Blood Clearance or ABC phenomenon), induce premature drug release, affect efficacy, and contribute to complement activation and other antibody-mediated toxicities [8–14]. The concerns about the contribution of anti-PEG antibodies to hypersensitivity reactions intensified with growing evidence of the induction of these antibodies by COVID-19 lipid nanoparticle-mRNA vaccines [15–20]. Alternative polymers receive increasing attention in attempts to overcome the negative consequences of PEG immunogenicity to PEGylated nanomedicines and PEG-containing products; however, many of these polymers were also found to be immunogenic warranting additional investigations [4]. Both structural aspects of the antibody binding to PEG and methodologies for the detection of induced and pre-existing antibodies have been discussed in the literature [8, 11, 12, 14, 21].

The purpose of the protocol described herein is to detect the presence of antibodies reactive to PEG2000, mPEG2000, and PEGylated liposomes used for delivery of the anti-cancer drug doxorubicin, also known as the Doxil formulation. The protocol can be used to assess the presence of both naturally occurring antibodies and antibodies induced as a result of exposure to PEGylated liposomes.

The three antigens—PEG2000, mPEG2000, and Doxebo—are coated on separate ELISA plates; one control (uncoated) plate is used to determine the assay threshold by accounting for a signal resulting from non-specific binding. After blocking, these plates are incubated with serial dilutions of the plasma or serum obtained from the donor's or patient's blood. When the antibodies specific to these antigens are present in the tested specimen, they are detected using anti-human IgG-HRP and anti-human IgM-HRP and visualized by the TMB substrate. The color change in each ELISA well after incubation with the substrate is proportional to the level of antibodies reactive to the antigens coated on the plate. The quantity of antibody is determined in terms of the titer, which is the highest dilution of the test plasma demonstrating an optical density reading above the assay threshold. The assay threshold is calculated as a mean O.D. value of the same plasma sample tested on the plate not coated with the antigen(s) plus 3 standard deviations (SD). Five donor plasma or sera samples can be analyzed in one batch (i.e., can be fit on one plate), and four plates are required for each batch.

This protocol can also be useful in assessing PEG and mPEG antibodies which may react with PEG present in other non-liposomal products. Other protocols, including ELISA and flow cytometry, are available and have been described by Roffler's team [10, 21].

2 Materials

1. PEG 2000.
2. mPEG2000.
3. Doxebo (Doxil blank liposome).
4. Instant nonfat dry milk.
5. Mouse anti-PEG monoclonal IgM antibody (*see* **Note 1**).
6. Biotinylated anti-PEG monoclonal IgM antibody (*see* **Note 1**).
7. Donkey anti-human IgG-HRP (*see* **Note 1**).
8. Rabbit anti-human IgM-HRP (*see* **Note 1**).
9. Goat anti-rat IgM-HRP (*see* **Note 1**).
10. Donkey anti-rabbit IgG-HRP (*see* **Note 1**).
11. Rabbit anti-PEG IgG (*see* **Note 1**).
12. Phosphate buffered saline (PBS).
13. Fetal bovine serum (FBS).
14. BupH carbonate-bicarbonate buffer packs (*see* **Note 1**).
15. Ultra TMB-ELISA substrate.
16. AGP6 rat anti-PEG antibody (*see* **Note 1**).
17. CHAPS, PlusOne (*see* **Note 1**).

3 Methods

3.1 Preparation of Reagents and Controls

1. Prepare 0.2 M carbonate-bicarbonate buffer with pH 9.4 by dissolving one pack of BupH carbonate-bicarbonate in 500 mL distilled water and mix well, filter through 0.2 μm filter and store at room temperature for up to 1 month.
2. Prepare Wash Buffer (1× PBS + 0.1% CHAPS) by adding 100 mL of 10× PBS to 900 mL distilled water. Then add 1 g CHAPS and mix well. Store at room temperature for 1 month.
3. Prepare Blocking Buffer (5% Nonfat Dry Milk in 1× PBS) by dissolving 25 g of nonfat dry milk in 500 mL of 1× PBS and mix well; store at 4 °C.
4. Prepare Dilution Buffer A (2% nonfat dry milk in 1× PBS) by dissolving 10 g of nonfat dry milk in 500 mL of 1× PBS and mix well; store at 4 °C.
5. Prepare Dilution Buffer B (4% FBS + 2% nonfat dry milk in 1× PBS) by adding 20 mL of FBS to 480 mL of Dilution Buffer A; store at 4 °C.
6. Prepare antigens as follows: Dilute PEG2000 in blocking buffer to a final concentration of 10 μg/mL; dilute

mPEG2000 in blocking buffer to a final concentration of 10 μg/mL; dilute Doxebo stock 300 times in blocking buffer (*see* **Note 2**).

7. Prepare Positive Control IgG by diluting rabbit anti-PEG IgG (PEG-B-47) in pooled normal human serum or plasma to a final concentration 10 μg/mL (*see* **Note 3**).
8. Prepare Positive Control IgM by diluting rat anti-PEG IgM (AGP6) in pooled normal human serum or plasma to a final concentration 10 μg/mL (*see* **Note 4**).
9. Prepare Stop Solution 2 N sulfuric acid (H_2SO_4) by slowly adding 27.7 mL H_2SO_4 into 200 mL of dH_2O water, mix the solution thoroughly, let it cool and bring the solution to 500 mL with dH_2O using a 1000 mL graduated cylinder; mix well and store in a bottle at room temperature.

3.2 Experimental Procedure

1. Coat 3 ELISA plates with 125 μL/well of 1 μg/mL of ANPEG-1 antibody in BupH buffer overnight at 4 °C. For the uncoated plate, add 125 μL/well of plain BupH buffer to it and incubate overnight at 4 °C (*see* **Note 5**).
2. Aspirate and discard coating solution, tap the plate dry on paper towels and add 250 μL/well of blocking buffer. Incubate 1 h at room temperature (RT) (uncoated plate incubates with blocking buffer 2 h at RT).
3. Aspirate blocking buffer from coated plates and tap the plates dry on paper towels.
4. Add 125 μL/well of antigen to coated plates and incubate at RT for 1 h (*see* **Note 6**).
5. Wash plates once with 1× PBS (300 μL/well).
6. Add 240 μL of Dilution Buffer A to wells in row A. Add 125 μL of Dilution Buffer B to wells in rows B-H (Fig. 1).
7. Add 10 μL (initial 1:25 dilution) of the test serum and controls to corresponding wells in row A.
8. Using a multichannel pipette, transfer 125 μL from row A to row B. Pipet up and down several times and transfer 125 μL from row B to row C. Repeat mixing and transferring to the next row until row H. After mixing, collect 125 μL from row H and discard it. This step results in a serial two-fold dilution of the test sera.
9. Incubate the plate at RT for 1 h.
10. Wash the plate two times with wash buffer. Rotate the plate after the first wash. Buffer volume per well is 300 μL.
11. Wash the plate one time with 1× PBS (300 μL/well). Invert and tap dry on a paper towel.

A

	1	2	3	4	5	6	7	8	9	10	11	12
A	PC	PC	TS 1	TS1	TS2	TS2	TS3	TS3	TS4	TS4	TS5	TS5
B	PC	PC	TS 1	TS1	TS2	TS2	TS3	TS3	TS4	TS4	TS5	TS5
C	PC	PC	TS 1	TS1	TS2	TS2	TS3	TS3	TS4	TS4	TS5	TS5
D	PC	PC	TS 1	TS1	TS2	TS2	TS3	TS3	TS4	TS4	TS5	TS5
E	PC	PC	TS 1	TS1	TS2	TS2	TS3	TS3	TS4	TS4	TS5	TS5
F	PC	PC	TS 1	TS1	TS2	TS2	TS3	TS3	TS4	TS4	TS5	TS5
G	PC	PC	TS 1	TS1	TS2	TS2	TS3	TS3	TS4	TS4	TS5	TS5
H	PC	PC	TS 1	TS1	TS2	TS2	TS3	TS3	TS4	TS4	TS5	TS5

B

	1	2	3	4	5	6	7	8	9	10	11	12
A	PC	PC	TS 1	TS1	TS2	TS2	TS3	TS3	TS4	TS4	TS5	TS5
B	PC	PC	TS 1	TS1	TS2	TS2	TS3	TS3	TS4	TS4	TS5	TS5
C	PC	PC	TS 1	TS1	TS2	TS2	TS3	TS3	TS4	TS4	TS5	TS5
D	PC	PC	TS 1	TS1	TS2	TS2	TS3	TS3	TS4	TS4	TS5	TS5
E	PC	PC	TS 1	TS1	TS2	TS2	TS3	TS3	TS4	TS4	TS5	TS5
F	PC	PC	TS 1	TS1	TS2	TS2	TS3	TS3	TS4	TS4	TS5	TS5
G	PC	PC	TS 1	TS1	TS2	TS2	TS3	TS3	TS4	TS4	TS5	TS5
H	PC	PC	TS 1	TS1	TS2	TS2	TS3	TS3	TS4	TS4	TS5	TS5

Fig. 1 Example plate map. There are two categories of plates. (**a**) Category A includes four plates: uncoated, coated with PEG2000, coated with mPEG2000, and coated with Doxebo. This group is used for the detection of anti-PEG IgG. The uncoated plate receives the same treatments as coated plates except for the initial coating with capture anti-PEG antibody. The titers of test sera in rows A to H for all test plates are 25, 50, 100, 200, 400, 800, 1600, 3200. Columns 1 and 2 receive donkey anti-rabbit-IgG HRP. All other wells receive donkey anti-human IgG-HRP. (**b**) Category B also includes four plates: uncoated, coated with PEG2000, coated with mPEG2000, and coated with Doxebo. This group is used for the detection of anti-PEG IgM. The uncoated plate receives the same treatments as coated plates except for the initial coating with capture anti-PEG antibody. The titers of test sera in rows A to H for all test plates are 25, 50, 100, 200, 400, 800, 1600, 3200. Columns 1 and 2 receive goat anti-rat IgM-HRP. All other wells get rabbit anti-human IgM-HRP

12. Add 125 μL per well of conjugate and incubate at RT for 1 h (*see* **Note** 7).
13. Repeat washes in **steps 10** and **11**. Buffer volume per well is 300 μL.
14. Add 125 μL of substrate to each well and develop the plate for 10–30 min.
15. Add 30 μL of stop solution, tap the plate to mix, and read the absorbance at 450 nm (*see* **Note 8**).

3.3 Data Analysis and Interpretation

1. A percent coefficient of variation should be calculated for each control or test according to the following formula:

$$\%CV = \frac{\text{Standard deviation}}{\text{Mean}} \times 100\%$$

2. %CV for each control and test sample should be less than 25%.
3. Review **Note 9**.

4 Notes

1. Details about antibodies used to develop this protocol are available at https://www.cancer.gov/nano/research/ncl/protocols-capabilities/ncl-method-ita-36.pdf. Equivalent supplies from alternate vendors can be used.
2. This dilution brings the mPEG concentration in the sample to 11 μg/mL.
3. This antibody will not recognize PEG2000 because, according to the manufacturer's info, they are specific to the methoxy group.
4. This antibody should recognize all PEGs because, according to the vendor's info, they are specific to the PEG backbone.
5. There are two categories of plates. Category A is used to assess the levels of PEG-specific IgG, and Category B is used for the detection of anti-PEG IgM. You will need one uncoated plate and three plates coated with individual antigens for each category. Examples of plate maps are shown in Fig. 1.
6. There are three antigens in total (PEG2000, mPEG2000, and Doxebo). For each group of five donors, prepare four plates in **step 1** (three coated with ANPEG-1 and one uncoated).
7. IgG plates will receive anti-human IgG-HRP. IgM plates will receive anti-human IgM-HRP. Wells containing positive control IgG serum will receive anti-rabbit IgG-HRP. Wells containing positive control IgM serum will receive anti-rat-IgM-HRP.
8. Examples of data generated using this assay can be found in reference [22].
9. It is important to understand that the detection of IgM and/or IgG reactive to PEG per se does not mean these antibodies have any functional consequences such as ABC, drug release, or complement activation-related toxicities [22]. Functional assays (e.g., complement activation) should be used to establish a correlation, and mechanistic studies are required to provide a further causation link between the anti-PEG antibodies and toxicity, clearance, or other affected biological properties of PEGylated products.

Acknowledgment

This project has been funded in whole or in part by federal funds from the National Cancer Institute, National Institutes of Health, under contract 75N91019D00024. The content of this publication does not necessarily reflect the views or policies of the

Department of Health and Human Services, nor does mention of trade names, commercial products, or organizations imply endorsement by the US Government.

References

1. Andrianov AK (2023) Noncovalent PEGylation of protein and peptide therapeutics. Wiley Interdiscip Rev Nanomed Nanobiotechnol 15(5):e1897
2. Chapa González C, Martínez Saráoz JV, Roacho Pérez JA, Olivas Armendáriz I (2023) Lipid nanoparticles for gene therapy in ocular diseases. Daru 31(1):75–82
3. Ibrahim M, Ramadan E, Elsadek NE, Emam SE, Shimizu T, Ando H, Ishima Y, Elgarhy OH, Sarhan HA, Hussein AK, Ishida T (2022) Polyethylene glycol (PEG): the nature, immunogenicity, and role in the hypersensitivity of PEGylated products. J Control Release 351:215–230
4. Shi D, Beasock D, Fessler A, Szebeni J, Ljubimova JY, Afonin KA, Dobrovolskaia MA (2022) To PEGylate or not to PEGylate: immunological properties of nanomedicine's most popular component, polyethylene glycol and its alternatives. Adv Drug Deliv Rev 180: 114079
5. Ibrahim M, Shimizu T, Ando H, Ishima Y, Elgarhy OH, Sarhan HA, Hussein AK, Ishida T (2023) Investigation of anti-PEG antibody response to PEG-containing cosmetic products in mice. J Control Release 354:260–267
6. Okada N, Taro S, Ando H, Nakamura S, Goda M, Abe M, Kitahara T, Ishida T, Ishizawa K (2023) Clinical impact of antipolyethylene glycol (PEG) antibody in hematological patients administered PEGylated-granulocyte colony-stimulating factor. Clin Pharmacol Drug Dev 12(8):826–831
7. Takata H, Shimizu T, Yamade R, Elsadek NE, Emam SE, Ando H, Ishima Y, Ishida T (2023) Anti-PEG IgM production induced by PEGylated liposomes as a function of administration route. J Control Release 360:285–292
8. Chen BM, Cheng TL, Roffler SR (2021) Polyethylene glycol immunogenicity: theoretical, clinical, and practical aspects of anti-polyethylene glycol antibodies. ACS Nano 15(9):14022–14048
9. Bavli Y, Winkler I, Chen BM, Roffler S, Cohen R, Szebeni J, Barenholz Y (2019) Doxebo (doxorubicin-free Doxil-like liposomes) is safe to use as a pre-treatment to prevent infusion reactions to PEGylated nanodrugs. J Control Release 306:138–148
10. Chen E, Chen BM, Su YC, Chang YC, Cheng TL, Barenholz Y, Roffler SR (2020) Premature drug release from polyethylene glycol (PEG)-coated liposomal doxorubicin via formation of the membrane attack complex. ACS Nano 14(7):7808–7822
11. Chen WA, Chang DY, Chen BM, Lin YC, Barenholz Y, Roffler SR (2023) Antibodies against poly(ethylene glycol) activate innate immune cells and induce hypersensitivity reactions to PEGylated nanomedicines. ACS Nano 17(6):5757–5772
12. Lee CC, Su YC, Ko TP, Lin LL, Yang CY, Chang SS, Roffler SR, Wang AH (2020) Structural basis of polyethylene glycol recognition by antibody. J Biomed Sci 27(1):12
13. Lin YC, Chen BM, Tran TTM, Chang TC, Al-Qaisi TS, Roffler SR (2023) Accelerated clearance by antibodies against methoxy PEG depends on pegylation architecture. J Control Release 354:354–367
14. Nguyen MT, Shih YC, Lin MH, Roffler SR, Hsiao CY, Cheng TL, Lin WW, Lin EC, Jong YJ, Chang CY, Su YC (2022) Structural determination of an antibody that specifically recognizes polyethylene glycol with a terminal methoxy group. Commun Chem 5(1):88
15. Dézsi L, Mészáros T, Kozma G, H-Velkei M, Oláh CZ, Szabó M, Patkó Z, Fülöp T, Hennies M, Szebeni M, Barta BA, Merkely B, Radovits T, Szebeni J (2022) A naturally hypersensitive porcine model may help understand the mechanism of COVID-19 mRNA vaccine-induced rare (pseudo) allergic reactions: complement activation as a possible contributing factor. GeroScience 44(2):597–618
16. Kozma GT, Mészáros T, Berényi P, Facskó R, Patkó Z, Oláh CZ, Nagy A, Fülöp TG, Glatter KA, Radovits T, Merkely B, Szebeni J (2023) Role of anti-polyethylene glycol (PEG) antibodies in the allergic reactions to PEG-containing Covid-19 vaccines: evidence for immunogenicity of PEG. Vaccine. https://doi.org/10.1016/j.vaccine.2023.06.009
17. Szebeni J, Kiss B, Bozó T, Turjeman K, Levi-Kalisman Y, Barenholz Y, Kellermayer M (2023) Insights into the structure of comirnaty covid-19 vaccine: a theory on soft, partially bilayer-covered nanoparticles with hydrogen

bond-stabilized mRNA-lipid complexes. ACS Nano 17(14):13147–13157

18. Szebeni J, Storm G, Ljubimova JY, Castells M, Phillips EJ, Turjeman K, Barenholz Y, Crommelin DJA, Dobrovolskaia MA (2022) Applying lessons learned from nanomedicines to understand rare hypersensitivity reactions to mRNA-based SARS-CoV-2 vaccines. Nat Nanotechnol 17(4):337–346
19. Lim XR, Chan GYL, Tan JWL, Ng CYL, Chua CG, Tan GB, Chan SSW, Ong KH, Tan YZ, Tan SHZ, Teo CML, Lee SSM, Thong BYH, Leung BPL (2023) Anaphylatoxin complement 5a in Pfizer BNT162b2-induced immediate-type vaccine hypersensitivity reactions. Vaccines (Basel) 11(6):1020
20. Zhou ZH, Cortese MM, Fang JL, Wood R, Hummell DS, Risma KA, Norton AE, KuKuruga M, Kirshner S, Rabin RL, Agarabi C, Staat MA, Halasa N, Ware RE, Stahl A, McMahon M, Browning P, Maniatis P, Bolcen S, Edwards KM, Su JR, Dharmarajan S, Forshee R, Broder KR, Anderson S, Kozlowski S (2023) Evaluation of association of anti-PEG antibodies with anaphylaxis after mRNA COVID-19 vaccination. Vaccine 41(28):4183–4189
21. Fang JL, Beland FA, Tang Y, Roffler SR (2021) Flow cytometry analysis of anti-polyethylene glycol antibodies in human plasma. Toxicol Rep 8:148–154
22. Neun BW, Barenholz Y, Szebeni J, Dobrovolskaia MA (2018) Understanding the role of anti-PEG antibodies in the complement activation by doxil in vitro. Molecules 23(7):1700

Chapter 20

In Vitro and In Vivo Methods for Analysis of Nanoparticles' Potential to Induce Delayed-Type Hypersensitivity Reactions

Timothy M. Potter, Barry W. Neun, and Marina A. Dobrovolskaia

Abstract

Delayed-type hypersensitivity (DTH) reactions are among the common reasons for drug withdrawal from clinical use during the post-marketing stage. Several in vivo methods have been developed to test DTH responses in animal models. They include the local lymph node assay (LLNA) and local lymph node proliferation assay (LLNP). While LLNA is instrumental in testing topically administered formulations (e.g., creams), the LLNP was proven to be predictive of drug-mediated DTH in response to small molecule pharmaceuticals. Global efforts in reducing the use of research animals lead to the development of in vitro models to predict test-materials' mediated DTH. Two such models include the analysis of surface marker expression in human cell lines THP-1 and U-937. These tests are known as the human cell line activation test (hCLAT) and myeloid U937 skin sensitization test (MUSST or U-SENS), respectively. Here we describe experimental procedures for all these methods, discuss their in vitro-in vivo correlation, and suggest a strategy for applying these tests to analyze engineered nanomaterials and nanotechnology-formulated drug products.

Key words Nanoparticles, Hypersensitivity, Leukocyte proliferation, Sensitizer, Allergen, Irritant, LLNA, LLNP, hCLAT, MUSST

1 Introduction

According to several reports from industry, academia, and regulatory agencies, many drugs are withdrawn from clinical use due to their adverse effects on the immune system function [1–4]. Although nanotechnology-formulated drugs represent a relatively new category of pharmaceutical products, their safety assessment is conducted using the regulatory framework established for other products, such as small molecules, therapeutic nucleic acids, and biologics; and, among other tests, it includes the analysis of immunotoxicity.

Acute immunotoxicity commonly observed in response to nanomaterials is related to hemolysis, complement activation, opsonization, phagocytosis, and cytokine secretion, and can be

Jeffrey D. Clogston et al. (eds.), *Characterization of Nanoparticles Intended for Drug Delivery*, Methods in Molecular Biology, vol. 2789, https://doi.org/10.1007/978-1-0716-3786-9_20,

accurately tested using in vitro assays [1]. Identifying changes in immune system function is more complex and requires more sophisticated approaches. Since many low molecular weight compounds can induce allergic contact dermatitis, evaluation of skin sensitization is one of the key safety endpoints for materials used in cosmetics [2]. Delayed-type hypersensitivity (DTH) reactions to systemically administered drugs are also a safety concern which cannot be revealed by standard toxicity studies [3]. Estimation of a material's potential to cause allergic contact dermatitis in humans was historically done using the Guinea Pig Maximization Test, Buehler's Test, and more recently by the Local Lymph Node Assay (LLNA). Local Lymph Node Proliferation (LLNP) is a modification of LLNA which was shown to accurately predict DTH reaction to systemically administered pharmaceuticals [3]. The change in LLNP involves subcutaneous (s.c.) injection of a test material instead of topical application onto a skin surface like in LLNA. Both LLNA and LLNP require administration of a test material to mice once a day for 3 consecutive days, followed by 2 days of rest and then intravenous (i.v.) administration of ^{3}H-thymidine. Animals are sacrificed 5 h after thymidine injections and draining lymph nodes are analyzed by scintillation counting. ^{3}H-thymidine incorporation into DNA of lymph node resident leukocytes is used as an indication of T cell activation involved in allergic sensitization.

Several years ago, the European Cosmetics Association (COLIPA), through a number of inter-laboratory studies, validated two in vitro assays that are proposed to decrease and possibly eliminate animal use in analysis of DTH reactions. These tests are the human cell line activation test (hCLAT) and myeloid U937 skin sensitization test (MUSST or U-SENS) [2, 4–8]. For the purpose of this chapter, we will refer to the MUSST/U-SENS assay as MUSST. They were originally developed by several researchers in Japan [2, 9–13], and subsequently adopted by several European labs [14–17]. The principle of these methods is based on the fact that allergic sensitization requires formation of T cell memory, and the essential step in the process is the activation of antigen-presenting cells such as dendritic cells. Due to the complexity of isolation and high inter-donor variability between primary human dendritic cell cultures, the researchers have selected monocyte-macrophage cell lines THP-1 and U-937. They found that these cells possess one of the properties of activated primary dendritic cells, i.e., they express CD86 and/or CD54 in response to allergens. Despite original publications describing the expression of both CD54 and CD86 by THP-1 cells, both our laboratory and BD Biosciences found that THP-1 lack CD86 (NCL unpublished data and reference [18]). The hCLAT and MUSST tests described in this chapter are adaptations of previously published methods and

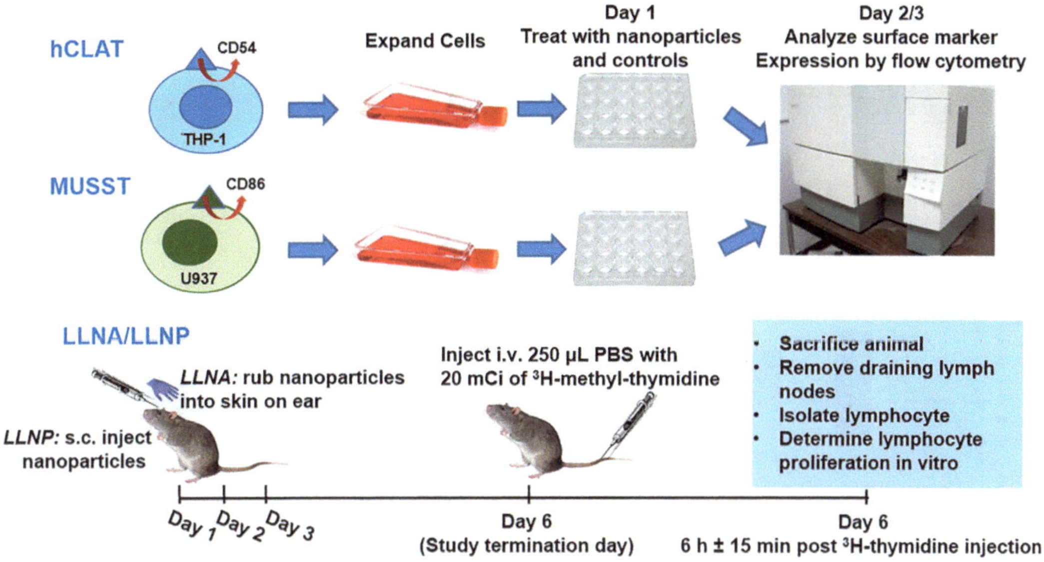

Fig. 1 Schematic of LLNA, LLNP, MUSST, and hCLAT methods. This figure depicts the major experimental steps of in vitro MUSST and hCLAT and in vivo LLNA and LLNP methods

evaluate the expression of surface markers CD54 and CD86 in THP-1 and U937 cell lines, respectively.

Both in vitro and in vivo DTH tests are schematically depicted in Fig. 1. These assays apply to engineered nanomaterials and nanotechnology-formulated drug products. However, due to the complexity of these novel materials, the in vitro-in vivo correlation between LLNA, LLNP, and hCLAT, MUSST assays may be observed for some but not all nanoparticles (Fig. 2). In vitro methods have a greater rate of positive response than in vivo tests; moreover, among in vitro assays, MUSST is more sensitive in identifying positive responses than hCLAT (Fig. 3). Therefore, the in vitro assays can be used for rapid screening of multiple nanoformulations. However, positive responders established in these in vitro tests require verification using a relevant in vivo method.

2 Materials (*See* Note 1)

2.1 hCLAT and MUSST

1. Phosphate buffered saline (PBS).
2. Heat-inactivated fetal bovine serum (FBS): Thaw a bottle of FBS at room temperature or overnight at 2–8 °C and allow to equilibrate to room temperature. Incubate 30 min at 56 °C in a water bath mixing every 5 min to heat-inactivate it. Single use aliquots may be stored at 2–8 °C for up to 1 month or at a nominal temperature of −20 °C for a longer duration.

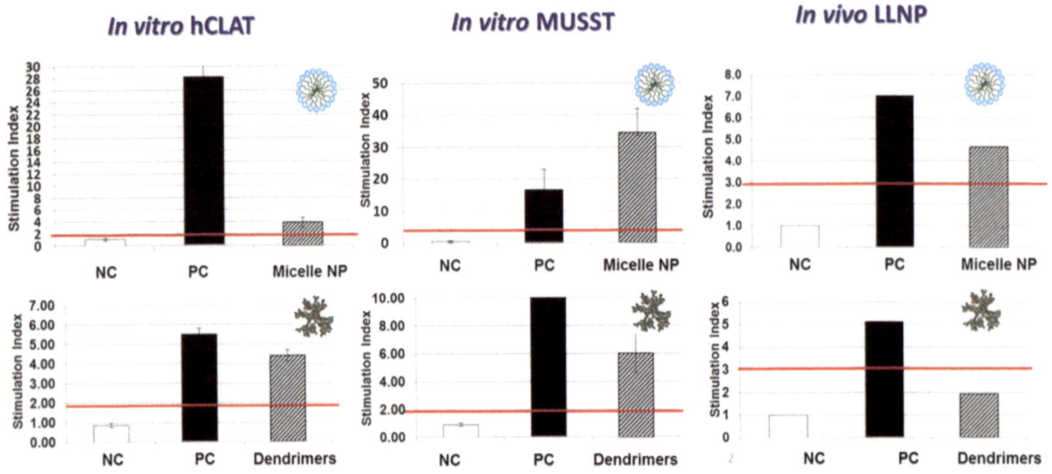

Fig. 2 Correlation between hCLAT, MUSST, and LLNP assays is observed for some but not all nanomaterials. Micelle nanoparticle (Cremophor EL) and dendrimers (generation 6 carboxy-terminated PAMAM dendrimers) were tested in vitro in hCLAT and MUSST assays and in vivo in the LLNP assay. The concentration of dendrimers tested in vitro is 100 μg/mL. That of Cremophor is 8.5 mg/mL and is equivalent to the cremophor concentration present in Taxol when the active pharmaceutical ingredient (paclitaxel) concentration is 25 μM. Taxol is the cremophor-based formulation of paclitaxel used in the clinic. It is associated with both acute and delayed type hypersensitivity. Animals were s.c. injected with 100 μL of nanoparticle stock solution. The concentration of the stock was 8.5 mg/mL and 0.1 mg/mL for cremophor and dendrimers, respectively. NC: negative control (1 mg/mL glycerol in hCLAT and MUSST and PBS in LLNP), PC: positive control (250 μg/mL nickel sulfate, 250 μM TNBS, and 50 mg/kg streptozotocin in hCLAT, MUSST, and LLNP, respectively). Shown is mean ± SD ($N = 3$). The red line depicts the minimum positive stimulation index of 3

3. Glycerol: 10 mL of a 10 mg/mL stock; filter through 0.2-micron filter. Make sure to use pyrogen-free water.
4. Reagent to distinguish between live and dead cells, such as trypan blue solution or acridine orange/propidium iodide (AO/PI) staining solution.
5. Bovine serum albumin (BSA) blocking buffer: Dissolve 50 mg of the reagent in 50 mL ice cold PBS. Mix well and place on ice. Prepare fresh before each use. Discard any leftover after the experiment is complete.
6. Globulins Cohn fraction II bocking buffer: Dissolve 5 mg of the reagent in 50 mL of ice cold PBS. Mix well and place on ice. Prepare fresh before each use. Discard any leftover after the experiment is complete.
7. 7-Amino-Actinomycin D (7-AAD).
8. Cellometer or hemacytometer to perform cell count.
9. Flow cytometer (*see* **Note 2**).

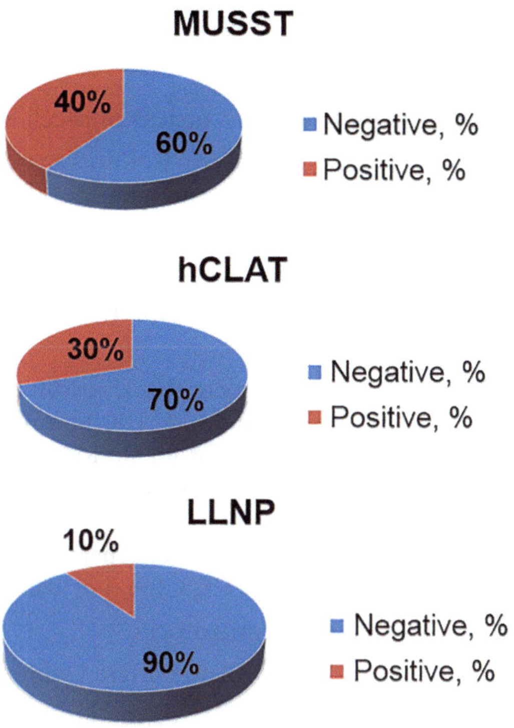

Fig. 3 Comparison of positive response rates between in vitro and in vivo tests. Twenty formulations including colloidal gold and colloidal silver nanoparticles with various sizes and surface coatings, generations 3, 4, 5, and 6 PAMAM dendrimers with different surface functionalities, iron oxide nanoparticles, liposomes, and emulsions were analyzed in vitro using hCLAT and MUSST assays and in vivo using the LLNP assay. Particles producing a stimulation index ≥3 were considered positive, while those with a stimulation index of less than 3 were considered negative. Percentage of positive and negative responders in each test is presented in the charts. Not all particles that were positive in vitro were positive in vivo, indicating that results correlated between in vitro and in vivo conditions for some but not all nanoparticles (Fig. 2, for example). The data show that in vitro assays have a greater rate of positive responses

2.2 hCLAT Only

1. Monocytic leukemia cell line THP-1 (ATCC).
2. R-phycoerythrin (PE) mouse IgG1 isotype control (BD Pharmingen, cat# 34076).
3. PE mouse anti-human CD54 (BD Pharmingen, cat# 555511).
4. Nickel(II) sulfate hexahydrate stock solution: 10 mL of a 10 mg/mL stock solution. Filter this stock through 0.2-micron filter.
5. Complete RPMI solution: 20% heat inactivated FBS, 4 mM L-glutamine, 2-mercaptoethanol to a final concentration of 0.05 mM, and 100 U/mL penicillin. Store at 2–8 °C protected

from light for no longer than 1 month. Before use, warm the medium in a water bath to 37 °C.

2.3 MUSST Only

1. Monocytic leukemia cell line U937 (ATCC).
2. Fluorescein isothiocyanate (FITC) mouse IgG1 isotype control (BD Pharmingen, cat# 555748).
3. FITC mouse anti-human CD86 (BD Pharmingen, cat# 555657).
4. 170 mM 2,4,6-Trinitrobenzenesulfonic acid solution (TNBS) stock.
5. Complete RPMI medium: 20% FBS (heat inactivated), 4 mM L-glutamine, 100 U/mL penicillin. Store at 2–8 °C protected from light for no longer than 1 month. Before use, warm the medium in a water bath to 37 °C.

2.4 LLNP and LLNA

1. 7–12 week-old male and female mice of the following strains: CBA/Ca, CBA/j or Balb/c.
2. Streptozotocin.
3. Phosphate buffered saline (PBS).
4. 1 mCi ^{3}H-thymidine.
5. 5% trichloroacetic acid (TCA) solution. Prepare 5% solution by dissolving 5 g of TCA in 100 mL of distilled water.
6. Tuberculin syringe.
7. 12 × 75 mm tube with cell strainer cap.
8. Scintillation liquid.
9. Scintillation counter.
10. 20 mL scintillation vials.

3 Methods

3.1 Preparation of Controls and Cell Lines

1. For *hCLAT and MUSST*, prepare glycerol as negative control. Use 100 μL of 10 mg/mL glycerol stock per 900 μL of cell suspension to achieve a final concentration of 1 mg/mL.
2. For *hCLAT*, prepare THP-1 cell line. Cultures can be maintained by the addition of fresh medium or replacement of culture medium. Alternatively, cultures can be established by centrifugation with subsequent resuspension at 2–4 × 10^5 viable cells/mL. Subculture is needed when cell concentration reaches 8 × 10^5 cells/mL. It is important not to allow cell concentration to exceed 1 × 10^6 cells/mL. Maintain cell density between 1 × 10^5 and 1 × 10^6 viable cells/mL. Do not use cells exceeding 22 passages.

3. For *hCLAT*, prepare nickel(II) sulfate hexahydrate as a positive control. Dilute the 10 mg/mL stock solution by adding 1 mL of the stock to 3 mL of media to obtain a 2.5 mg/mL intermediate solution. Add 100 μL of the 2.5 mg/mL intermediate solution to 900 μL of cell suspension to obtain a final concentration of 250 μg/mL.
4. For *MUSST*, prepare U937 cell line. Cultures can be maintained by the addition of fresh medium or replacement of culture medium. Alternatively, cultures can be established by centrifugation with subsequent resuspension at 1–2 × 10^5 viable cells/mL. Subculture is needed when cell concentration reaches 8 × 10^5 cells/mL. It is important not to allow cell concentration to exceed 1 × 10^6 cells/mL. Maintain cell density between 1 × 10^5 and 2 × 10^6 viable cells/mL. Do not use cells exceeding 22 passages.
5. For *MUSST*, prepare TNBS solution as a positive control. Dilute the 170 mM TNBS stock solution by adding 18 μL of the stock to 982 μL of the media to obtain a 3 mM working stock. Add 100 μL of the 3 mM working stock solution to 900 μL of the cell suspension to obtain a final concentration of 300 μM.
6. For *LLNP*, prepare ^{3}H-thymidine dosing solution. Add 11.5 mL of PBS to a dosing vial. Remove stock of the radiolabeled material from storage and add 1 mL of this stock to the dosing vial containing PBS. Mix well to prepare a dosing solution. To confirm the concentration of this dosing solution, remove two 8 μL aliquots of the solution and mix with 20 mL of water, then use 1 mL aliquots of these diluted dosing solutions and mix each with 10 mL of the scintillation liquid. These aliquots are used to confirm that each milliliter of the dosing solution contains ~80 μCi of ^{3}H-thymidine. Dispose of all radioactive material and perform post-work contamination survey according to the protocol approved by your institutional committee.
7. Prepare streptozotocin by dissolving lyophilized powder in sterile PBS or saline at concentration 1.5–2 mg/mL. The final dose of streptozotocin is 50 mg/kg and the injection volume is 50–100 μL per animal.

3.2 Preparation of Study Samples

1. The *hCLAT and MUSST* require 0.4 mL of nanoparticles dissolved/resuspended in complete culture medium to a concentration of 10× the highest tested concentration. The concentration is selected based on the plasma concentration of the nanoparticle at the intended therapeutic dose. For the purpose of this protocol, this concentration is called "theoretical plasma concentration." Considerations for estimating theoretical plasma concentration were reviewed elsewhere [1] and

are summarized in Box 1. The assay will evaluate four concentrations: 10× (or when feasible 100×, 30×, or 5×) of theoretical plasma concentration, and two 1:5 serial dilutions of the theoretical plasma concentration. When the intended therapeutic concentration is unknown, the highest final concentration is 1 mg/mL or the highest reasonably achievable concentration. For example, if the final theoretical plasma concentration to be tested is 0.1 mg/mL, then a stock of 10 mg/mL will be prepared and diluted ten-fold (1 mg/mL), followed by two five-fold serial dilutions (0.2 and 0.04 mg/mL, respectively). When 0.1 mL of each sample is added to the plate and mixed with 0.9 mL of cell suspension, the final nanoparticle concentrations tested in the assay are 1.0, 0.1, 0.02, and 0.004 mg/mL.

2. For *LLNP and LLNA*, prepare test nanoparticles in an appropriate vehicle which may be peculiar to each formulation. Nanomaterials are tested at three doses, selected based on the existing information about the test nanomaterial from previous toxicological studies: (1) at the maximum tolerated dose (MTD) (e.g., a dose which demonstrated a moderate decrease in leukocyte count and was not accompanied by life-threatening abnormalities); (2) half of MTD; and, (3) quarter of MTD.

Box 1
Estimation of nanomaterial concentration for in vitro assay. The box shows an example of calculating nanoparticle concentrations for in vitro assays. The information presented in the box has been summarized previously [1].

Box 1 Example Calculation of Nanoparticle Concentration for In Vitro Test

In this example, assume the mouse dose is known to be 123 mg/kg.

$$\text{Human dose} = \frac{\text{Mouse dose}}{12.3} = \frac{123\ \text{mg/kg}}{12.3} = 10\ \text{mg/kg}$$

Blood volume constitutes approximately 8% of the body weight (e.g., a 70 kg human has approximately 5.6 L of blood (or 8% of 70 kg)), allowing for a rough estimation of the blood concentration.

(continued)

$$\text{In vitro concentration}_{\text{human matrix}} = \frac{\text{Human dose}}{\text{Human blood volume}} = \frac{70\ \text{kg}^{*}10\ \text{mg/kg}}{5.6\ \text{L}} = 0.125\ \text{mg/mL}$$

3.3 hCLAT Experimental Procedure

1. Adjust THP-1 cell number to 1×10^6 viable cells/mL and aliquot 900 μL of the cell suspension to each well on the 24-well plate.
2. Add 100 μL of test nanoparticles or controls to appropriate wells. Prepare duplicate wells for each sample. Incubate for 24 h at 37 °C and 5% CO_2.
3. After 24 h of incubation with test materials and controls, harvest the cells and determine cell viability by trypan blue or another relevant method.
4. Harvest cells into round bottom 12 × 75 mm tubes and centrifuge at 400 × *g* for 5 min.
5. Aspirate supernatant and wash cells with 2 mL of ice cold PBS.
6. Resuspend cells in 400 μL of ice cold Cohn fraction blocking buffer and incubate on ice for 10 min.
7. Wash cells two times with 1 mL of BSA blocking buffer using centrifugation at 400 × *g* at 4 °C for 5 min to pellet the cells. Discard wash supernatant.
8. After final wash, resuspend cells in 100 μL of ice cold BSA blocking buffer.
9. Add 20 μL of monoclonal antibody against CD54 and isotype control antibody to appropriate tubes (final concentration of antibody is 5 μg/mL).
10. Incubate at 4 °C for 30 min.
11. Wash cells twice with 1 mL of BSA blocking buffer and centrifugation at 400 × *g* for 5 min to pellet the cells and discard the wash supernatant.
12. After final wash, resuspend the cells in 400 μL of BSA blocking buffer.
13. Add 5 μL of 7AAD to each sample.
14. Analyze the samples using flow cytometer (*see* **Note 2**; Fig. 4).

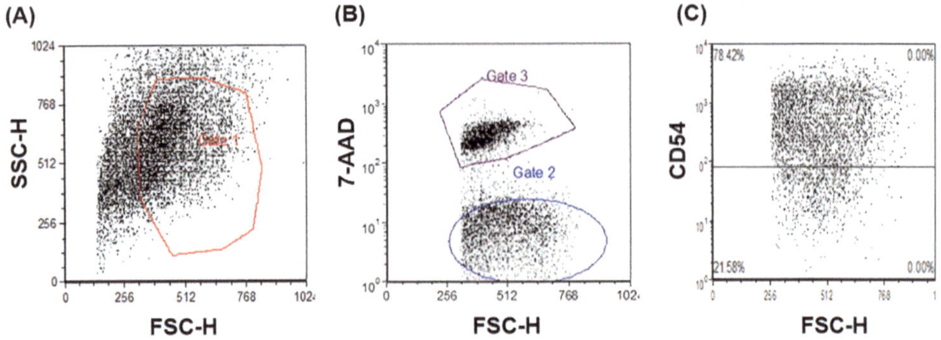

Fig. 4 An example of flow cytometry analysis of CD54 expression in THP-1 cells. (**a**) Forward scatter (FSC) vs. side scatter (SSC) dot plot gated on cells of interest. (**b**) FSC vs. FL3 cells stained with 7AAD dot plot showing 7AAD negative cells (gate 2). Gate 3 shows dead cells. (**c**) FSC vs. FL2 profile of cells stained with mouse anti-human CD54-PE dot plot gated to show only cells from gate 2

3.4 MUSST Experimental Procedure

1. Adjust U937 cell count to 1.25×10^5 cells/mL and aliquot 900 μL of the cell suspension to each well on the 24-well plate.
2. Add 100 μL of test nanoparticles or controls to appropriate wells. Prepare duplicate wells for each sample.
3. After 48 h of incubation with test materials and controls, harvest the cells and determine cell viability using trypan blue or another similar approach.
4. Harvest cells into polystyrene round bottom 12 × 75 mm tubes and centrifuge at 400 × *g* for 5 min.
5. Aspirate supernatant and wash cells with 2 mL of ice cold PBS.
6. Resuspend cells in 400 μL of ice cold Cohn fraction blocking buffer and incubate on ice for 10 min.
7. Wash cells two times with 1 mL of BSA blocking buffer using centrifugation at 400 × *g* at 4 °C for 5 min to pellet the cells and discard wash supernatant.
8. After final wash, resuspend cells in 100 μL of ice cold BSA blocking buffer.
9. Add 20 μL of monoclonal antibody against CD86 and isotype control antibody to appropriate tubes (final concentration of the antibody is 5 μg/mL).
10. Incubate at 4 °C for 30 min.
11. Wash cells twice with 1 mL of BSA blocking buffer and centrifuge at 400 × *g* for 5 min to pellet the cells. Discard the wash supernatant.
12. After final wash, resuspend the cells in 400 μL of BSA blocking buffer.
13. Add 5 μL of 7AAD to each sample.
14. Analyze the samples using flow cytometer (*see* **Note 2**; Fig. 5).

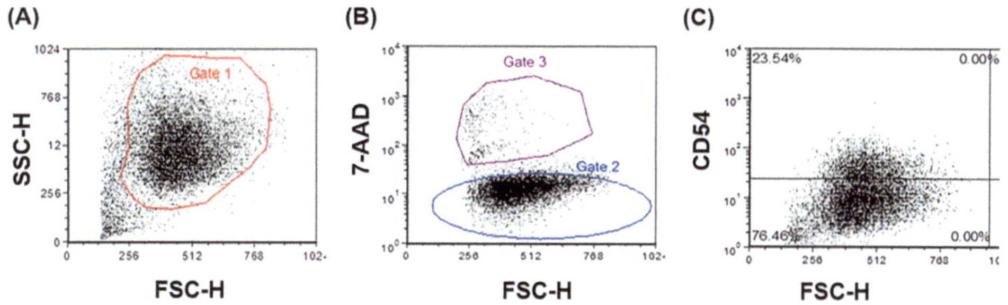

Fig. 5 An example of flow cytometry analysis of CD86 expression in U937 cells. (**a**) Forward scatter (FSC) vs. side scatter (SSC) dot plot gated on cells of interest. (**b**) FSC vs. FL3 cells stained with 7AAD dot plot showing 7AAD negative cells (gate 2). Gate 3 shows dead cells. (**c**) FSC vs. FL1 profile of cells stained with mouse anti-human CD86 FITC dot plot gated to show only cells from gate 2

3.5 LLNA and LLNP Experimental Procedure

1. On Day 1, weigh each mouse before applying treatment (*see* **Note 3**). If using LLNA, apply 25 μL of the test solution (either negative control [PBS], positive control [10% neutral buffered formalin] or nanoparticles) topically to the dorsal surface of both ears (*see* **Note 4**). Just before application of the treatment, wipe the dorsal surface of each ear with acetone to aid in absorption. If using LLNP, inject 100 μL of the test solution (either negative control [PBS], positive control [streptozotocin] or nanoparticles) s.c. in the midline of the head (*see* **Notes 5** and **6**).
2. Repeat Day 1 treatment on 2 subsequent days (Days 2 and 3). Observe each animal for signs of local irritation.
3. On Days 4, 5, and 6 of the study, observe each animal for signs of local irritation and record animal weights.
4. On Day 6 (72 ± 3 h) after the first treatment, inject each animal intravenously (i.v.) with 250 μL of sterile PBS containing 20 μCi of ^{3}H-thymidine via the lateral tail vein. Use a 1 mL syringe with a needle size no larger than 25 gauge (*see* **Note 7**).
5. Sacrifice the animals 5 h ± 45 min after injection of the thymidine using carbon dioxide asphyxiation or another technique approved by your institute's animal care committee. Isolate the draining lymph nodes of each ear and collect them into vials containing sterile PBS (*see* **Note 8**).
6. Place lymph nodes into the top compartment of the cell strainer cap. Using the plunger from the 1 mL syringe, press the lymph node into the 5 mL round bottom tube. Prepare one tube per individual animal.
7. Rinse the cap with 5 mL of PBS to move all single cells into the tube. Discard the strainer cap and remaining tissue debris.

8. Centrifuge for 10 min at 190 × *g* and 4 °C. Remove supernatant and leave 1 mL of solution above the pellet.
9. Repeat **step 8** two more times.
10. After the final wash, remove all PBS and gently agitate to break up the pellet.
11. Add 3 mL of 5% TCA and incubate at 4 °C overnight.
12. Centrifuge for 10 min at 190 × *g* and 4 °C. Remove supernatant and resuspend the pellet in 1 mL of 5% TCA.
13. Transfer each sample to a scintillation vial and add 10–15 mL of scintillation fluid.
14. Shake well. Dark adapt the samples for 30 min.
15. Analyze the samples using a scintillation counter.

3.6 Calculations and Data interpretation: hCLAT and MUSST

1. A percent coefficient of variation should be calculated for each control or test according to the following formula:

$$\%CV = \frac{\text{Standard deviation}}{\text{Mean}} \times 100\%$$

2. Measure the stimulation index (SI) using the following formula:

$$SI = \frac{\text{Mean\%positive cells}_{\text{test sample}}}{\text{Mean\%positive cells}_{\text{negative control}}}$$

An SI ≥3 is considered a positive response.

3. %CV for each control and test sample should be less than 30%.
4. If a positive control or negative control fails to meet acceptance criterion described above, the assay should be repeated.
5. Within the acceptable assay, if two of three replicates of unknown sample fail to meet acceptance criterion described in **step 3** above, this unknown sample should be re-analyzed.

3.7 Calculations and Data Interpretation: LLNA and LLNP

1. The proliferative response (PR) in each sample is determined by the number of radioactive disintegrations per minute per lymph node, using the following formula:

$$PR = \frac{\left(DPM_{\text{test sample}} - DPM_{\text{background}}\right)}{LN}$$

where, DPM is the result reported by the scintillation counter and LN is the number of the lymph nodes in the sample. The background DPM ($DPM_{\text{background}}$) is determined by analyzing 10 mL of the scintillation liquid without any sample.

2. Measure SI according to the formula:

$$SI = \frac{PR_{\text{test sample}}}{PR_{\text{negative control}}}$$

An SI of ≥3 is considered a positive response.

4 Notes

1. When using reagents and instruments from sources other than that used in our protocols, the assay performance may change. We recommend performing the assay qualification to verify the assay functionality before proceeding with test samples.
2. To set up the instrument for data acquisition and analysis, please refer to the operation manual of your flow cytometer. The protocol described in this chapter was developed using a FACSCalibur. With this instrument, we first optimize the instrument settings to bring the cells in focus using forward and side scatter. Next, we create a gate around the cell population and analyze cells from this gate on the fluorescent channel 3 (FL-3) to distinguish between dead and live cells. We create a gate around the live cell population in FL-3. Cells negative for 7-AAD staining detected in this channel are the live cells, while cells positive in 7-AAD staining are dead cells and are excluded from the analysis. We acquire $CD54^{+}$ cells and $CD86^{+}$ cells from the 7-AAD negative population using fluorescent channels 2 and 1 (FL-2 and FL-1), respectively. The FL-1 and FL-2 on the FACSCalibur instrument detect the FITC and PE labels of the antibodies, respectively. Each test sample is also stained using the isotype control antibody. The percent positive cells in each sample stained with CD54 or CD86 antibody is determined against respective isotype controls of the same sample. If antibodies with different fluorophores are used, one would need to select the appropriate isotype control and adjust both the antibody concentration and the acquisition channel.
3. This procedure requires an animal study protocol and radiation program approved by relevant institutional committees at your organization.
4. Consider using a collar to prevent animals from ingesting the treatments during the grooming process.
5. The point of injection is very important. It should be in the midline, between the eyes on the top of the head. Point the needle toward the back of the animal so that the end of the

needle is just forward of the point between the ears. Injection should be performed slowly so that the bulk of the liquid is placed either between or just forward of the ears. Injecting too far to the rear will result in reduced response in the draining lymph nodes.

6. Typical injection volume is 100 μL, but it may be lower and vary between 50 and 100 μL. The volume should not exceed 5 mL/kg, and the needle size should be 25 G or smaller.
7. Placing animals on a warm plate may be needed to dilate veins for easier i.v. injections.
8. The lymph nodes should be processed as soon as possible within 2 h after isolation from the animals.

Acknowledgement

This project has been funded in whole or in part with federal funds from the National Cancer Institute, National Institutes of Health, under contract 75N91010D00024. The content of this publication does not necessarily reflect the views or policies of the Department of Health and Human Services, nor does mention of trade names, commercial products, or organizations imply endorsement by the US Government.

References

1. Dobrovolskaia MA, McNeil SE (2013) Understanding the correlation between in vitro and in vivo immunotoxicity tests for nanomedicines. J Control Release 172(2):456–466
2. Sakaguchi H, Ryan C, Ovigne JM, Schroeder KR, Ashikaga T (2010) Predicting skin sensitization potential and inter-laboratory reproducibility of a human Cell Line Activation Test (h-CLAT) in the European Cosmetics Association (COLIPA) ring trials. Toxicol In Vitro 24(6):1810–1820
3. Weaver JL, Chapdelaine JM, Descotes J, Germolec D, Holsapple M, House R, Lebrec H, Meade J, Pieters R, Hastings KL, Dean JH (2005) Evaluation of a lymph node proliferation assay for its ability to detect pharmaceuticals with potential to cause immune-mediated drug reactions. J Immunotoxicol 2(1):11–20
4. Bauch C, Kolle SN, Fabian E, Pachel C, Ramirez T, Wiench B, Wruck CJ, van Ravenzwaay B, Landsiedel R (2011) Intralaboratory validation of four in vitro assays for the prediction of the skin sensitizing potential of chemicals. Toxicol In Vitro 25(6):1162–1168
5. Nukada Y, Ashikaga T, Miyazawa M, Hirota M, Sakaguchi H, Sasa H, Nishiyama N (2012) Prediction of skin sensitization potency of chemicals by human Cell Line Activation Test (h-CLAT) and an attempt at classifying skin sensitization potency. Toxicol In Vitro 26(7):1150–1160
6. Natsch A, Ryan CA, Foertsch L, Emter R, Jaworska J, Gerberick F, Kern P (2013) A dataset on 145 chemicals tested in alternative assays for skin sensitization undergoing prevalidation. J Appl Toxicol 33(11):1337–1352
7. Piroird C, Ovigne JM, Rousset F, Martinozzi-Teissier S, Gomes C, Cotovio J, Alepee N (2015) The Myeloid U937 Skin Sensitization Test (U-SENS) addresses the activation of dendritic cell event in the adverse outcome pathway for skin sensitization. Toxicol In Vitro 29(5):901–916
8. Urbisch D, Mehling A, Guth K, Ramirez T, Honarvar N, Kolle S, Landsiedel R, Jaworska J, Kern PS, Gerberick F, Natsch A, Emter R, Ashikaga T, Miyazawa M, Sakaguchi

H (2015) Assessing skin sensitization hazard in mice and men using non-animal test methods. Regul Toxicol Pharmacol 71(2):337–351

9. Sakaguchi H, Ashikaga T, Miyazawa M, Yoshida Y, Ito Y, Yoneyama K, Hirota M, Itagaki H, Toyoda H, Suzuki H (2006) Development of an in vitro skin sensitization test using human cell lines; human Cell Line Activation Test (h-CLAT). II. An inter-laboratory study of the h-CLAT. Toxicol In Vitro 20(5): 774–784
10. Ashikaga T, Yoshida Y, Hirota M, Yoneyama K, Itagaki H, Sakaguchi H, Miyazawa M, Ito Y, Suzuki H, Toyoda H (2006) Development of an in vitro skin sensitization test using human cell lines: the human Cell Line Activation Test (h-CLAT). I. Optimization of the h-CLAT protocol. Toxicol In Vitro 20(5):767–773
11. Sakaguchi H, Ashikaga T, Miyazawa M, Kosaka N, Ito Y, Yoneyama K, Sono S, Itagaki H, Toyoda H, Suzuki H (2009) The relationship between CD86/CD54 expression and THP-1 cell viability in an in vitro skin sensitization test--human cell line activation test (h-CLAT). Cell Biol Toxicol 25(2): 109–126
12. Miyazawa M, Ito Y, Yoshida Y, Sakaguchi H, Suzuki H (2007) Phenotypic alterations and cytokine production in THP-1 cells in response to allergens. Toxicol In Vitro 21(3):428–437
13. Yoshida Y, Sakaguchi H, Ito Y, Okuda M, Suzuki H (2003) Evaluation of the skin sensitization potential of chemicals using expression of co-stimulatory molecules, CD54 and CD86, on the naive THP-1 cell line. Toxicol In Vitro 17(2):221–228
14. Ball N, Gordon N, Casal E, Parish J (2011) Evaluation of auto bi-level algorithm to treat pressure intolerance in obstructive sleep apnea. Sleep Breath 15(3):301–309
15. Aeby P, Ashikaga T, Bessou-Touya S, Schepky A, Gerberick F, Kern P, Marrec-Fairley M, Maxwell G, Ovigne JM, Sakaguchi H, Reisinger K, Tailhardat M, Martinozzi-Teissier S, Winkler P (2010) Identifying and characterizing chemical skin sensitizers without animal testing: Colipa's research and method development program. Toxicol In Vitro 24(6):1465–1473
16. Maxwell G, Aeby P, Ashikaga T, Bessou-Touya S, Diembeck W, Gerberick F, Kern P, Marrec-Fairley M, Ovigne JM, Sakaguchi H, Schroeder K, Tailhardat M, Teissier S, Winkler P (2011) Skin sensitisation: the Colipa strategy for developing and evaluating non-animal test methods for risk assessment. ALTEX 28(1): 50–55
17. Luebke R (2012) Immunotoxicant screening and prioritization in the twenty-first century. Toxicol Pathol 40(2):294–299
18. Mittar T, Parambar R, McIntyre C (2011) Flow cytometry and high-content imaging to identify markers of monocyte-macrophage differentiation. Application note

Chapter 21

Analysis of Nanoparticle Adjuvant Properties

Barry W. Neun, Edward Cedrone, and Marina A. Dobrovolskaia

Abstract

Nanoparticles can be engineered for targeted antigen delivery to immune cells and for stimulating an immune response to improve the antigen immunogenicity. This approach is commonly used to develop nanotechnology-based vaccines. In addition, some nanotechnology platforms may be initially designed for drug delivery, but in the course of subsequent characterization, additional immunomodulatory functions may be discovered that can potentially benefit vaccine efficacy. In both of these scenarios, an in vivo proof of concept study to verify the utility of the nanocarrier for improving vaccine efficacy is needed. Here we describe an experimental approach and considerations for designing an animal study to test adjuvant properties of engineered nanomaterials in vivo.

Key words Nanoparticles, Vaccines, Adjuvant, Antigen, Antibody

1 Introduction

Nanoparticles can provide a wide variety of advantages over conventional adjuvants. Some benefits include improved solubility of hydrophobic antigens, option to control antigen release, lower frequency and severity of side effects, protection of the antigen from degradation, and concurrent delivery of multiple antigens, as reviewed elsewhere [1, 2]. Different nanomaterials, including but not limited to polymeric, chitosan, magnetic, latex, gold, silica, and polystyrene nanoparticles, have been described as successful antigen carriers and vaccine adjuvants [3–10]. The most common target cells for nanoparticle-based vaccines are antigen presenting cells (APC) such as dendritic cells (DC). DC can internalize nanoparticles loaded with antigens via multiple routes [6, 11]. Macrophages have also been shown to utilize multiple pathways for nanoparticle internalization [12].

Nanoparticle physicochemical properties, such as size, zeta potential, and surface functionalities, play a fundamental role in particle interaction with the immune cells. Several studies demonstrated that smaller particles generate stronger immune responses

Jeffrey D. Clogston et al. (eds.), *Characterization of Nanoparticles Intended for Drug Delivery*, Methods in Molecular Biology, vol. 2789, https://doi.org/10.1007/978-1-0716-3786-9_21,

than their larger counterparts [3, 6–8, 11, 13]. This property can be explained by the ability of smaller particles to travel to the draining lymph nodes (LN) where they target LN-resident DC, B cells, and macrophages [13, 14]. Particles with sizes of 100 nm or larger tend to stay at the site of injection where they stimulate resident APC. A nanoparticle size of about 50 nm is suggested to be the most optimal for particle uptake by DC [15–17]. It has been further demonstrated that smaller particles are more potent in inducing IFN-α responses, while larger particles preferentially stimulate the production of TNF-α [18]. Moreover, smaller particles promote Th1 and $CD8^+$ T cell responses, while their larger counterparts preferentially stimulate Th2 responses [3, 6, 8]. Engineering the particle surface is another approach commonly used in vaccine design. For example, positively charged particles demonstrate greater uptake by DC [19–21], stronger induction of DC maturation [22–26] and higher cytotoxic T lymphocyte responses [27]. A comprehensive review of nanotechnology-based vaccines is available elsewhere [28, 29].

Nanotechnology platforms are also used to formulate drugs and imaging agents. Sometimes, in the course of characterization of such particles, one may discover immunomodulatory functions, which may potentially benefit vaccine efficacy. In this case, the carrier may be re-purposed for delivery of an antigen. In both situations, when the adjuvant property was intentionally created and when it was unexpectedly discovered, a proof of concept experiment is needed to verify the nanoparticles' adjuvant property. Here we describe an experimental approach, which can be used for such a proof of concept study, along with considerations about route of administration and dosing regimen (Fig. 1).

2 Materials

1. Six week old male and female mice of C57BL6 strain.
2. Phosphate buffered saline (PBS).
3. Model antigen, Ovalbumin-Endofit. Ovalbumin is supplied as a lyophilized powder. Reconstitute in sterile PBS or saline before injection to a final concentration of 1 mg/mL. Use freshly prepared material, discard any leftovers.
4. ELISA to detect antibody to the model antigen: Mouse anti-OVA ELISA kit.
5. ELISA to detect antibody to target antigen.
6. Test nanomaterial and its formulation buffer (vehicle control).

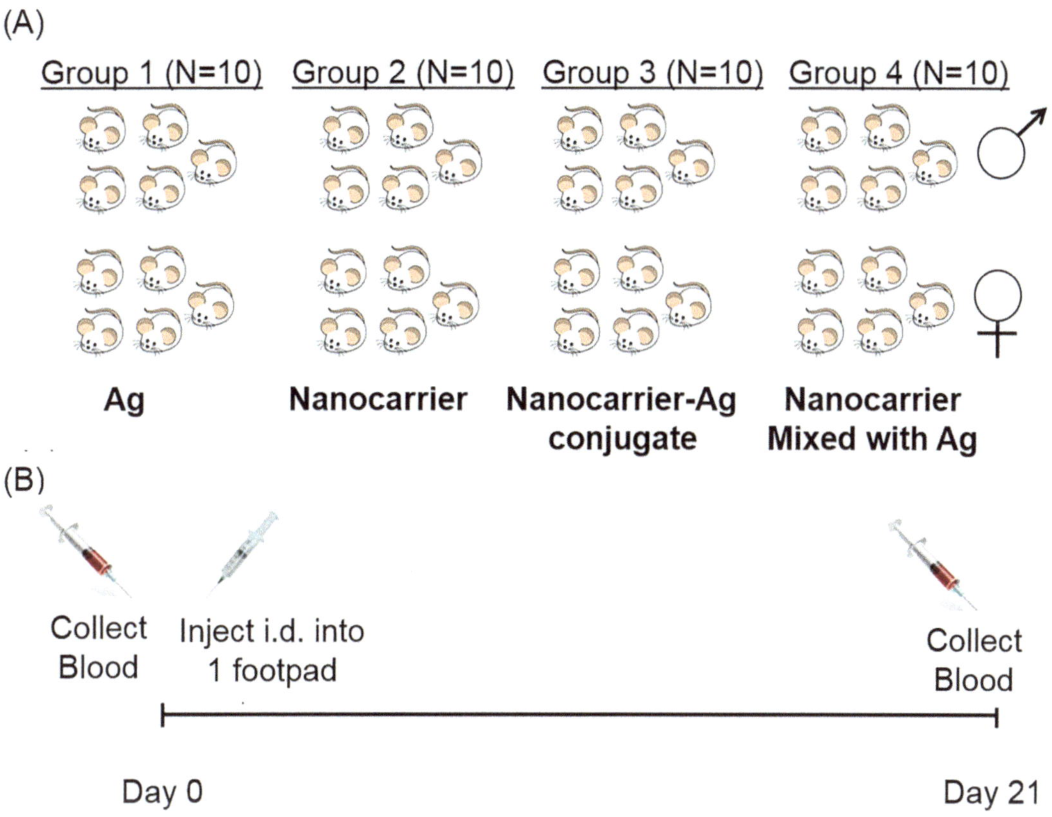

Fig. 1 Schematic of the study protocol. This figure shows the main experimental steps of the method used to assess adjuvanticity of nanoparticles. (**a**) Treatment groups and number of animals. (**b**) Study outline. *Ag* antigen, *i.d.* intradermal

3 Methods

3.1 Preparation of Control, Test Antigen, and Test Nanoparticle

1. Prepare vehicle control. Vehicle control is the buffer or media used to formulate test nanomaterials. Typical excipients employed in nanoformulations are trehalose, sucrose, and albumin. However, other reagents and materials are also used alone or in combination. Vehicle control should match the formulation buffer of the test nanomaterial by both composition and concentration. This control can be skipped if nanoparticles are stored in PBS.
2. Prepare test antigen of interest according to the manufacturer's instructions or based on the information unique to this antigen.
3. Prepare test nanoparticles in an appropriate vehicle, which is specific to each formulation. Nanomaterials are tested at three doses, selected based on the existing information about the test nanomaterial from previous immunological studies.

3.2 In Vivo Assay

1. On Day 0, determine the weight of each animal and collect 100 μL of blood via the tail vein to establish antibody background for each animal (*see* **Notes 1–4**).
2. Inject test samples and controls intradermally into the dorsal foot skin as a 30 μL total volume bolus (*see* **Notes 2** and **5**).
3. On Day 21, determine the weight of each animal, then euthanize the animals via CO_2 asphyxiation. Perform terminal blood collection via cardiac puncture (*see* **Note 6**).
4. Allow the blood to clot, centrifuge the tubes for 5 min at 2500 × *g*, and transfer serum to fresh tubes. The serum can be analyzed immediately or stored frozen at −20 °C.
5. On the day of analysis, thaw sera at room temperature and analyze by ELISA for the presence of antibodies.

3.3 Calculations and Result Interpretation

1. Percent Coefficient of Variation (%CV) should be calculated for each sample analyzed on ELISA plate according to the following formula:

$$\frac{\text{Standard Deviation}}{\text{Mean}} \times 100\%$$

2. The %CV for each sample should be less than 25%.
3. Compare IgM and IgG levels in the treatment groups with that in the control group immunized with antigen alone. Perform statistical analysis using traditional methods.

4 Notes

1. Conduct thorough physicochemical, immunological, and toxicological characterization of a nanomaterial-based formulation before its testing in the animal study. Ensure that the test nanoparticle is endotoxin-free. Endotoxin is a potent immunostimulant and can also act as a vaccine adjuvant.
2. Check your Institutional Animal Care and Use Committee requirements for animal husbandry and needle size. We do not recommend using needles larger than 29 gauge for injections. Perform injections into only one hind foot per animal. Provide soft bedding and ensure the animals are able to reach food and water. Monitor animals daily for pain/distress or complications at the injection site for the duration of the experiment or until there is no evidence of pain or discomfort. Consider humane euthanasia for mice exhibiting signs of severe pain or distress.

3. The following test groups are included in a typical study: (1) nanoparticle formulated antigen [10 animals: 5 males and 5 females]; (2) antigen alone [10 animals: 5 males and 5 females]; (3) nanoparticle carrier only [10 animals: 5 males and 5 females]; and (4) nanoparticle carrier mixed with model antigen (i.e., antigen is not bound or conjugated to the particle surface or entrapped into the particle) [10 animals: 5 males and 5 females]. Group 1 can be skipped if the test antigen is not conjugated to the nanoparticle carrier. The inclusion of the standard adjuvant (e.g., alum or complete Freund adjuvant [CFA]) should be considered to estimate the potency of the nanoparticle carrier.
4. When designing a study, one may also consider a traditional immunization protocol, which includes prime and boost dosing regimen on Days 0 and 7, respectively. The traditional protocol will require s.c. vaccination on two flanks of each animal, and therefore will require larger material quantity. However, if the test nanoparticle hydrodynamic size is less than 100 nm, nanoparticles can deliver the antigen directly into the draining lymph nodes. Intradermal injection into the footpad described in this protocol is optimal and provides better efficacy for such carriers (Fig. 2). If, however, the test nanoparticle size is over 100 nm, nanoparticles will stimulate the immune cells at the site of injection, and a traditional immunization regimen including s.c. vaccination should be considered.
5. Nanoparticle dose per injection typically does not exceed 500 μg (i.e., per 30 μL volume). The dose of the model antigen (OVA) is 30 μg per injection (i.e., per 30 μL volume). The concentration of the target antigen may be different and should be determined experimentally. Using 30 μg per injection can be considered as a starting point.
6. Additional blood collection can be scheduled on Day 7 if monitoring of antigen-specific IgM generation is of interest.

Acknowledgement

This project has been funded in whole or in part with federal funds from the National Cancer Institute, National Institutes of Health, under contract 75N91010D00024. The content of this publication does not necessarily reflect the views or policies of the Department of Health and Human Services, nor does mention of trade names, commercial products, or organizations imply endorsement by the US Government.

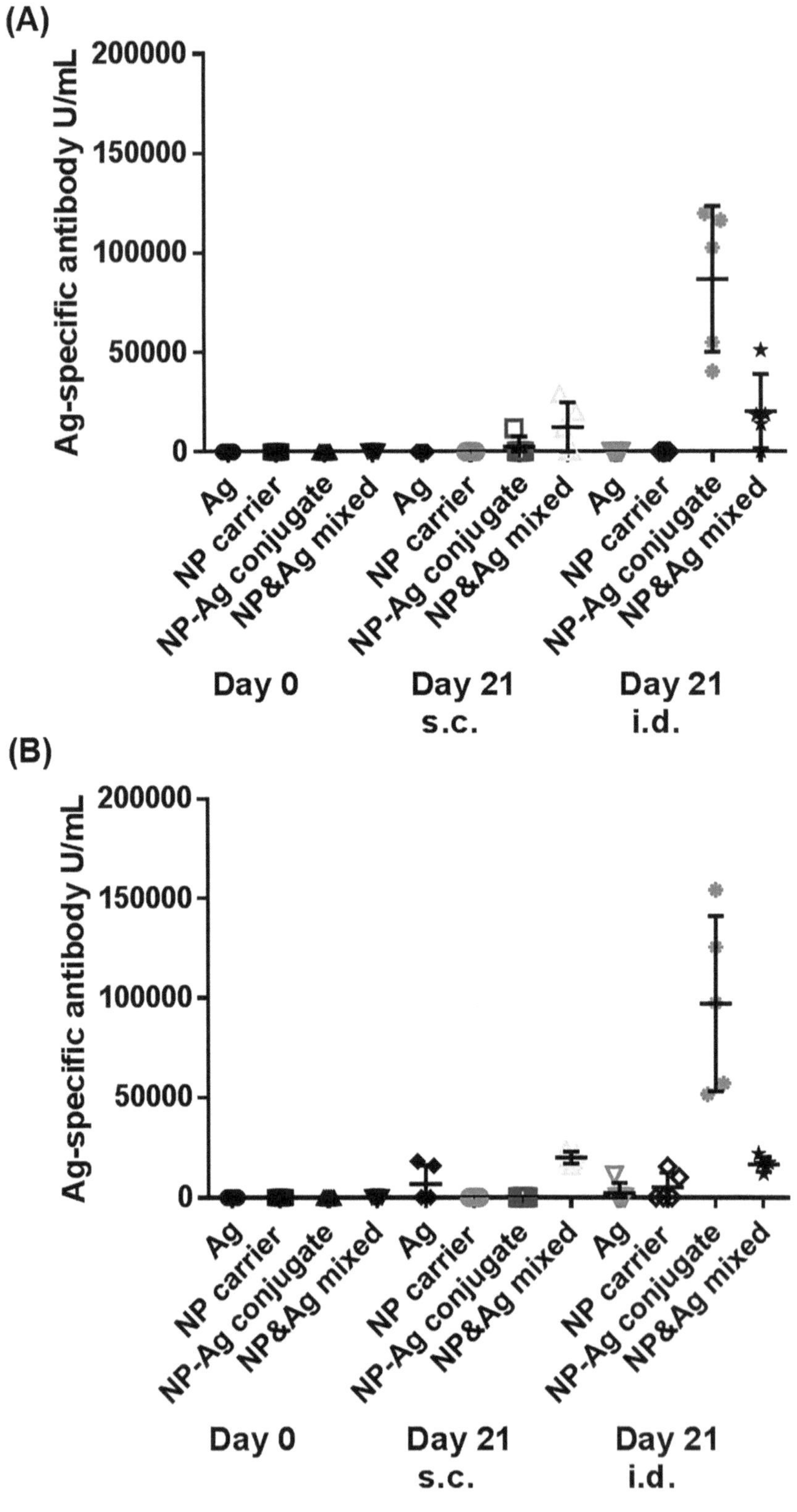

Fig. 2 An example of data generated using this protocol. The hydrodynamic size of the test nanoparticle was <100 nm. Therefore, it was expected that lymphatic drainage would be essential for the efficacy of the

References

1. Xiang SD, Scalzo-Inguanti K, Minigo G, Park A, Hardy CL, Plebanski M (2008) Promising particle-based vaccines in cancer therapy. Expert Rev Vaccines 7(7):1103–1119
2. Xiang SD, Selomulya C, Ho J, Apostolopoulos V, Plebanski M (2010) Delivery of DNA vaccines: an overview on the use of biodegradable polymeric and magnetic nanoparticles. Wiley Interdiscip Rev Nanomed Nanobiotechnol 2(3):205–218
3. Fifis T, Mottram P, Bogdanoska V, Hanley J, Plebanski M (2004) Short peptide sequences containing MHC class I and/or class II epitopes linked to nano-beads induce strong immunity and inhibition of growth of antigen-specific tumour challenge in mice. Vaccine 23(2):258–266
4. Pavelic K, Hadzija M, Bedrica L, Pavelic J, Dikic I, Katic M, Kralj M, Bosnar MH, Kapitanovic S, Poljak-Blazi M, Krizanac S, Stojkovic R, Jurin M, Subotic B, Colic M (2001) Natural zeolite clinoptilolite: new adjuvant in anticancer therapy. J Mol Med (Berl) 78(12):708–720
5. Walsh MC, Banas JA, Mudzinski SP, Preissler MT, Graziano RF, Gosselin EJ (2003) A two-component modular approach for enhancing T-cell activation utilizing a unique anti-FcgammaRI-streptavidin construct and microspheres coated with biotinylated-antigen. Biomol Eng 20(1):21–33
6. Fifis T, Gamvrellis A, Crimeen-Irwin B, Pietersz GA, Li J, Mottram PL, McKenzie IF, Plebanski M (2004) Size-dependent immunogenicity: therapeutic and protective properties of nano-vaccines against tumors. J Immunol 173(5):3148–3154
7. Minigo G, Scholzen A, Tang CK, Hanley JC, Kalkanidis M, Pietersz GA, Apostolopoulos V, Plebanski M (2007) Poly-L-lysine-coated nanoparticles: a potent delivery system to enhance DNA vaccine efficacy. Vaccine 25(7): 1316–1327
8. Mottram P, Leong D, Crimeen-Irwin B, Gloster S, Xiang SD, Meanger J, Ghildyal R, Vardaxis N, Plebanski M (2007) Type 1 and type 2 immunity following vaccination is influenced by nanoparticle size: formulation of a model vaccine for respiratory syncytial virus. Mol Pharm 4(1):73–84
9. Tighe H, Corr M, Roman M, Raz E (1998) Gene vaccination: plasmid DNA is more than just a blueprint. Immunol Today 19(2):89–97
10. Weiss R, Scheiblhofer S, Freund J, Ferreira F, Livey I, Thalhamer J (2002) Gene gun bombardment with gold particles displays a particular Th2-promoting signal that over-rules the Th1-inducing effect of immunostimulatory CpG motifs in DNA vaccines. Vaccine 20(25–26):3148–3154
11. O'Hagan DT, MacKichan ML, Singh M (2001) Recent developments in adjuvants for vaccines against infectious diseases. Biomol Eng 18(3):69–85
12. Franca A, Aggarwal P, Barsov EV, Kozlov SV, Dobrovolskaia MA, Gonzalez-Fernandez A (2011) Macrophage scavenger receptor a mediates the uptake of gold colloids by macrophages in vitro. Nanomedicine (Lond) 6(7): 1175–1188
13. Manolova V, Flace A, Bauer M, Schwarz K, Saudan P, Bachmann MF (2008) Nanoparticles target distinct dendritic cell populations according to their size. Eur J Immunol 38(5): 1404–1413
14. Reddy ST, van der Vlies AJ, Simeoni E, Angeli V, Randolph GJ, O'Neil CP, Lee LK, Swartz MA, Hubbell JA (2007) Exploiting lymphatic transport and complement activation in nanoparticle vaccines. Nat Biotech 25(10):1159–1164
15. Aoyama Y, Kanamori T, Nakai T, Sasaki T, Horiuchi S, Sando S, Niidome T (2003) Artificial viruses and their application to gene delivery. Size-controlled gene coating with Glycocluster nanoparticles. J Am Chem Soc 125(12):3455–3457

Fig. 2 (continued) formulation. The intradermal (i.d.) injection was performed according to the protocol described in this chapter. In parallel, the same particle was also administered to animals via the traditional subcutaneous (s.c.) route and dosed using a traditional prime and boost regimen on Days 0 and 7, respectively. Individual animal responses are shown for (**a**) males and (**b**) females. The data demonstrate that immunization using a single footpad injection described in this protocol resulted in greater antibody levels than repeated administration of the same nanoparticles via s.c. route. The total dose of antigen delivered via the s.c. route was four times greater than that delivered via i.d. route. Antibody levels for all animals at Day 0 were the same. *Ag* antigen; *NP* nanoparticle

16. Nakai T, Kanamori T, Sando S, Aoyama Y (2003) Remarkably size-regulated cell invasion by artificial viruses. Saccharide-dependent self-aggregation of glycoviruses and its consequences in glycoviral gene delivery. J Am Chem Soc 125(28):8465–8475
17. Wang J, Fu L, Gu F, Ma Y (2011) Notch1 is involved in migration and invasion of human breast cancer cells. Oncol Rep 26:1295–1303
18. Rettig L, Haen SP, Bittermann AG, von Boehmer L, Curioni A, Krämer SD, Knuth A, Pascolo S (2010) Particle size and activation threshold: a new dimension of danger signaling. Blood 115(22):4533–4541
19. Foged C, Brodin B, Frokjaer S, Sundblad A (2005) Particle size and surface charge affect particle uptake by human dendritic cells in an in vitro model. Int J Pharm 298:315–322
20. Villanueva A, Cañete M, Roca AG, Calero M, Veintemillas-Verdaguer S, Serna CJ, MdP M, Miranda R (2009) The influence of surface functionalization on the enhanced internalization of magnetic nanoparticles in cancer cells. Nanotechnology 20(11):115103
21. Thiele L, Merkle HP, Walter E (2003) Phagocytosis and phagosomal fate of surface-modified microparticles in dendritic cells and macrophages. Pharm Res 20(2):221–228
22. Little SR, Lynn DM, Ge Q, Anderson DG, Puram SV, Chen J, Eisen HN, Langer R (2004) Poly-beta amino ester-containing microparticles enhance the activity of nonviral genetic vaccines. Proc Natl Acad Sci U S A 101(26):9534–9539
23. Thiele L, Rothen-Rutishauser B, Jilek S, Wunderli-Allenspach H, Merkle HP, Walter E (2001) Evaluation of particle uptake in human blood monocyte-derived cells in vitro. Does phagocytosis activity of dendritic cells measure up with macrophages? J Control Release 76(1-2):59–71
24. Reddy ST, van der Vlies AJ, Simeoni E, Angeli V, Randolph GJ, O'Neil CP, Lee LK, Swartz MA, Hubbell JA (2007) Exploiting lymphatic transport and complement activation in nanoparticle vaccines. Nat Biotechnol 25(10):1159–1164
25. Jilek S, Merkle HP, Walter E (2005) DNA-loaded biodegradable microparticles as vaccine delivery systems and their interaction with dendritic cells. Adv Drug Deliv Rev 57(3): 377–390
26. Jilek S, Ulrich M, Merkle HP, Walter E (2004) Composition and surface charge of DNA-loaded microparticles determine maturation and cytokine secretion in human dendritic cells. Pharm Res 21(7):1240–1247
27. Singh M, Briones M, Ott G, O'Hagan D (2000) Cationic microparticles: a potent delivery system for DNA vaccines. Proc Natl Acad Sci U S A 97(2):811–816
28. Fesenkova V (2013) Nanoparticles and dendritic cells. In: Dobrovolskaia MA, McNeil SE (eds) Handbook of immunological properties of engineered nanomaterials. World Scientific Publishing Ltd, Singapore
29. Xiang SD, Fuchsberger M, Jkarlson TDL, Hardy CL, Selomulya C, Plebanski M (2013) Nanoparticles, immunomodulation and vaccine delivery. In: Dobrovolskaia MA, McNeil SE (eds) Handbook of immunological properties of engineered nanomaterials. World Scientific Publishing Ltd, Singapore

Chapter 22

Methods for Analysis of Nanoparticle Immunosuppressive Properties

Edward Cedrone, Timothy M. Potter, Barry W. Neun, and Marina A. Dobrovolskaia

Abstract

Adverse drug effects on immune system function represent a significant concern in the pharmaceutical industry, because 10–20% of drug withdrawal from the market is attributed to immunotoxicity. Immunosuppression is one such adverse effect. The traditional immune function test used to estimate materials' immunosuppression is T cell dependent antibody response (TDAR). This method involves a 28-day in vivo study evaluating the animal's antibody titer to a known antigen (Keyhole Limpet Hemocyanin; KLH) with and without challenge. Due to the limited quantities of novel drug candidates, an in vitro method called human lymphocyte activation (HuLA) assay has been developed to substitute the traditional TDAR assay during early preclinical development. In this test, leukocytes isolated from healthy donors vaccinated with the current year's flu vaccine are incubated with Fluzone in the presence or absence of nanoparticles. The antigen-specific lymphocyte proliferation is then measured by ELISA analyzing incorporation of BrdU into DNA of the proliferating cells. Here we describe the experimental procedures for investigating immunosuppressive properties of nanoparticles by both TDAR and HuLA assays, discuss the in vitro-in vivo correlation of these methods, and show a case study using the iron oxide nanoparticle formulation, Feraheme.

Key words Nanoparticles, Immunosuppression, Leukocyte proliferation, Antigen, TDAR, HuLa

1 Introduction

Approximately 10–20% of drugs have been withdrawn from clinical use between 1969 and 2005 due to their adverse effects on immune system function [1–4]. These concerns were described in several reports from academia [1], the pharmaceutical industry [2], and the US Food and Drug Administration [3, 4]. The range of adverse immune reactions responsible for drug withdrawal included anaphylaxis, allergy, hypersensitivity, idiosyncratic reactions, and immunosuppression [1–4].

Jeffrey D. Clogston et al. (eds.), *Characterization of Nanoparticles Intended for Drug Delivery*, Methods in Molecular Biology, vol. 2789, https://doi.org/10.1007/978-1-0716-3786-9_22,

Nanotechnology-formulated complex drug formulations are a relatively new class of therapeutics. Investigation of nanoparticle immunotoxicity is performed according to the framework established for other therapeutics and includes analysis of immunosuppressive properties [5–7]. The traditional immune function test used to estimate materials' immunosuppression is T-cell dependent antibody response (TDAR) [8]. Although the likelihood of identifying immunotoxicity increases with progression from preclinical in vitro to in vivo studies, in vitro methods are often the first choice in the early phases of the drug's safety investigation. In addition to the higher throughput and lower cost, in vitro methods are also helpful when quantities of novel drug candidates are limited. To study immunosuppressive properties of novel drugs, an in vitro method called human lymphocyte activation (HuLA) assay has been developed to substitute the traditional in vivo TDAR assay [9]. The schematics of both methods are shown in Fig. 1. The HuLA assay was validated across a wide range of immunosuppressive agents with different mechanisms of action (cyclosporine, dexamethasone, rapamycin, mycophenolic acid, and methotrexate) and demonstrated sensitivity to these drugs at their respective therapeutic concentrations. The primary endpoints of TDAR and HuLA assays are antigen-specific antibody response and lymphocyte proliferation, respectively. Collinge, et al. demonstrated that cell proliferation in the HuLA assay is specific to the flu antigen and involves $CD4^+$ and $CD8^+$ T cells and B lymphocytes [9]. They further confirmed that cell proliferation is accompanied by an increase in flu-specific, antibody-secreting cells.

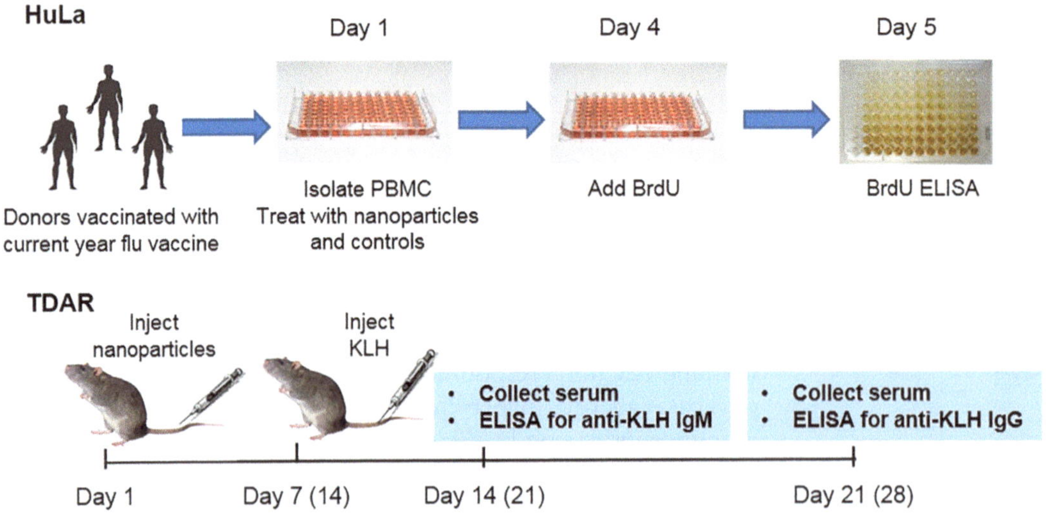

Fig. 1 Schematic of TDAR and HuLA assays. This figure shows the major experimental steps of the HuLA and TDAR methods. For the TDAR assay, the day number for KLH injection and IgM/IgG blood collection may change, as it depends on the test nanomaterial's clinically intended dosing regimen. However, there is always a 7-day interval between the KLH injection and IgM blood collection. The IgG blood collection can be done at day 14 or 21 after the KLH injection

In this chapter, we describe experimental procedures for analyzing nanoparticle immunosuppressive properties by both TDAR and HuLA assays. Furthermore, using the iron oxide nanoparticle formulation Feraheme, we show a correlation between HuLA and TDAR assays (Fig. 2a, b). We chose Feraheme as a model particle for the in vitro-in vivo comparison because several literature reports demonstrated immunosuppressive properties of another commercial iron oxide formulation, Resovist [10–12], and because immunosuppression was a commonly observed toxicity of investigational iron oxide formulations that we analyzed in our laboratory. The main limitation of this immunosuppressive assay pair is that in vitro assays cannot account for nanoparticle biodistribution, potential inter-sex differences in particle metabolism, and sensitivity to particle-mediated toxicity (Fig. 2c). Therefore, an in vitro assay is useful for identifying potential immunosuppressive nanoparticle candidates, while an in vivo method is still needed to confirm the results of the in vitro test.

2 Materials

2.1 HuLA

1. Human blood from at least three prescreened donors, anticoagulated with Li-heparin. The blood from at least three donor volunteers vaccinated with the current season flu vaccine should be drawn in vacutainers containing Li-heparin as anticoagulant. The first 10 cc collected during phlebotomy should be discarded. Cells from each donors should be tested separately (*see* **Note 1**).
2. Ficoll-Paque Premium.
3. Phosphate buffered saline (PBS).
4. 4 mg/mL clinical-grade dexamethasone (DXM) stock.
5. Fluzone (Sanofi Pasteur). This is a prescription medication that is seasonal and may not be available to all research laboratories. Fluzone is supplied as a stock with a final concentration of 90 μg/mL of influenza hemagglutinin. Dilute the commercial stock 1:50 with complete culture media. The vaccine is available in both single and multidose vials. Mutidose vials usually contain preservatives (e.g., mercury) which are contraindicated in this assay. Use only preservative-free versions.
6. Heat-inactivated fetal bovine serum (FBS). Thaw a bottle of FBS at room temperature or overnight at 2–8 °C and allow to equilibrate to room temperature. Incubate 30 min at 56 °C in a water bath mixing every 5 min to heat-inactivate it. Single use aliquots may be stored at 2–8 °C for up to 1 month or at a nominal temperature of −20 °C.
7. Complete RPMI medium: RPMI 1640, 10% FBS (heat inactivated), 2 mM L-glutamine, 100 U/mL penicillin, 100 μg/mL

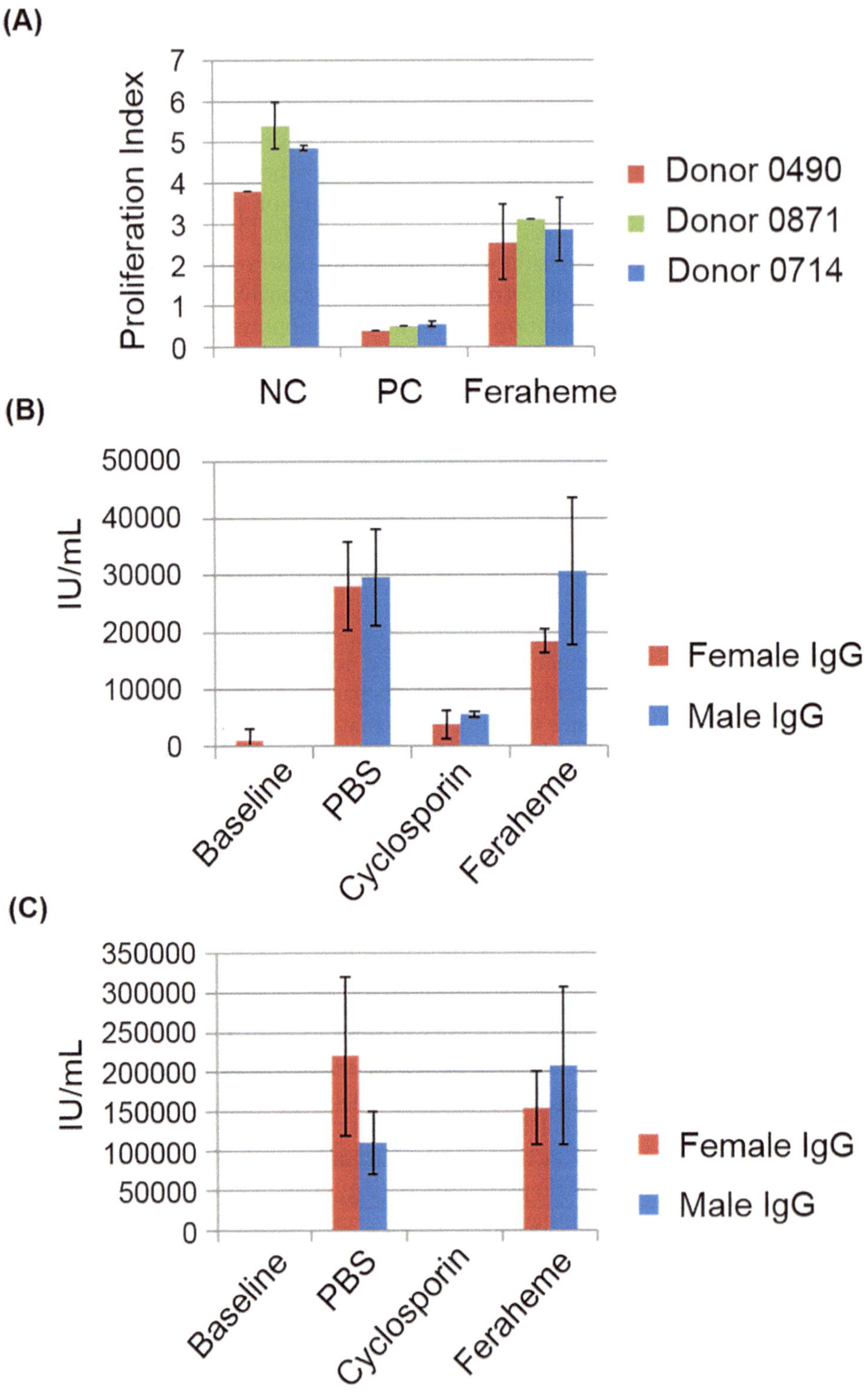

Fig. 2 Correlation between HuLA and TDAR assays. The iron oxide nanoparticle Feraheme was analyzed in vitro and in vivo. (**a**) The results of the HuLA assay show a statistically significant ($p < 0.05$) decrease in

streptomycin sulfate. Store at 2–8 °C protected from light for no longer than 1 month. Before use, warm the medium in a water bath to 37 °C.

8. Hank's balanced salt solution (HBSS).
9. Trypan Blue solution.
10. (3-(4, 5-dimethyl-2-thiazolyl)-2,5-diphenyl-2H-tetrazolium bromide) (MTT).
12. Glycine.
13. Sodium chloride.
14. Commercial BrdU cell proliferation assay (e.g., Calbiochem, QIA58). After initial thaw of the commercially supplied material, divide it into small aliquots and store at −20 °C. On the day of the experiment, thaw the required number of aliquots and dilute BrdU label 1:2000 in fresh complete media. Prepare immediately before use. Dilute BrdU-specific antibody by 1: 100 in Antibody Diluent. Prepare immediately before use. After initial thaw, divide into small aliquots and store at −20 °C. Reconstitute Peridoxase Goat Anti-Mouse IgG HRP Conjugate in 250 μL of PBS and let incubate at room temperature for 10 min. Once reconstituted, divide into small aliquots and store at −20 °C. For use, dilute Peridoxase Goat Anti-Mouse IgG HRP Conjugate in Conjugate Diluent according to the dilution instructions on the vial as the dilution factor is lot specific. Prepare immediately before use. Allow the Fixative/ Denaturing Solution to sit at room temperature for 4 h prior to use. Thaw the Conjugate Diluent, Substrate, Plate Wash Concentrate, and Stop Solution overnight at 4 °C. Once thawed, these components can be stored at 4 °C. Dilute Plate Wash Concentrate (20X) to 1X by adding 25 mL of Concentrate to 475 mL of deionized water. Store at 4 °C.
15. 96-well flat bottom plates for BrdU plate.
16. Hemocytometer.
17. Test nanomaterial and the buffer or media it was formulated in.

Fig. 2 (continued) lymphocyte proliferation in response to 0.3 mg/mL of Feraheme. *NC* Negative Control (PBS), *PC* Positive Control (Dexamethasome). (**b**) TDAR investigation was conducted in 6-week-old CD1 mice. Each treatment group included 5 males and 5 females. The animals received i.v. injection of PBS as a negative control or were fed daily with a clinical formulation of cyclosporine in oil for the positive control. Feraheme treatment groups received a single injection of 30 mg/kg Feraheme (the dose is provided in terms of iron) 1 h prior to KLH injection. The results show a decrease ($p < 0.05$) in anti-KLH IgM titer in females but not in males. (**c**) The same animals as in (**b**) were sacrificed and terminal blood was collected at day 21 of the study for analysis of serum IgG. The results show no significant change in the IgG levels or difference between males and females

2.2 TDAR

1. 6-week old male and female mice of CD-1 strain.
2. Keyhole Limpet Hemocyanin (KLH): 8 mg/mL in PBS; filter through sterile 0.2 μm filter and keep at room temperature for no longer than 2 h. Discard any leftovers. Each animal should receive 0.25 mL of this stock solution. The dose per animal is 2 mg (*see* **Note 2**).
3. 100 mg/mL Cyclosporin A (Neoral). Cyclosporin is supplied as an oral solution with a concentration of 100 mg/mL. The dose, 100 mg/kg/day, is administered to mice via oral gavage. The maximum volume of the gavage is 10 mL/kg or 0.2 mL per 20 g animal. To deliver a 100 mg/kg dose by gavage, dilute the commercial 100 mg/mL stock of cyclosporine to 10 mg/mL for a 20 g animal for example. To adjust the stock solution concentration, perform dilutions of the commercial stock in corn oil (*see* **Note 3**).
4. Phosphate buffered saline (PBS).
5. Corn oil, any brand available in a grocery store.
6. Mouse anti-KLH IgM ELISA commercial kit (e.g., Life Diagnostics Inc., 4000-1).
7. Mouse anti-KLH IgG ELISA commercial kit (e.g., Life Diagnostics Inc., 4010-1).
8. Test nanomaterial.

3 Methods

3.1 Preparation of Controls and Study Sample for HuLA

1. Prepare negative control. Use PBS and process this control in the same way as the test samples.
2. Prepare positive control of stock DXM. Clinical grade DXM is provided at a stock concentration of 4 mg/mL. Dilute the commercial stock in culture medium for a final concentration of 250 μg/mL.
3. Prepare vehicle control sample. Vehicle control is the buffer or media used to formulate the test nanomaterials. Common excipients used in nanoformulations are trehalose, sucrose, and albumin. However, other reagents and materials are also used alone or in combination. Vehicle control should match the formulation buffer of the test nanomaterial by both composition and concentration. This control can be skipped if nanoparticles are stored in PBS.
4. This assay requires 0.5 mL of nanoparticles dissolved/resuspended in complete culture medium to a concentration of 16X the highest tested concentration. The concentration is selected based on the plasma concentration of the nanoparticle at the

intended therapeutic dose. For the purpose of this protocol, this concentration is called the "theoretical plasma concentration." Considerations for estimating theoretical plasma concentration were reviewed elsewhere [13] and are summarized in Box 1. The assay will evaluate four concentrations: 10X (or when feasible 100X, 30X, or 5X) of the theoretical plasma concentration and two 1:5 serial dilutions of the theoretical plasma concentration. When the intended therapeutic concentration is unknown, the highest final concentration is 1 mg/mL or the highest reasonably achievable concentration. For example, if the final theoretical plasma concentration to be tested is 0.2 mg/mL, then a stock of 32 mg/mL will be prepared and diluted ten-fold (3.2 mg/mL), followed by two 1:5 serial dilutions (0.64 and 0.13 mg/mL, respectively). When 0.01 mL of each sample is added to the plate and mixed with 0.05 mL of Fluzone and 0.1 mL of the cell suspension, the final nanoparticle concentrations tested in the assay are: 2.0, 0.2, 0.04, and 0.008 mg/mL.

Box 1 Estimation of nanomaterial concentration for in vitro assay. The box shows an example of calculating the nanoparticle concentration for the in vitro HuLA assay (adopted from [13]).

Box 1: Example Calculation of Nanoparticle Concentration for In Vitro Test

In this example, assume the mouse dose is known to be 123 mg/kg.

$$\text{Human dose} = \frac{\text{Mouse dose}}{12.3} = \frac{123\ \text{mg/kg}}{12.3} = 10\ \text{mg/kg}$$

Blood volume constitutes approximately 8% of the body weight (e.g., a 70 kg human has approximately 5.6 L (8% of 70) of blood), allowing for a very rough estimation of the maximum blood concentration.

$$\text{in vitro concentration}_{\text{human matrix}} = \frac{\text{human dose}}{\text{human blood volume}}$$

$$\frac{70\ \text{kg} * 10\ \text{mg/kg}}{5.6\ \text{L}} = 0.125\ \text{mg/mL}$$

3.2 Preparation of Study Sample for TDAR

1. Prepare test nanoparticles in appropriate vehicle, which is specific to each formulation. Nanomaterials are tested at three doses, selected based on the existing information about the test nanomaterial from previous toxicological studies, at the: (1) no observable adverse effect level (NOAEL); (2) maximum tolerated dose (MTD) (e.g., a dose which demonstrated a

moderate decrease in leukocyte count and was not accompanied by life-threatening abnormalities); and (3) dose between dose levels 1 and 2. The frequency and the route of administration for each test nanomaterial is selected based on its intended clinical use (e.g., once a week, every day, every other day, etc., and i.v., s.c., oral, etc.).

3.3 Human Lymphocyte Activation (HuLA) Assay

1. Place freshly drawn blood into 15 mL or 50 mL conical centrifuge tubes, add an equal volume of room temperature PBS, and mix well.
2. Slowly layer the Ficoll-Paque solution underneath the blood/PBS mixture by placing the tip of the pipet containing Ficoll-Paque at the bottom of the blood sample tube. Alternatively, the blood/PBS mixture may be slowly layered over the Ficoll-Paque solution. Use 3 mL of Ficoll-Paque solution per 4 mL of blood/PBS mixture. For example, use 15 mL of Ficoll-Paque solution per 20 mL of diluted blood.
3. Centrifuge 30 min at 900 × *g* at 18–20 °C, without brake.
4. Using a sterile pipet, remove the upper layer containing plasma and platelets and discard it.
5. Using a fresh sterile pipet, transfer the mononuclear cell layer into another centrifuge tube.
6. Wash cells by adding an excess of HBSS and centrifuging for 10 min at 400 × *g* at 18–20 °C. The HBSS volume should be about three times the volume of the mononuclear layer.
7. Discard the supernatant and repeat wash **step 6** one more time.
8. Resuspend cells in complete RPMI 1640 medium. Dilute cells 1:5 or 1:10 with trypan blue, count cells, and determine viability using trypan blue exclusion. If viability is at least 90%, proceed to the next step.
9. Adjust cell concentration to 1×10^6/mL with complete medium.
10. Aliquot 100 μL of cell suspension to the appropriate wells of a 96-well plate (Fig. 3). Repeat this step for each individual donor.
11. Add 10 μL of test nanoparticle, positive control, and negative control to the respective wells. Prepare no-cell control wells containing nanoparticles only (Fig. 3).
12. Incubate at 37 °C for 1 h.
13. Add 50 μL of media or Fluzone vaccine to appropriate wells. Do not add Fluzone to "No Fluzone" wells.
14. Incubate at 37 °C for 72 h.
15. Add 20 μL of BrdU label to appropriate wells on BrdU plate, taking care not to add BrdU to "No BrdU" wells.

	1	2	3	4	5	6	7	8	9	10	11	12
A	Medium (baseline)	NC	VC	PC		TS1 (0.008 mg/mL)	TS1 (0.008 mg/mL)	TS1 (0.008 mg/mL)	TS1 (0.04 mg/mL)	TS1 (0.04 mg/mL)	TS1 (0.04 mg/mL)	TS1 (0.2 mg/mL)
B	Medium (baseline)	NC	VC	PC		TS1 (0.008 mg/mL)	TS1 (0.008 mg/mL)	TS1 (0.008 mg/mL)	TS1 (0.04 mg/mL)	TS1 (0.04 mg/mL)	TS1 (0.04 mg/mL)	TS1 (0.2 mg/mL)
C	TS1 (0.2 mg/mL)	TS1 (0.2 mg/mL)	TS1 (2.0 mg/mL)	TS1 (2.0 mg/mL)	TS1 (2.0 mg/mL)	NC	VC	PC	Medium (baseline)	Cells, No Fluzone	Cells, No Fluzone	Cells, No BrdU
D	TS1 (0.2 mg/mL)	TS1 (0.2 mg/mL)	TS1 (2.0 mg/mL)	TS1 (2.0 mg/mL)	TS1 (2.0 mg/mL)	NC	VC	PC	Medium (baseline)	Cells, No Fluzone	Cells, No Fluzone	Cells, No BrdU
E	Cells, No BrdU											
F	Cells, No BrdU											
G	TS1 (0.008 mg/mL)	TS1 (0.04 mg/mL)	TS1 (0.2 mg/mL)	TS1 (2.0 mg/mL)								
H	TS1 (0.008 mg/mL)	TS1 (0.04 mg/mL)	TS1 (0.2 mg/mL)	TS1 (2.0 mg/mL)								

Fig. 3 Example plate map for the HuLA assay. Wells 1–4 in Rows G and H are the cell-free test samples; they do not receive cells. *NC* Negative Control, *PC* Positive Control; *TS* Test Sample, *VC* Vehicle Control

16. Incubate at 37 °C for 24 h.
17. Spin BrdU plate for 4 min at 400 × *g*.
18. Aspirate media from BrdU plate, add 200 μL of fixative per well, and incubate at room temperature for 30 min. Remove fixative and tap the plate on paper towel.
19. Add 100 μL of diluted BrdU-specific antibody to all wells.
20. Incubate at room temperature for 1 h.
21. Aspirate plate.
22. Wash plate 3 times using 250 μL/well of 1× Wash Buffer. Blot the plate on paper towels to remove excess buffer.
23. Add 100 μL of Peroxidase-conjugated Goat Anti Mouse IgG to all wells.
24. Incubate at room temperature for 30 min.
25. Aspirate plate.
26. Wash plate 3 times using 250 μL/well of 1× Wash Buffer. Blot the plate on paper towels to remove excess buffer.
27. Fill wells completely with distilled water.
28. Aspirate plate. Blot the plate on paper towels to remove excess buffer.

29. Add 100 μL of Substrate Solution to all wells.
30. Incubate at room temperature for 30 min in the dark.
31. Add 100 μL of Stop Solution.
32. Read plate at dual wavelengths of 450 nm (Lm1) and 540 nm (Lm2) within 30 min of addition of Stop Solution. The final result is recorded as OD = Lm1-Lm2.

3.4 TDAR Assay

1. On Day 1 of the study, weigh each animal and administer nanoparticles at dose level and route of administration, as discussed in Subheading 3.2. The typical treatment groups, each containing five males and five females, include: Untreated control, vehicle alone, KLH alone, cyclosporin/KLH (positive control), and nanoparticles/KLH (test sample).
2. On Day 11, administer KLH (*see* **Note 4**).
3. On Day 17 weigh each animal, then collect 100 μL of blood from each mouse via tail vein. Allow blood to clot. Collect and freeze serum in a fresh tube for the subsequent IgM analysis (*see* **Note 4**).
4. On Day 25 weigh each animal, then collect 100 μL of blood from each mouse via tail vein. Allow blood to clot. Collect and freeze serum in a fresh tube for the subsequent IgG analysis. Humanely euthanize the animals according to your institute's animal care and use policies, for example via carbon dioxide asphyxiation (*see* **Note 4**).
5. Analyze serum using a commercial ELISA kit for the presence of KLH-specific IgM and IgG.

3.5 Calculations and Data Interpretation for HuLA

1. A percent coefficient of variation should be calculated for each control or test according to the following formula:

$$\frac{\text{Standard Deviation}}{\text{Mean}} \times 100\%$$

%CV for each control and test sample should be less than 30%.

2. Calculate the Stimulation Index (SI) using the following formula:

$$\text{Mean OD} - \text{test sample}\frac{\text{Mean OD}_{\text{cells, no BrdU}}}{\text{Mean OD}_{\text{No Fluzone sample}}} - \text{Mean OD}_{\text{cells,no BrdU}}$$

A good SI in the absence of an immunosuppressive agent is ≥3. Preferable SI is ≥6–7.

3. If positive control or negative control fails to meet acceptance criterion in **step 1** above, the assay should be repeated.

4. Within the acceptable assay, if two of three replicates of unknown sample fail to meet acceptance criterion described in **step 1**, this unknown sample should be re-analyzed.
5. If significant variability is observed in results obtained using leukocytes from three initial donors, the experiment needs to be repeated with additional donor cells.
6. The positive control is considered positive if it results in at least a two-fold reduction in SI, when compared to the baseline sample.
7. Statistical analysis is used to identify the nanoparticle candidate with potential immunosuppressive properties.

3.6 Calculations and Data Interpretation for TDAR

1. %CV for each sample analyzed on ELISA plate should be calculated as described in **step 1** of Subheading 3.5 for the HuLA assay. %CV should be no more than 25%.
2. Compare IgM and IgG levels in the treatment groups with that in the control group immunized with KLH alone. Perform statistical analysis using traditional methods.

4 Notes

1. Although we do not endorse suppliers, and materials from other sources can be used, we found that a new user benefits from knowing catalog information of reagents used in our assays. If you need ideas of what reagents we have used, please review NCL Method ITA-18 available at https://www.cancer.gov/nano/research/ncl/protocols-capabilities. When other reagents are used, the assay performance may change. When using reagents and instruments from sources other than that used in our protocols, assay performance qualification is needed to verify the assay functionality and validity.
2. Check your institutional Animal Care and Use Committee (ACUC) requirements for the needle size. We do not recommend using needles larger than 25 gauge for i.v. injections.
3. The commercial formulation of cyclosporine A contains castor oil. Castor oil will cause diarrhea in animals which may lead to dehydration. To avoid diarrhea in the positive control group, corn oil should be used for the preparation of the working stock of cyclosporine, which is administered to animals via oral gavage.
4. The timing between nanoparticle administration and that of KLH may be modified to select the most relevant schedule mimicking the clinically intended use. For example, nanoparticles and KLH can be administered on Day 1 with an hour interval. This dosing regimen is appropriate for nanoparticles

intended for a single dose administration. Such a change will result in a different number of days for the study. Blood collection for IgM and IgG assessment is done 7 and 14–21 days after KLH administration, respectively.

Acknowledgement

This project has been funded in whole or in part with federal funds from the National Cancer Institute, National Institutes of Health, under contract 75N91010D00024. The content of this publication does not necessarily reflect the views or policies of the Department of Health and Human Services, nor does mention of trade names, commercial products, or organizations imply endorsement by the US Government.

References

1. Wilke RA, Lin DW, Roden DM, Watkins PB, Flockhart D, Zineh I, Giacomini KM, Krauss RM (2007) Identifying genetic risk factors for serious adverse drug reactions: current progress and challenges. Nat Rev Drug Discov 6(11):904–916
2. Smith DA, Schmid EF (2006) Drug withdrawals and the lessons within. Curr Opin Drug Discov Devel 9(1):38–46
3. Wysowski DK, Nourjah P (2004) Analyzing prescription drugs as causes of death on death certificates. Public Health Rep 119(6):520
4. Wysowski DK, Swartz L (2005) Adverse drug event surveillance and drug withdrawals in the United States, 1969–2002: the importance of reporting suspected reactions. Arch Intern Med 165(12):1363–1369
5. Tyner K, Sadrieh N (2011) Considerations when submitting nanotherapeutics to FDA/CDER for regulatory review. Methods Mol Biol 697:17–31
6. Tyner KM, Zou P, Yang X, Zhang H, Cruz CN, Lee SL (2015) Product quality for nanomaterials: current U.S. experience and perspective. Wiley Interdiscip Rev Nanomed Nanobiotechnol 7(5):640–654
7. Bancos S, Tyner K, Weaver JL (2013) Immunotoxicity testing for drug–nanoparticle conjugates: regulatory considerations. In: Dobrovolskaia MA, McNeil SE (eds) Handbook of immunological properties of engineered nanomaterials. World Scientififc Publishing, Singapore
8. Dobrovolskaia MA, Germolec DR, Weaver JL (2009) Evaluation of nanoparticle immunotoxicity. Nat Nanotechnol 4(7):411–414
9. Collinge M, Cole SH, Schneider PA, Donovan CB, Kamperschroer C, Kawabata TT (2010) Human lymphocyte activation assay: an in vitro method for predictive immunotoxicity testing. J Immunotoxicol 7(4):357–366
10. Shen CC, Liang HJ, Wang CC, Liao MH, Jan TR (2011) A role of cellular glutathione in the differential effects of iron oxide nanoparticles on antigen-specific T cell cytokine expression. Int J Nanomedicine 6:2791–2798
11. Shen CC, Liang HJ, Wang CC, Liao MH, Jan TR (2012) Iron oxide nanoparticles suppressed T helper 1 cell-mediated immunity in a murine model of delayed-type hypersensitivity. Int J Nanomedicine 7:2729–2737
12. Shen CC, Wang CC, Liao MH, Jan TR (2011) A single exposure to iron oxide nanoparticles attenuates antigen-specific antibody production and T-cell reactivity in ovalbumin-sensitized BALB/c mice. Int J Nanomedicine 6:1229–1235
13. Dobrovolskaia MA, McNeil SE (2013) Understanding the correlation between in vitro and in vivo immunotoxicity tests for nanomedicines. J Control Release 172(2): 456–466. PMID: 23742883; PMCID: PMC5831149. https://doi.org/10.1016/j.jconrel.2013.05.025

Chapter 23

Evaluation of the Acute Anaphylactoid Reactogenicity of Nanoparticle-Containing Medicines and Vaccines Using the Porcine CARPA Model

Janos Szebeni

Abstract

A small fraction, up to 10%, of people treated intravenously with state-of-the-art nanoparticulate drugs or diagnostic agents develop an acute infusion reaction which can be severe or even lethal. Activation of the complement (C) system can play a causal, or contributing role in these atypical, "pseudoallergic" reactions, hence their name, C activation-related pseudoallergy (CARPA). Intravenous (i.v.) administration of the human reaction-triggering (very small) dose of a test sample in pigs triggers a symptom tetrad (characteristic hemodynamic, hematological, skin, and laboratory changes) that correspond to the major human symptoms. Quantitating these changes provides a highly sensitive and reproducible method for assessing the risk of CARPA, enabling the implementation of appropriate preventive measures. Accordingly, the porcine CARPA model has been increasingly used for the safety evaluation of therapeutic and diagnostic nanomedicines and, recently, mRNA-lipid nanoparticle vaccines. This chapter provides details of the experimental procedure followed upon using the model.

Key words Nanoparticles, CARPA, Complement, Hemodynamic changes, Pig

1 Introduction

Severe manifestations of nanomedicine (NM)-induced hypersensitivity reactions (HSRs), also known as infusion, or anaphylactoid reactions, are usually rare (occur in less than 10% of patients). Despite decades of research showing that the incidence of such HSRs depends on both drug and patient features, it is largely unknown why some NMs and patients are prone to these reactions and why others are not. For this reason, assessing the reactogenicity of NMs in an animal model of reactive, i.e., hypersensitive people, is an important safety measure.

It is known that complement activation can play a causal or contributing role to HSRs to many NMs, micellar solvents, polymers, biologicals, contrast agents, conjugated macromolecules

Jeffrey D. Clogston et al. (eds.), *Characterization of Nanoparticles Intended for Drug Delivery*, Methods in Molecular Biology, vol. 2789, https://doi.org/10.1007/978-1-0716-3786-9_23,

[1–3] and vaccines that get in direct contact with blood [4, 5]. The phenomenon has been referred to as complement activation-related pseudo-allergy (CARPA) [1–3, 6, 7]; thus, measurement of complement activation by reactogenic nanoparticles (NPs) has become an established approach to evaluate the CARPAgenic potential of NPs.

However, assessing complement activation in vitro informs only about the afferent arm of HSRs, one of more possible pathways of reaction induction. CARPA is a complex in vivo phenomenon, a chain reaction involving many molecular and cellular interactions and checkpoints. The relationship between complement activation and HSR is not always straightforward; for example, not all people develop HSR in whom a reactogenic drug induces complement activation [8], and not all HSRs involve complement activation [9]. It is hypothesized that the mechanism underlying HSRs is simultaneous double (or multiple) "hits" on immune cells [10]. Thus, there is a need for in vivo assays to provide a clinically relevant, overall estimate of the CARPA risk of NMs that measures the combined impacts of all factors involved. Importantly, since CARPA develops in hypersensitive individuals, a small fraction in the population, the in vivo CARPA test must reproduce the symptoms in this small fraction, rather than the lack of response in the overwhelming majority. The porcine CARPA model, outlined in Fig. 1, is a disease model that satisfies the above goals and preconditions for many reasons. One is the identity, or similarity of symptoms to those typical of human reactions. Table 1 summarizes these symptoms in pigs and humans and categorizes them in four groups: (1) hemodynamic/cardiopulmonary, (2) hematologic, (3) dermatologic, and (4) immune changes. Most specific symptoms in these categories are identical, or very similar in pigs and humans, although some have no equivalent in the other species. For example, pulmonary hypertension, the most quantitative and reproducible symptom in pigs, has not been documented in humans, since pulmonary arterial pressure is not measured during HSRs. On the other hand, the characteristic psychosomatic and vegetative symptoms in man is not included in the pigs' clinical picture, since they cannot be measured.

Another reason why the porcine CARPA model is clinically relevant is the sensitivity of the test. It was established that the amount of drug (phospholipid in liposomes) in the minimally reactive i.v. bolus injected i.v. in pigs is approximately the same that reaches the blood of reactive patients during infusion [11]. It is usually 1–2 orders of magnitude lower than the therapeutic dose. Yet a further essential feature of the model is the intra- and inter-animal reproducibility of reactions, particularly the rise of pulmonary arterial pressure (PAP). The reactions to a certain test agent

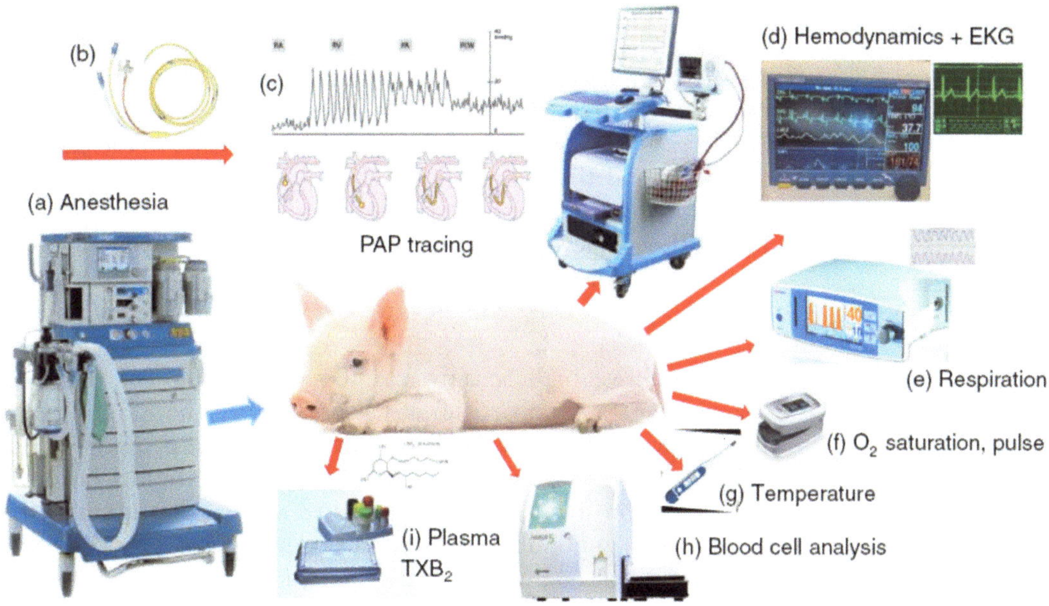

Fig. 1 Instruments and endpoints measured in the porcine CARPA model. (**a**) Anesthesia machine; (**b**) Swan-Ganz catheter; (**c**) blood pressure wave forms directing the passage of the tip of the Swan-Ganz catheter via the right atrium (RA), right ventricle (RV), and pulmonary artery (PA) until being wedged into the pulmonary capillary bed; (**d**) computerized multiple parameter hemodynamic monitoring system (1000 Hz sampling rate). From the continuous recording of systemic arterial pressure (SAP) and pulmonary arterial pressure (PAP) values, online averaging is performed and recorded, together with the heart rate, derived from SAP signal; (**e**) capnograph connected to the tracheal tube to measure respiratory rate, $etCO_2$, and $inCO_2$; (**f**) pulse oximeter (fixed on the tail) measures O_2 saturation in blood and pulse rate; (**g**) temperature is measured with a thermometer placed in the rectum; (**h**) veterinary hematology analyzer measuring all blood cell counts and WBC differential; (**i**) ELISA or other techiques for measuring biomarkers of allergic/inflammatory reactions, e.g., TxB2, histamine, prostaglandins, leukotrienes, adenosine, tryptase, PAF, cytokines, chemokines, anaphylatoxins (C3a, C5a), sC5b-9, and other immune activation endpoints, plasma protein, drug and metabolite levels, antigens and toxins, as well as natural (preformed) and/or induced antibodies as measures of the status of innate and adaptive immunity

Table 1
Symptoms of porcine CARPA and their human equivalents

Organ system	Porcine	Human	Refs
Hemodynamic/ cardiopulmonary	Hypo- or hypertension, tachy- or bradycardia, arrhythmia, ventricular fibrillation, cardiac arrest		[6]
	Pulmonary hypertension reduction of exhaled CO_2	Dyspnoea, tachypnoea	
Hematological	Leukopenia, granulopenia, compensatory leukocytosis, thrombocytopenia	Leukocytosis	[28]
Dermatological	Skin flushing, rash		
Immune	Rises of C3a, SC5b-9, and thromboxane B2 (TxB2) Fever	Rise of SC5b-9	[28]

can be reproduced in essence in all pigs, with small individual variation [12, 13]. On the other hand, real-time recordings of the initial systemic and pulmonary arterial blood pressure and heart rate changes following bolus administration of different reactive NPs present different patterns that are characteristic to the particle, the dose, and the administration conditions. This fact raises the possibility that after appropriate standardization, the model could be used for fingerprinting the adverse immune reactivity of different NPs. This specific variability of immuno-hemodynamic response reflect the complexity of the underlying mechanism [12, 13].

The above high sensitivity and reproducibility of the pigs' reaction to i.v. nanoparticles are essential since the model has to mimic the reaction in a small percentage (<5%) of people who are hypersensitive to a nanomedicine [14]. Had the model have the same reaction incidence and individual variation as seen in the general human population, hundreds of animals would need to be used to model the human situation [12, 14].

Last, but not least, the size of adolescent pigs makes it easy to measure the physiological and blood parameters detailed below. Also, justification of pig studies is usually easier than getting ethical approval for studies on dogs, other pet animals, or non-human primates.

Since its first description [6], the model has been successfully used for analyzing different aspects of CARPA caused by liposomes, liposomal drugs, emulsions, polymers, and various conjugated NPs [2, 3, 6, 12, 13, 15–26] and, recently, with the reactogenic mRNA-LNP vaccines [4, 5]. The goals and other details of these studies are summarized in Table 2. The model has also attracted regulatory recommendation as a sensitive preclinical in vivo test enabling nanotoxicological bioequivalence assays, such as needed for the approval of generic liposomal drugs [27].

2 Materials

1. InnoCare-T hemodynamic monitoring system (Innomed Medical Co. Hungary)
2. Capnograph CAP10 (Medlab, Medizinische Diagnosegerate GmbH, Karlsruhe, Germany)
3. Swan-Ganz catheter (Arrow International Inc., catalog number AI-07124) 5 Fr. 110 cm)
4. Ag/AgCl monitoring electrodes (3M Health Care, No: 2228)
5. Vetenirary hematology analyzer

Table 2
Chronological list of pig studies which included the analysis of hemodynamic changes and other endpoints of hypersensitivity reactions to i.v. drugs

Year	Tested drugs/agents[a]	Major findings[b]	Reference
1989	Cholesterol-containing liposomes used for immunizing pigs against hypercholesterolemia-induced arteriosclerosis	Liposomes caused major cardiopulmonary distress and TxA2 release via anti-cholesterol antibody-mediated C activation in pig blood	[29]
1992	Albunex microspheres used as ultrasound contrast agents	HSR involves TxA2-mediated pulmonary hypertension in pigs	[30]
1994	Albunex microspheres used as ultrasound contrast agents	NPs are cleared mainly by pulmonary intravascular macrophages (PIMs) in the lung of pigs	[31]
1998 1999	Liposome-encapsulated hemoglobin used for blood substitution and control MLV liposomes	Hemodynamic changes are due to C activation with subsequent secretion of TxA2. PAP is dose-dependent, highly reproducible endpoint of HSRs	[7] [6]
2000	Different liposomes applied to dissect the structural factors contributing to pulmonary hypertension in pigs	Vesicle size, lamellarity, charge and infusion speed are all critical determinants of the rise of PAP	[11]
	PEGylated liposomal doxorubicin (Doxil/Caelyx), a reactogenic anticancer drug	Doxil activates C in vitro and its dose-dependent hemodynamic effects in pigs mimic the human HSRs to this drug	[32]
2005	Negatively charged multilamellar vesicles applied as a model for reactogenic liposomes	The hemodynamic disturbance during HSRs is also manifested in cerebrovascular changes, explaining the psychic symptoms of HSRs	[17]
2006		The symptoms of HSRs reproduce those of cardiac anaphylaxis. The reaction can be reproduced only partially with injection of C5a	[23]
2008	Oversulfated chondroitin sulfate (OCS), a contaminant of heparin that caused a US nationwide outbreak of severe adverse reactions during 2007–2008	OCS induced contact and complement system activation and cardiopulmonary distress only in pigs but not in other species, mimicking the human symptoms of severe heparin reactions	[33]
2010	Liposomal bisphosphonates (LBPs) developed for the prevention of myocardial infarction via macrophage inhibition	LBPs triggered no or minor HSRs in pigs, which correlated with their C activating capability in vitro	[34]
2011	PEI-PEG block-copolymers used as models for polymeric drug carrier nanosystems	25K-PEI activated C in vitro and caused HSRs in pigs; its PEGylation decreased, but did not eliminate these effects	[21]
2012	PEGylated liposomal doxorubicin (Doxil, Caelyx) and liposomal amphotericin-B (AmBisome)	Doxil and AmBisome activated C in vitro and caused proportional HSRs in pigs. The effect of Doxil, but not of AmBisome, was tachyphylactic	[25]

(continued)

Table 2 (continued)

Year	Tested drugs/agents[a]	Major findings[b]	Reference
	Lipophilic prodrugs of methotrexate and melphalan	The products are reactogenic	[35]
	PEGylated liposomal doxorubicin (Doxil, Caelyx)	Prevention of infusion reactions to Doxil via tachyphylaxis induction by placebo vesicles called Doxebo	[26]
	Hemoglobin vesicles (HbVs) used as an oxygen carrier blood substitute	By optimizing the lipid composition of HbVs both C activation and the HSR of pigs could be attenuated. The reaction was tachyphylactic	[36]
	Review on porcine CARPA: methods, symptoms, quantification		[24]
2013	Review on the cardiopulmonary changes in porcine CARPA		[37]
2014	TRO40303, a cardioprotective sterane compound, inhibitor of transitional permeability pores in mitochondria	Consistent with the safety and tolerance in a phase I trial, TRO40303 did not activate C and caused HSR in pigs	[38]
	Different liposomes	Features of CARPA in pigs and rats	[18]
	Intralipid, used for reversing the symptoms of local anesthetic overdose	Intralipid caused major HSR in pigs, although C activation could not be detected in pig blood in vitro	[15]
	Review: Cardiovascular manifestations of complement activation-related pseudoallergy		[39]
2015	Review: methods, symptoms, quantification, variation		[13]
	Review on porcine PIM cells		[40]
	US patent on desensitization against PEGylated NPs with Doxebo in pigs		[41]
2016	siRNA-containing LNP formulations	A stepwise micro-dosing protocol is reaction-free in pigs, suggesting safety in patients	[42]
	Known reactogenic nanomedicines (Doxil, AmBisome, Cremophor EL)	The hemodynamic derangement and other changes caused by reactogenic nanomedicines were similar in pigs and Göttingen miniature pigs	[20]
	Nano-systems intended for cardiovascular applications (liposomes, LNPs, polymeric, and iron oxide NPs)	The non-reactive NPs were suggested to have the least risk for HSRs in man	[43]
2017	Nitroglycerin encapsulated shear-responsive liposomes against MI	Liposomes did not cause significant reaction	[44] [45]
	MRI contrast agent superparamagnetic iron oxide NPs (SPIONdex)	C activation and HSR in pigs can be eliminated by reducing the size of SPIONdex NPs	[46]
2018		Absence of HSRs	[47]
	Polystyrene NPs (PS-NPs) used as a model for reactogenic drug delivery nano-systems	PS-NP-induced cardiopulmonary distress depends on the shape of particles, spheres being more reactogenic than rods or disks. C activation was not measurable in pig whole blood	[48]
		Spherical PS-NP-induced cardiopulmonary distress in pigs	[22]

(continued)

Table 2 (continued)

Year	Tested drugs/agents[a]	Major findings[b]	Reference
		showed significant correlation with C activation in human serum. PS-NPs were opsonized in pig serum by C3 derivatives, indicating C activation	
	Hemostatic NPs based on PEG-PLGA-PLL-PEG-cRGD copolymers, developed to control traumatic blood loss	In a porcine liver injury model, these NPs led to massive vasodilation and exsanguination due to CARPA. This adverse effect could be attenuated by tailoring the zeta potential of NPs	[49]
	Review on porcine CARPA: The role of complement activation		[10]
	Review on porcine CARPA: Pros and cons of translation to a preclinical safety test		[12]
2019	Doxil and placebo Doxil (Doxebo) used to clarify the mechanism of HSRs to Doxil and other PEGylated NPs	C activation and the HSR caused by Doxil was greatly amplified in Doxebo-immunized animals in which the anti-PEG IgM levels were increased	[50]
	Dialysis membranes	No difference in CARPAgenicity (Abstract)	[51]
	Liposomal cortisol phosphate developed against chronic inflammatory diseases	Slow, stepwise infusion with micro-dosing minimizes the risk for HSRs	[52]
	TC^{99m}-Fucoidan, a sulfated fucose-rich polysaccharide developed for the detection of P-selectin expression in cardiovascular diseases	The drug did not cause C activation or HSR in pigs, suggesting safety for human use for the imaging of activated endothelium	[53]
2020	Pseudo-anaphylaxis to PEGylated nanomedicines in pigs: Use for safety testing (Abstracts)		[54, 55]
	Review on the clinical relevance of the porcine CARPA model		[14]
	Review on the rationale of using the pig model for nanomedicine safety assessment		[56]
2021	Hemodialysis membranes	Hemodialysis causes CARPA in pigs	[57]
2022	Thrombin-inhibitor nucleic acid-mRNA LNP	The LNP causes no reaction	[58]
	Hemodialysis membranes	Correspondence on the mechanism of dialysis-membrane-caused CARPA in pigs	[59, 60]
	Comirnaty (mRNA-LNP-containing Covid-19 vaccine)	I.v. injection of Comirnaty can cause CARPA in pigs	[5]
2023		Anti-PEG hyperimmune pigs develop anaphylactic shock to Comirnaty injection i.v.	[4]
	Different iron-oxide contrast agents	Intravenous administration of ferucarbotran or ferumoxytol induced strong CARPA in pigs, while SPION (Dex) did not elicit any HSR	[61]
	Zymosan	Zymosan-induced cytokine release in pigs provides a model for cytokine storm	[62]

Table modified and extended from reference [14]

[a] Test agents injected in pigs as i.v. bolus and/or infusion over hours

[b] Selected from many other observations

6. Reference substances known to cause hemodynamic changes in CARPA (e.g., zymosan)
7. Commercial or home-made ELISA kits

3 Methods

3.1 Animal Preparation and Surgical Procedure

1. Group animals according to the study plan and allow the time to acclimate (*see* **Notes 1–5** and Table 3).
2. Sedate animals by intramuscular injection of Calypsol/Xilazine at 2–4/1.5–2 mL dose based on body weight (*see* **Note 6**).
3. Perform intubation with 6.5–7 French endotracheal tubes to maintain free airways and to enable controlled ventilation if necessary (*see* **Note 7**).
4. Maintain anesthesia using isoflurane inhalation narcosis (2–2.5%) with oxygen.
5. Monitor respiration using a capnograph (connected to the tracheal tube to monitor respiratory rate and $etCO_2$) and a pulse-oximeter (fixed on the tail) to monitor tissue oxygen saturation (*see* **Notes 8** and **9**).
6. Measure the body temperature in the rectum.
7. Introduce a Swan-Ganz catheter into the pulmonary artery via the right external jugular vein, passing through the right atrium and right ventricle (*see* **Note 10**).
8. Record the pulmonary arterial pressure (PAP) by pressure transducer connected to the cannula in the right pulmonary artery.
9. Introduce another cannula in the left femoral artery (*see* **Note 10**).

Table 3
Husbandry, food, and water supply parameters

Husbandry	Group caging
Cage type	Steel with plastic grated floor
Cage size	2 m × 3 m
Temperature	18–22 °C
Relative humidity	40–70%
Diet name and type	Normal, powder format (No. 521-102-84)
Fasting	Prior to experiment pigs are fasted for 12 h
Water supply	Tap water ad libitum

10. Record the systemic arterial pressure (SAP) by the second pressure transducer connected to the cannula in the left femoral artery (*see* **Note 11**).
11. Cannulate the left femoral vein for blood sampling and the left external jugular vein for administration of reference and test items (*see* **Note 10**).
12. Record ECG through foam type Ag/AgCl monitoring electrodes (*see* **Note 12**, Tables 2 and 3).

3.2 Drug Dosing and Animal Monitoring

1. Prepare test and reference samples by reconstituting them in physiologically relevant buffers (*see* **Note 13**).
2. Administer test and reference samples by bolus injection or infusion.
3. Record hemodynamic changes and collect blood samples for subsequent analysis (*see* **Note 14**, Tables 4 and 5).

Table 4
Standard experimental endpoints to be measured and documented before and after administration of test and reference substances[a]

Pulmonary Arterial Pressure (PAP)
Systemic Arterial Pressure (SAP) Heart rate (HR)
ECG (Einthoven's lead II) Respiratory rate
End-tidal CO_2 (et CO_2) Tissue oxygen saturation (SpO_2 %)
Body temperature (°C)

[a]Optional additional parameters to be measured or calculated: cardiac output, pulse pressure (pulse amplitude), central venous pressure, left ventricular end-diastolic pressure, systemic vascular resistance, pulmonary vascular resistance

Table 5
Hematology parameters and sampling guide

Sample site	Left femoral vein
Time of sampling	Before each injection at 0′ and after 1′, 3′, 5′, 10′, 15′, and 30′
Anticoagulant/sample volume	EDTA spray coated mini-tube/0.2 mL blood

3.3 Analysis of Collected Blood Samples

1. Using hematology analyzer, examine the first set of collected blood samples for the following parameters: WBC, RBC, Hgb, PLT, Ly, Gr, Mi (*see* **Note 15**).
2. Centrifuge a second set of whole blood at 2500 RPM for 10 min at 4 °C; transfer each of the resultant plasma samples into three Eppendorf tubes (*see* **Note 16**).
3. Analyze plasma samples for the different analytes to be tested (*see* **Notes 17–19**).

3.4 Data Analysis and Reporting

1. Twenty (20) second intervals are evaluated from the PAP, SAP, and HR curves at every minute; multiple measurements are averaged and compared between the treatment groups.
2. AUC values (min*mmHg) are calculated for PAP and SAP curves, summarized, and compared between the treatment groups.
3. ECG signals are evaluated by an ECG analysis software. Time intervals of 20 s are selected and analyzed as follows: All heart cycles from the selected periods are plotted in a group of 5, and the "pig" attributes are chosen. The start/end of ECG waves marked by the software can be either accepted or modified by the investigator if necessary. The average time intervals and voltages of waves are then compared between treatment groups.
4. For statistical analysis, data from the test item and reference item treated animals are compared by a two tailed independent sample t-test. In the case of composite scores, if the distribution of the data is asymmetric, a non-parametric Mann-Whitney U test is performed to compare the groups. Statistical difference is declared significant if the *p*-value is less than the significance level ($p < 0.05$) (*see* **Note 20**).

4 Notes

1. Consult your institution to obtain Animal Care and Use Committee (ACUC) approval for the animal husbandry and experiments.
2. Animals are identified by way of ear notching performed soon after their birth. Notching provides a permanent identification system with information about the pig parentage, birth weight, medication, etc.
3. In the course of the study the individual identification of the animals is performed by plastic ear tags with animal identity number.

4. The cages are marked with identification cards, with information about study, sex, and individual animal numbers.
5. The drinking water is analyzed twice yearly and is considered not to contain any contaminants that could affect the health of the animals and consequently the integrity of the study.
6. It is recommended to perform the sedative injections in the stalls to avoid stress to animals, then transport them to the operating room.
7. Animals are expected to breath spontaneously.
8. Both parameters are monitored by an anesthesia monitor and the data are written in a lab notebook.
9. A capnograph is connected to the tracheal tube to monitor $etCO_2$ and the respiratory rate.
10. All incision areas must be shaved and disinfected by liberal application of povidone-iodine 10% and closed after fixation of cannulas.
11. The left femoral vein is cannulated for blood sampling and the left external jugular vein is cannulated for administration of reference and test items.
12. The two recording electrodes are positioned on shaved, alcohol-cleaned skin: One electrode on the right side of the chest wall under the clavicula, while the other on the chest wall in the middle underarm line, at the height of the eighth rib. The reference electrode is positioned on the ear.
13. Commonly used buffers include phosphate-buffered saline (PBS), normal saline (0.9% sodium chloride), or 5% dextrose.
14. Hemodynamic parameters (e.g., PAP, SAP) should be recorded at the baseline (i.e., before the administration of test and reference materials) and after dosing with test substances. Remember to collect the blood into more than one tube to allow the analysis of whole blood by a hematology analyzer and preparation of plasma for analysis by ELISA.
15. Blood samples are usually analyzed by a hematology analyzer immediately after collection.
16. These plasma samples may be analyzed immediately after collection or stored at −80 °C for future analysis. Storing in single-use aliquots avoids multiple freeze/thaw cycles which may affect the accuracy of unstable parameters.
17. These analytes can be accurately measured using commercial ELISA kits. Each vendor supplies a detailed procedure for performing ELISA assays.
18. In an average pig, thromboxane levels vary between 1000–2000 pg/mL and that of C3a between 400–600 U/

mL. These baseline levels usually elevate many times after the injection of a reactogenic test material.

19. Blood samples remaining from the study should be discarded according to institutional rules for the treatment of biohazardous materials.
20. Standardized principal component analysis may also be performed and added to the study results in order to identify different groups of objects, according to the data structure.

Acknowledgement

This project has been funded by the European Union Horizon 2020 projects 825828 (Expert) and 952520 (Biosafety) and grants from the National Research, Development, and Innovation Office (NKFIH) of Hungary (2020-1.1.6-JÖVŐ-2021-00013 and 2020-1.1.6-JÖVŐ-2021-00010.

References

1. Szebeni J (2001) Complement activation-related pseudoallergy caused by liposomes, micellar carriers of intravenous drugs, and radiocontrast agents. Crit Rev Ther Drug Carrier Syst 18(6):567–606
2. Szebeni J (2005) Complement activation-related pseudoallergy: a new class of drug-induced acute immune toxicity. Toxicology 216(2–3):106–121
3. Szebeni J (2014) Complement activation-related pseudoallergy: a stress reaction in blood triggered by nanomedicines and biologicals. Mol Immunol 61(2):163–173
4. Barta BA, Radovits T, Dobos AB, Kozma GT, Meszaros T, Berenyi P, Facsko R, Fulop TG, Merkely B, Szebeni J (2023) The COVID-19 mRNA vaccine Comirnaty induces anaphylactic shock in an anti-PEG hyperimmune large animal model: role of complement activation in cardiovascular, hematological and inflammatory mediator changes. bioRxiv. https://doi.org/10.1101/2023.05.19.541479
5. Dezsi L, Meszaros T, Kozma G, H-Velkei M, Olah CZ, Szabo M, Patko Z, Fulop T, Hennies M, Szebeni M, Barta BA, Merkely B, Radovits T, Szebeni J (2022) A naturally hypersensitive porcine model may help understand the mechanism of COVID-19 mRNA vaccine-induced rare (pseudo) allergic reactions: complement activation as a possible contributing factor. Geroscience 44(2):597–618
6. Szebeni J, Fontana JL, Wassef NM, Mongan PD, Morse DS, Dobbins DE, Stahl GL, Bunger R, Alving CR (1999) Hemodynamic changes induced by liposomes and liposome-encapsulated hemoglobin in pigs – a model for pseudoallergic cardiopulmonary reactions to liposomes: role of complement and inhibition by soluble CR1 and anti-C5a antibody. Circulation 99(17):2302–2309
7. Szebeni J, Fontana JL, Wassef NM, Mongan PD, Morse DS, Stahl GL, Bunger R, Alving CR (1998) Liposome induced and complement-mediated cardiopulmonary distress in pigs as a model of pseudo-allergic reactions to liposomal drugs. Mol Immunol 35(6–7):401–401
8. Chanan-Khan A, Szebeni J, Savay S, Liebes L, Rafique NM, Alving CR, Muggia FM (2003) Complement activation following first exposure to pegylated liposomal doxorubicin (Doxil): possible role in hypersensitivity reactions. Ann Oncol 14(9):1430–1437
9. Orfi E, Hricisak L, Dezsi L, Hamar P, Benyo Z, Szebeni J, Szenasi G (2022) The hypertensive effect of amphotericin B-containing liposomes (Abelcet) in mice: dissecting the roles of C3a and C5a anaphylatoxins, macrophages and thromboxane. Biomedicines 10(7):1764
10. Szebeni J (2018) Mechanism of nanoparticle-induced hypersensitivity in pigs: complement or not complement? Drug Discov Today 23(3):487–492
11. Szebeni J, Baranyi L, Savay S, Bodo M, Morse DS, Basta M, Stahl GL, Bunger R, Alving CR (2000) Liposome-induced pulmonary hypertension: properties and mechanism of a

complement-mediated pseudoallergic reaction. Am J Physiol Heart Circ Physiol 279(3): H1319–H1328
12. Szebeni J, Bedocs P, Dezsi L, Urbanics R (2018) A porcine model of complement activation-related pseudoallergy to nano-pharmaceuticals: pros and cons of translation to a preclinical safety test. Precis Nanomed 1(1):63–73
13. Urbanics R, Bedocs P, Szebeni J (2015) Lessons learned from the porcine CARPA model: constant and variable responses to different nanomedicines and administration protocols. Eur J Nanomed 7(3):219–231
14. Szebeni J, Bawa R (2020) Human clinical relevance of the porcine model of pseudoallergic infusion reactions. Biomedicines 8(4):82
15. Bedocs P, Capacchione J, Potts L, Chugani R, Weiszhar Z, Szebeni J, Buckenmaier CC (2014) Hypersensitivity reactions to intravenous lipid emulsion in swine: relevance for lipid resuscitation studies. Anesth Analg 119(5):1094–1101
16. Bodo M, Szebeni J, Baranyi L, Savay S, Pearce FJ, Alving CR, Bunger R (2005) Cerebrovascular involvement in liposome – induced cardiopulmonary distress in pigs. J Liposome Res 15(1–2):3–14
17. Bodo M, Szebeni J, Baranyi L, Savay S, Pearce FJ, Alving CR, Bunger R (2005) Rheoencephalographic evidence of complement activation-related cerebrovascular changes in pigs. J Cereb Blood Flow Metab 25(1_suppl): S550
18. Dezsi L, Fulop T, Meszaros T, Szenasi G, Urbanics R, Vazsonyi C, Orfi E, Rosivall L, Nemes R, Kok RJ, Metselaar JM, Storm G, Szebeni J (2014) Features of complement activation-related pseudoallergy to liposomes with different surface charge and PEGylation: comparison of the porcine and rat responses. J Control Release 195:2–10
19. Fulop T, Nemes R, Meszaros T, Urbanics R, Kok RJ, Jackman JA, Cho NJ, Storm G, Szebeni J (2018) Complement activation in vitro and reactogenicity of low-molecular weight dextran-coated SPIONs in the pig CARPA model: correlation with physicochemical features and clinical information. J Control Release 270:268–274
20. Jackman JA, Meszaros T, Fulop T, Urbanics R, Szebeni J, Cho NJ (2016) Comparison of complement activation-related pseudoallergy in miniature and domestic pigs: foundation of a validatable immune toxicity model. Nanomedicine 12(4):933–943
21. Merkel OM, Urbanics R, Bedocs P, Rozsnyay Z, Rosivall L, Toth M, Kissel T, Szebeni J (2011) In vitro and in vivo complement activation and related anaphylactic effects associated with polyethylenimine and polyethylenimine-graft-poly(ethylene glycol) block copolymers. Biomaterials 32(21):4936–4942
22. Meszaros T, Kozma GT, Shimizu T, Miyahara K, Turjeman K, Ishida T, Barenholz Y, Urbanics R, Szebeni J (2018) Involvement of complement activation in the pulmonary vasoactivity of polystyrene nanoparticles in pigs: unique surface properties underlying alternative pathway activation and instant opsonization. Int J Nanomedicine 13:6345–6357
23. Szebeni J, Baranyi L, Savay S, Bodo M, Milosevits J, Alving CR, Bunger R (2006) Complement activation-related cardiac anaphylaxis in pigs: role of C5a anaphylatoxin and adenosine in liposome-induced abnormalities in ECG and heart function. Am J Physiol-Heart C 290(3):H1050–H1058
24. Szebeni J, Bedocs P, Csukas D, Rosivall L, Bunger R, Urbanics R (2012) A porcine model of complement-mediated infusion reactions to drug carrier nanosystems and other medicines. Adv Drug Deliver Rev 64(15): 1706–1716
25. Szebeni J, Bedocs P, Rozsnyay Z, Weiszhar Z, Urbanics R, Rosivall L, Cohen R, Garbuzenko O, Bathori G, Toth M, Bunger R, Barenholz Y (2012) Liposome-induced complement activation and related cardiopulmonary distress in pigs: factors promoting reactogenicity of Doxil and AmBisome. Nanomedicine 8(2):176–184
26. Szebeni J, Bedocs P, Urbanics R, Bunger R, Rosivall L, Toth M, Barenholz Y (2012) Prevention of infusion reactions to PEGylated liposomal doxorubicin via tachyphylaxis induction by placebo vesicles: a porcine model. J Control Release 160(2):382–387
27. European Medicines Agency (2013) Reflection paper on the data requirements for intravenous liposomal products developed with reference to an innovator liposomal product, 21 Feb 2013
28. Milosevits G, Meszaros T, Orfi E, Bakos T, Garami M, Kovacs G, Dezsi L, Hamar P, Gyorffy B, Szabo A, Szenasi G, Szebeni J (2021) Complement-mediated hypersensitivity reactions to an amphotericin B-containing lipid complex (Abelcet) in pediatric patients and anesthetized rats: benefits of slow infusion. Nanomedicine 34:102366
29. Wassef NM, Johnson SH, Graeber GM, Swartz GM Jr, Schultz CL, Hailey JR, Johnson AJ, Taylor DG, Ridgway RL, Alving CR (1989)

Anaphylactoid reactions mediated by autoantibodies to cholesterol in miniature pigs. J Immunol 143(9):2990–2995

30. Ostensen J, Hede R, Myreng Y, Ege T, Holtz E (1992) Intravenous injection of Albunex microspheres causes thromboxane mediated pulmonary hypertension in pigs, but not in monkeys or rabbits. Acta Physiol Scand 144(3):307–315
31. Walday P, Tolleshaug H, Gjoen T, Kindberg GM, Berg T, Skotland T, Holtz E (1994) Biodistributions of air-filled albumin microspheres in rats and pigs. Biochem J 299(Pt 2):437–443
32. Szebeni J, Baranyi L, Savay S, Lutz HU, Jelezarova E, Bunger R, Alving CR (2000) The role of complement activation in hypersensitivity to pegylated liposomal doxorubicin (DOXIL (R)). J Liposome Res 10(4):467–481
33. Kishimoto TK, Viswanathan K, Ganguly T, Elankumaran S, Smith S, Pelzer K, Lansing JC, Sriranganathan N, Zhao G, Galcheva-Gargova Z, Al-Hakim A, Bailey GS, Fraser B, Roy S, Rogers-Cotrone T, Buhse L, Whary M, Fox J, Nasr M, Dal Pan GJ, Shriver Z, Langer RS, Venkataraman G, Austen KF, Woodcock J, Sasisekharan R (2008) Contaminated heparin associated with adverse clinical events and activation of the contact system. N Engl J Med 358(23):2457–2467
34. Epstein-Barash H, Gutman D, Markovsky E, Mishan-Eisenberg G, Koroukhov N, Szebeni J, Golomb G (2010) Physicochemical parameters affecting liposomal bisphosphonates bioactivity for restenosis therapy: internalization, cell inhibition, activation of cytokines and complement, and mechanism of cell death. J Control Release 146(2):182–195
35. Kuznetsova NR, Sevrin C, Lespineux D, Bovin NV, Vodovozova EL, Meszaros T, Szebeni J, Grandfils C (2012) Hemocompatibility of liposomes loaded with lipophilic prodrugs of methotrexate and melphalan in the lipid bilayer. J Control Release 160(2):394–400
36. Sakai H, Suzuki Y, Sou K, Kano M (2012) Cardiopulmonary hemodynamic responses to the small injection of hemoglobin vesicles (artificial oxygen carriers) in miniature pigs. J Biomed Mater Res A 100(10):2668–2677
37. Dezsi L, Szenasi G, Urbanics R, Rosivall L, Szebeni J (2013) Cardiopulmonary and hemodynamic changes in complement activation-related pseudoallergy. Health 5(6):1032–1038
38. Le Lamer S, Paradis S, Rahmouni H, Chaimbault C, Michaud M, Culcasi M, Afxantidis J, Latreille M, Berna P, Berdeaux A, Pietri S, Morin D, Donazzolo Y, Abitbol JL, Pruss RM, Schaller S (2014) Translation of TRO40303 from myocardial infarction models to demonstration of safety and tolerance in a randomized Phase I trial. J Transl Med 12:38
39. Dezsi L, Urbanics R, Meszaros T, Vazsonyi C, Fulop T, Orfi E, Rosivall L, Szebeni J, Szanasi G (2014) Cardiovascular manifestations of complement activation-related pseudoallergy following administration of liposomal nanomedicines. Acta Physiol 211:94–94
40. Csukas D, Urbanics R, Weber G, Rosivall L, Szebeni J (2015) Pulmonary intravascular macrophages: prime suspects as cellular mediators of porcine CARPA. Eur J Nanomed 7(1): 27–36
41. Barenholz Y, Szebeni J, Toth M, Rosivall L (2015) Particulate drug carriers as desensitizing agents. U.S. Patent 9,078,812 B2
42. Kasperovic P (2016) Dosages and methods for delivering lipid formulated nucleic acid molecules. US 2016/0122759 A1
43. Matuszak J, Baumgartner J, Zaloga J, Juenet M, da Silva AE, Franke D, Almer G, Texier I, Faivre D, Metselaar JM, Navarro FP, Chauvierre C, Prassl R, Dezsi L, Urbanics R, Alexiou C, Mangge H, Szebeni J, Letourneur D, Cicha I (2016) Nanoparticles for intravascular applications: physicochemical characterization and cytotoxicity testing. Nanomedicine 11(6):597–616
44. Bugna S, Buscema M, Matviykiv S, Urbanics R, Weinberger A, Meszaros T, Szebeni J, Zumbuehl A, Saxer T, Muller B (2016) Surprising lack of liposome-induced complement activation by artificial 1,3-diamidophospholipids in vitro. Nanomed-Nanotechnol 12(3):845–849
45. Buscema M, Matviykiv S, Meszaros T, Gerganova G, Weinberger A, Mettal U, Mueller D, Neuhaus F, Stalder E, Ishikawa T, Urbanics R, Saxer T, Pfohl T, Szebeni J, Zumbuehl A, Muller B (2017) Immunological response to nitroglycerin-loaded shear-responsive liposomes in vitro and in vivo. J Control Release 264:14–23
46. Unterweger H, Janko C, Schwarz M, Dezsi L, Urbanics R, Matuszak J, Orfi E, Fulop T, Bauerle T, Szebeni J, Journe C, Boccaccini AR, Alexiou C, Lyer S, Cicha I (2017) Non-immunogenic dextran-coated superparamagnetic iron oxide nanoparticles: a biocompatible, size-tunable contrast agent for magnetic resonance imaging. Int J Nanomedicine 12:5223–5238
47. Unterweger H, Dézsi L, Matuszak J, Janko C, Poettler M, Jordan J, Bäuerle T, Szebeni J, Fey T, Boccaccini AR, Alexiou C, Cicha I

(2018) Dextran-coated superparamagnetic iron oxide nanoparticles for magnetic resonance imaging: evaluation of size-dependent imaging properties, storage stability and safety. Int J Nanomedicine 13:1899–1915

48. Wibroe PP, Anselmo AC, Nilsson PH, Sarode A, Gupta V, Urbanics R, Szebeni J, Hunter AC, Mitragotri S, Mollnes TE, Moghimi SM (2017) Bypassing adverse injection reactions to nanoparticles through shape modification and attachment to erythrocytes. Nat Nanotechnol 12(6):589–594
49. Onwukwe C, Maisha N, Holland M, Varley M, Groynom R, Hickman D, Uppal N, Shoffstall A, Ustin J, Lavik E (2018) Engineering intravenously administered nanoparticles to reduce infusion reaction and stop bleeding in a large animal model of trauma. Bioconjug Chem 29(7):2436–2447
50. Kozma GT, Meszaros T, Vashegyi I, Fulop T, Orfi E, Dezsi L, Rosivall L, Bavli Y, Urbanics R, Mollnes TE, Barenholz Y, Szebeni J (2019) Pseudo-anaphylaxis to polyethylene glycol (PEG)-coated liposomes: roles of anti-PEG IgM and complement activation in a porcine model of human infusion reactions. ACS Nano 13(8):9315–9324
51. Piecha D, Dezsi L, Moore C, Petho A, Rosivall L, Canaud B, Szebeni J, Steppan S (2019) SP539 porcine CARPA model to study hypersensitivity-like reactions during dialysis. Nephrol Dial Transplant 34(Supplement 1):gfz103.SP539
52. Fulop T, Kozma GT, Vashegyi I, Meszaros T, Rosivall L, Urbanics R, Storm G, Metselaar JM, Szebeni J (2019) Liposome-induced hypersensitivity reactions: risk reduction by design of safe infusion protocols in pigs. J Control Release 309:333–338
53. Chauvierre C, Aid-Launais R, Aerts J, Chaubet F, Maire M, Chollet L, Rolland L, Bonafe R, Rossi S, Bussi S, Cabella C, Dezsi L, Fulop T, Szebeni J, Chahid Y, Zheng KH, Stroes ESG, Le Guludec D, Rouzet F, Letourneur D (2019) Pharmaceutical development and safety evaluation of a GMP-grade Fucoidan for molecular diagnosis of cardiovascular diseases. Mar Drugs 17(12):699
54. Szebeni J (2020) Testing of drug-induced pseudo-allergies in a porcine model. World Allergy Organ J 13(8):100250
55. Szebeni J, Kozma G, Meszaros T, Urbanics R (2020) Pseudo-anaphylaxis to PEGylated nanomedicines in a porcine model. World Allergy Organ J 13(8):100178
56. Bedocs P, Szebeni J (2020) The critical choice of animal models in nanomedicine safety assessment: a lesson learned from hemoglobin-based oxygen carriers. Front Immunol 11:584966
57. Petho A, Piecha D, Meszaros T, Urbanics R, Moore C, Canaud B, Rosivall L, Mollnes TE, Steppan S, Szenasi G, Szebeni J, Dezsi L (2021) A porcine model of hemodialyzer reactions: roles of complement activation and rinsing back of extracorporeal blood. Ren Fail 43(1):1609–1620
58. Ke W, Chandler M, Cedrone E, Saito RF, Rangel MC, de Souza Junqueira M, Wang J, Shi D, Truong N, Richardson M, Rolband LA, Dreau D, Bedocs P, Chammas R, Dokholyan NV, Dobrovolskaia MA, Afonin KA (2022) Locking and unlocking thrombin function using immunoquiescent nucleic acid nanoparticles with regulated retention in vivo. Nano Lett 22(14):5961–5972
59. Petho A, Szebeni J (2022) Pulmonary hypertension in a hemodialysis porcine model: possible unforeseen causes? Ren Fail 44(1):693
60. Soliman KM, Rao V, Daoud A (2022) Pulmonary hypertension in a hemodialysis porcine model: possible unforeseen causes. Ren Fail 44(1):945–945
61. Unterweger H, Janko C, Folk T, Cicha I, Kovacs N, Gyebnar G, Horvath I, Mathe D, Zheng KH, Coolen BF, Stroes E, Szebeni J, Alexiou C, Dezsi L, Lyer S (2023) Comparative in vitro and in vivo evaluation of different iron oxide-based contrast agents to promote clinical translation in compliance with patient safety. Int J Nanomedicine 18:2071–2086
62. Kokeny G, Bakos T, Barta BA, Nagy GV, Meszaros T, Kozma GT, Szabo A, Szebeni J, Merkely B, Radovits T (2023) Zymosan particle-induced hemodynamic, cytokine and blood cell changes in pigs: an innate immune stimulation model with relevance to cytokine storm syndrome and severe COVID-19. Int J Mol Sci 24(2):1138

Chapter 24

Immunophenotyping, Part I: Instrument Calibration and Reagent Qualification for Immunophenotyping Analysis of Human Peripheral Blood Mononuclear Cell Cultures

Hannah S. Newton, Jenny Zhang, and Marina A. Dobrovolskaia

Abstract

Nanoparticles are increasingly used in biomedical applications to influence the way the immune system reacts to tumors and infectious disease-causing agents. Nanoparticles not-intended for immunomodulation can also influence immune responses by affecting immune cell subsets' viability and/or activity. While immunophenotyping is commonly used to assess the effects of drugs and nanoparticles on immune cell subsets, no standardized approach exists due to the breadth of available cell models and instrumentation. In this chapter, we describe a protocol for flow cytometer calibration and reagent qualification prior to its use in the immunophenotyping experiment. The strategies described herein can be adapted to other instruments. The subsequent chapter—immunophenotyping part II (Chap. 25)—provides detailed instructions for applying this methodology to analyze nanoparticle effects on subsets of immune cells present in peripheral blood.

Key words Nanoparticles, Immunophenotyping, PBMC, Instrument setup, Calibration, Reagent qualification

1 Introduction

Immunophenotyping is the use of antigen expression for the identification of distinct immune cell subsets (and their activation statuses) [1–4]. This technique can detect minute changes in cell populations and thus is used to characterize the cell makeup in many diseases as well as determine effects of treatments, such as nanoparticles [5]. It is important to develop a method that allows for immunological evaluation of nanoparticles because some nanoparticles are designed to modify the immune system while others cause immunotoxicity [4, 5]. Currently, the most common technique used to perform immunophenotyping is multicolor flow cytometry [2, 3].

Jeffrey D. Clogston et al. (eds.), *Characterization of Nanoparticles Intended for Drug Delivery*, Methods in Molecular Biology, vol. 2789, https://doi.org/10.1007/978-1-0716-3786-9_24,

This protocol covers two separate immunophenotyping panels utilizing 11–12 antibody-fluorophore conjugates:

Immunophenotyping Panel #1 (or Lymphocyte Panel): This panel includes antibody-fluorophore conjugates that allow for analysis of different lymphocyte populations including B cells and T cells ($CD8^+$ T cells, $CD4^+$ T cells, regulatory T (Treg) cells, naïve T cells, and γδ TCR T cells). This panel also determines cellular CD25 and CD154 expression, which are markers of proliferation and co-stimulation/presentation, respectively.

Immunophenotyping Panel #2 (or Monocyte, Dendritic cell (DC), Natural Killer (NK) Cell Panel): This panel includes antibody-fluorophore conjugates that allow for analysis of different cell populations including $CD14^+$ monocytes, DCs (plasmacytoid (p) and myeloid (m) DCs), and NK cells along with NK T cells. This panel also examines cellular CD69 and CD54 expression which are markers of early activation and adhesion, respectively.

When used in conjunction with other immunoassays, this protocol aids in establishing efficacy and safety profiles of engineered nanoparticles used for vaccine or drug delivery. The protocol has two parts: part I described in this chapter is intended for instrument calibration, and part II, described in Chap. 25, is intended for the analysis of nanoparticle-treated cells.

The protocol described herein uses compensation beads and peripheral blood mononuclear cells (PBMC) derived from the blood of healthy donor volunteers for instrument calibration and reagent qualification. This protocol is required before proceeding to the second part of the protocol (Chap. 25) for the immunophenotyping of nanoparticle-treated cells.

Experimental steps described herein, allow for proper optimization and compensation of the antibody-fluorophore conjugates used in the panels. The antibody titrations allow for the optimization of antibody concentrations in which the optimal concentration/dilution is defined as the concentration that leads to the greatest difference between the positive and negative mean fluorescence intensities. Single stain control experiments performed with compensation beads allow for the necessary compensation of the optimized antibody-fluorophores used in each panel. Lastly, fluorescence minus one (FMO) controls serve as a way to fine-tune the compensation obtained from the single stain control experiment to account for any minute changes in the protocol, including the presence of PBMC rather than compensation beads and fixation.

The variety and complexity of immune cell phenotypes that can be assessed during immunophenotyping depends on the available flow cytometry instrumentation. This protocol is optimized for the flow cytometer NovoCyte 3005 by Acea Biosciences Inc. (part of Agilent Technologies), which has the capacity to acquire data from

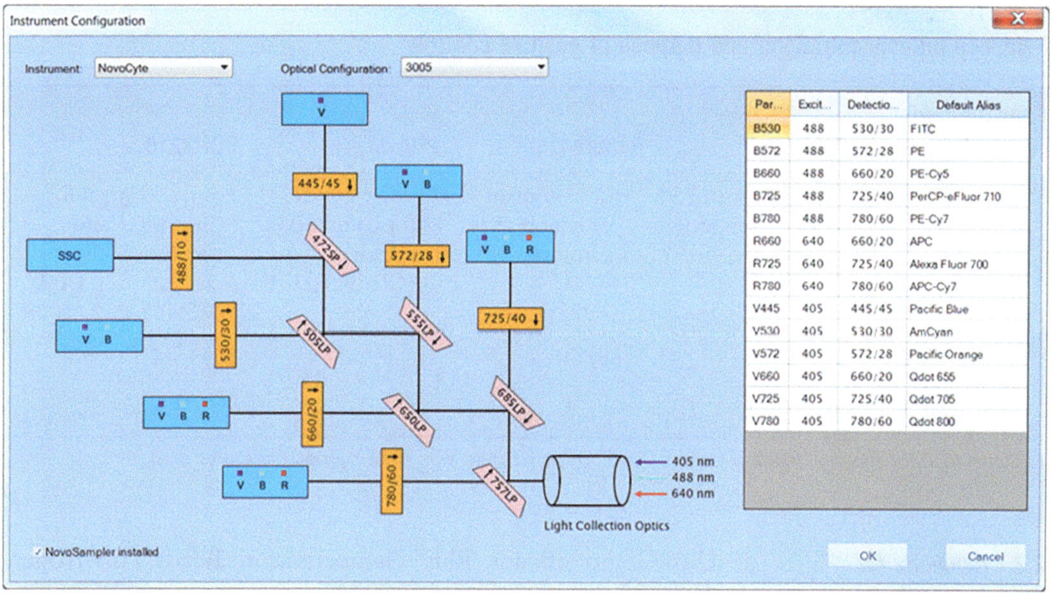

Fig. 1 Instrument configuration and settings. Parameters used on a NovoCyte 3005 instrument are shown from reference [7] (*see* **Note 1**)

14 distinct fluorophores simultaneously. NovoCyte 3005 is equipped with three lasers (405 nm, 488 nm, and 640 nm) and six detectors (445/45 nm, 530/30 nm, 572/28 nm, 660/20 nm, 725/40 nm, and 780/60 nm) [6] (Fig. 1, *see* **Note 1**).

2 Materials

2.1 Reagents for PBMC Cultures

1. Human blood anti-coagulated with Li-heparin and obtained from healthy donors (*see* **Notes 2–4**).
2. Phosphate buffered saline (PBS).
3. Roswell park memorial institute (RPMI) 1640.
4. Fetal bovine serum (FBS).
5. Penicillin streptomycin solution.
6. L-glutamine.
7. Ficoll Paque Premium.
8. Hank's balanced salt solution (HBSS).
9. Any reagent for live and dead cell counting.
10. Phytohemagglutinin M (PHA-M) (Table 1).
11. Phorbol 12-myristate 13-acetate (PMA).
12. Ionomycin.
13. Oligodeoxyribonucleotide, Human TLR9 ligand (ODN2216).

Table 1
Guidance on concentrations and purpose of positive controls

Description/ control	PMA	Ionomycin	PHA-M	ODN2216
Primary purpose	Positive control for lymphocyte activation (CD154; CD25)	Positive control for lymphocyte activation (CD154; CD25)	Positive control for lymphocyte activation (CD69; CD54)	Positive control for monocyte and dendritic cell activation (CD69; CD54)
Final concentration in assay	50 ng/mL	1 μg/mL	10 μg/mL	5 μg/mL

Reagents used for assay validation are summarized. Other reagents with similar properties can be used

2.2 Reagents for Flow Cytometry

1. UltraComp eBeads Plus Compensation Beads (Invitrogen, 01-3333-42).
2. eBioscience Flow Cytometry Staining Buffer (Invitrogen, 00-4222-26).
3. Paraformaldehyde (PFA), 20% solution.
4. NovoRinse (Agilent Technologies, 872B603).
5. NovoClean (Agilent Technologies, 872B602).
6. NovoFlow (Agilent Technologies, 871B607).
7. Antibodies/Dyes (Tables 2 and 3).

2.3 General Supplies and Equipment

1. Pipettes covering a range of 0.05 to 10 mL.
2. 96-well U-bottom plates.
3. 24-well flat bottom plates.
4. Polypropylene tubes, 15 and 50 mL.
5. Microcentrifuge tubes.
6. 5 mL Polystyrene round bottom tubes, 12 × 75 mm.
7. Centrifuge.
8. Cell culture incubator with 5% CO_2 and 95% humidity.
9. Biohazard safety cabinet approved for level II handling of biological material.
10. Water bath.
11. Vortex.
12. Instrument for cell counting.
13. Flow Cytometer (Acea Novocyte 3005).

Table 2
Anti-human labeling antibodies used in the immunophenotyping panel 1 (lymphocyte panel)

Fluorochrome	Marker	Cell Type	Purpose	Company	Concentration (μg/mL)	Catalog #
FITC	CD8a	T cells	Cytotoxic T cells	BioLegend	200	300905 or 300906
PE	CD4	T cells	T-helper	BioLegend	100	317409 or 317409
PE-Cy7	CD19	B cells	LPS co-receptor	BioLegend	100	302215 or 302216
APC	CCR4	T-reg	CD194	BioLegend	50	359407 or 359408
AF700	CD45RA	Naïve cells	Memory status	Invitrogen	50	56-0458-42
APC-Fire 750	TCR-γδ	TCR- γδ T cells	TCR- γδ receptor	BioLegend	400	331227 or 331228
PacBlue	CD45	All leukocytes	Pan-leukocyte	BioLegend	100	368539 or 368540
Zombie Aqua	Live/dead	Dead cells	Excludes dead cells	BioLegend	Not Available	423101 or 423102
BV570	CD3	T cells	T cells	BioLegend	80	300435 or 300436
BV650	CD25	T-reg and activation status of cells	IL-2 receptor α	BioLegend	100	302633 or 302634
BV711	CD154	lymphocytes	CD40L	BioLegend	100	310837 or 310838
BV785	CD127	T-reg	IL-7 receptor α	BioLegend	50	351329 or 351330

3 Methods

3.1 Preparation of Reagents and Controls

1. Prepare complete RPMI 1640 medium by supplementing stock RPMI medium with the following reagents: 10% heat inactivated FBS (*see* **Note 5**); 2 mM L-glutamine; 100 U/mL penicillin; 100 μg/mL streptomycin (*see* **Note 6**).
2. Prepare PMA stock by reconstituting the dry powder in DMSO to a stock concentration of 1 mg/mL (*see* **Notes 7** and **8**).
3. Ionomycin stock is supplied as 1% free acid and 99% ethyl alcohol and a concentration of 10 mg/mL (*see* **Note 9**).

Table 3
Anti-human labeling antibodies for immunophenotyping panel 2 (monocyte, DC, NK cell panel)

Fluorochrome	Marker	Cell Type	Purpose	Company	Concentration (μg/mL)	Catalog #
FITC	CD56	NK cells	Neural cell adhesion molecule	BioLegend	200	362545 or 362546
PE	CD14	Monocytes	LPS co-receptor	BioLegend	200	301805 or 301806
PE-Cy7	CD19	B cells	LPS co-receptor	BioLegend	100	302215 or 302216
APC	CD123	pDCs	Interleukin-3 receptor	BioLegend	100	306011 or 306012
AF700	CD54	Lymphocytes and monocytes	Adhesion	BioLegend	400	353125 or 353126
APC-Fire 750	CD20	B cells	B cells	BioLegend	200	302357 or 302358
PacBlue	CD45	All leukocytes	Pan-leukocyte	BioLegend	100	368539 or 368540
Zombie Aqua	Live/ dead	Dead cells	Excludes dead cells	BioLegend	Not Available	423101 or 423102
BV570	CD3	T cells	T cells	BioLegend	80	300435 or 300436
BV650	CD69	Neutrophils/ monocytes	Early activation	BioLegend	50	310933 or 310934
BV785	CD11c	mDCs	Integrin α x	BioLegend	160	301643 or 301644

4. Prepare 1 mg/mL ODN 2216 stock by reconstituting the supplied powder in pyrogen-free, nuclease-free water to a stock concentration of 1 mg/mL (*see* **Note 10**).
5. Prepare 1 mg/mL PHA-M stock by adding 1 mL of sterile PBS or cell culture medium per 1 mg of PHA-M to the vial and gently rotating the vial to mix (*see* **Note 11**).

3.2 PBMC Isolation

1. Place freshly drawn blood into 15 or 50 mL conical centrifuge tubes, add an equal volume of room-temperature 1X PBS, and mix well (*see* **Notes 2–4**).
2. Slowly layer the Ficoll-Paque solution underneath the blood/PBS mixture by placing the tip of the pipet containing Ficoll-Paque at the bottom of the blood sample tube (*see* **Notes 12** and **13**).

3. Centrifuge 30 min at 900 × *g* and 18–20 °C (*see* **Note 14**).
4. Using a sterile pipet, remove the upper layer containing plasma and platelets and discard it.
5. Using a fresh sterile pipet, transfer the mononuclear cell layer into another centrifuge tube.
6. Wash cells by adding an excess of HBSS and centrifuging for 10 min at 400 × *g*, 18–20 °C. The HBSS volume should be ~3 times the volume of mononuclear layer (*see* **Note 15**).
7. Discard supernatant and repeat wash step one more time.
8. Resuspend cells in complete RPMI 1640 medium and determine cell viability using any available live-dead stain reagent (*see* **Notes 16** and **17**).

3.3 Optional Antibody Titration for Flow Cytometry (see Notes 18–20)

3.3.1 Bead and Live/Dead PBMC Plate Preparation (Without Antibodies or Dyes)

1. Make a tenfold dilution of UltraComp eBeads Plus Compensation Beads using staining buffer as the diluent (*see* **Note 21**).
2. Add 50 μL of tenfold diluted beads to each of the necessary 96-well U-bottom wells of the plate labeled "Bead and Live/Dead PBMC Plate" (*see* **Notes 22** and **23**).
3. Add 50 μL of PBMC suspended in 1X PBS to each of the necessary wells of the 96-well plate labeled "Bead and Live/Dead PBMC Plate" (*see* **Notes 24–26**).

3.3.2 Antibody/Dye Titration Plate Preparation

1. Add 20 μL of staining buffer to each appropriate well in a 96-well U-bottom plate labeled "Antibody/Dye Titration Plate" for preparation of the antibody dilutions (*see* **Note 27**).
2. Add 20 μL of 1X PBS to each appropriate well in the 96-well U-bottom plate labeled "Antibody/Dye Titration Plate" for preparation of the Live/Dead (L/D)-Zombie Aqua dye dilutions (*see* **Note 27**).
3. Prepare each antibody/dye titration down a single column of the plate as follows by adding 5 μL of appropriate antibody/dye to well B1, mixing well and transferring 5 μL of diluted antibody/dye from B1 to well C1, mixing well and repeating this procedure for wells C1, D1, E1, and F1 (*see* **Note 28** and Fig. 3).

3.3.3 Combination of Antibody/Dye Titrations to Bead and Live/Dead PBMC Plate

1. Add 5 μL of appropriate stock antibody/dye to the corresponding well in row A of the "Bead and Live/Dead PBMC Plate."
2. Transfer 5 μL of appropriate antibody/dye dilution from "Antibody/Dye Titration Plate" to the corresponding well of the "Bead and Live/Dead PBMC Plate" (*see* **Note 29**).

3. Incubate the "Bead and Live: Dead PBMC Plate" (now with beads + antibody and PBMC + dye) in the dark at room temperature for 30 min.
4. Centrifuge plate for 1–5 min at 300 × *g*.
5. Manually aspirate 40 μL of supernatant from each well.
6. Add 40 μL of staining buffer (or 1X PBS for dye samples) to each well. Resuspend pellets (*see* **Note 30**).

3.3.4 Plate Reading with NovoCyte 3005

1. Turn on computer > NovoExpress > username and password.
2. Turn on flow cytometer.
3. Check default parameters: All parameters.
4. Set up the experimental parameters as follows:
   (a) Mode: custom; 96 well U-bottom.
   (b) Click on appropriate squares to create sample. Rename each well accordingly in work list.
   (c) Indicate to run samples vertically.
   (d) Mix: 1 cycle, every 12 wells; 1500 rpm, 10 s; no rinse.

Table 4
Cytometer parameters. Left: Panel #1; Right: Panel #2

Parameter	Alias	Gain	A	H	Parameter	Alias	Gain	A	H
FSC	FSC	–	☒	☒	FSC	FSC	–	☒	☒
SSC	SSC	–	☒	☒	SSC	SSC	–	☒	☒
B530	CD8-FITC	460	☒	☒	B530	CD56-FITC	460	☒	☒
B572	CD4-PE	397	☒	☒	B572	CD14-PE	397	☒	☒
B660	PE-Cy5	537	☐	☐	B660	PE-Cy5	537	☐	☐
B725	PerCP-eFluor 710	648	☐	☐	B725	PerCP-eFluor 710	648	☐	☐
B780	CD19-PE-Cy7	478	☒	☒	B780	CD19-PE-Cy7	478	☒	☒
R660	CCR4-APC	537	☒	☒	R660	CD123-APC	537	☒	☒
R725	CD45RA-Alexa Fluor 700	648	☒	☒	R725	CD54-Alexa Fluor 700	648	☒	☒
R780	TCR-g/d-APC-Fire750	478	☒	☒	R780	CD20-APC-Fire750	478	☒	☒
V445	CD45-Pacific blue	425	☒	☒	V445	CD45-Pacific blue	425	☒	☒
V530	L/D-zombie aqua	460	☒	☒	V530	L/D-zombie aqua	460	☒	☒
V572	CD3-BV570	397	☒	☒	V572	CD3-BV570	397	☒	☒
V660	CD25-BV650	537	☒	☒	V660	CD69-BV650	537	☒	☒
V725	CD154-BV711	648	☒	☒	V725	Qdot 705	648	☐	☐
V780	CD127-BV785	478	☒	☒	V780	CD11c-BV785	478	☒	☒

Parameters used on the Novocyte 3005 flow cytometer (*see* **Note 31**)

(e) Experimental Run. Stop conditions: Events 12,000; 30 μL. Flow rate: Fast.

(f) Cytometer Settings: Parameters (Table 4).

5. Insert plate.
6. Click run plate. Highlight all boxes that are to be run (all). Click run; Click okay.
7. Save file.

3.3.5 Data Analysis in NovoExpress

1. Make a forward scatter (FSC) vs side scatter (SSC) plot. Draw an region of interest (ROI) around the beads.
2. Make a histogram by double-clicking on the bead population.
3. Change the axes of the histogram (right click): X-axis: Appropriate fluorophore; Y-axis: Count.
4. Drag the analysis tab to each sample of corresponding fluorophore.
5. To overlay graphs, duplicate histogram by right clicking. Right click on the duplicated histogram → Edit overlay → Add → Select the diluted samples to add → Add and close.
6. Right click on graph → style → partial overlap.
7. Copy image and make a PowerPoint with serial dilution.
8. Optimal dilution is determined by the antibody/dye concentration that has the largest difference between positive and negative stain (*see* **Note 32**).
9. Repeat analysis for each serial dilution.
10. Save files.

3.4 Single Stain Controls with Compensation Beads (see Notes 18 and 33)

3.4.1 Plate Preparation with Beads and Cells

1. Make a tenfold dilution of UltraComp eBeads Plus Compensation Beads using staining buffer as the diluent (*see* **Note 21**).
2. Add 50 μL of diluted beads to appropriate wells of a 96-well U-bottom plate (*see* **Note 34**).
3. Add 50 μL of cells prepared in 1X PBS (1:1 Live: Dead PBMC; approximately 5×10^5 cells/mL) to 2 wells (*see* **Notes 35** and **36**).

3.4.2 Antibody Dilution Preparation

1. Prepare at least 5 μL of each antibody dilution with staining buffer (Table 5, *see* **Note 37**).
2. Transfer 5 μL of antibody dilution/dye to the corresponding well of the plate with the beads/cells.
3. Incubate the plate in the dark for about 30 min.
4. Centrifuge plate for 5 min at 300 × *g*.
5. Manually aspirate 30 μL of supernatant from the well.

Table 5
Human immunophenotyping panel antibody dilutions

Immunophenotyping panel #1		Immunophenotyping panel #2	
Antibody	**Dilution**	**Antibody**	**Dilution**
CD8-FITC	Dil 25	CD56-FITC	Dil 125
CD4-PE	Dil 125	CD14-PE	Dil 125
CD19-PE/Cy7	Dil 125	CD19-PE/Cy7	Dil 125
CCR4-APC	Dil 125	CD123-APC	Dil 25
CD45RA-AF700	Dil 25	CD54-AF700	Dil 125
TCR-gamma/delta APC-Fire750	Dil 50	CD20-APC/Fire750	Dil 25
CD45-PacBlue	Dil 25	CD45-PacBlue	Dil 25
CD3-BV570	Dil 25	CD3-BV570	Dil 25
CD25-BV650	Dil 100	CD69-BV650	Dil 25
CD154-BV711	Dil 50	CD11c-BV785	Dil 25
CD127-BV785	Dil 25	Zombie aqua	Dil 125
Zombie aqua	Dil 125		

6. Add 50 μL of staining buffer (or 1X PBS for dye samples) to the well and resuspend the pellet.
7. Read the plate on the cytometer.

3.4.3 Plate Reading

1. Turn on the computer → NovoExpress → username and password.
2. Turn on the flow cytometer.
3. Save file.
4. Instrument → Auto compensation.
5. In pop-up, de-select unused parameters. Rename other channels to the fluorophores used → Okay (*see* **Note 38**).
6. Set up the experimental parameters.
   (a) Mode: custom; 96-well U-bottom.
   (b) Select wells and rename and reposition each well accordingly in work list.
   (c) Indicate to run samples vertically.
   (d) Mix: 1 cycle, every 2 wells; 1500 rpm, 10 s; Rinse: 1 cycle, every 3 wells.

(e) Experimental Run. Stop conditions: Events 25,000; 50 μL. Flow rate: Fast.

(f) Cytometer settings: Parameters (Table 4, *see* **Note 39**).

7. Insert plate into the flow cytometer.
8. Click run plate. Highlight all boxes that are to be run (all) → run → Okay.
9. Can adjust the main cell gate for cell samples.
10. Save file.
11. Export FCS files and save the compensation matrix.

3.5 Flow Cytometry Control Experiments with PBMC and FMO Controls: Day 1, PBMC (see Note 40)

1. Adjust concentration of PBMC from Subheading 3.2, **step 8** to 1.25×10^6 viable cells/mL using complete RPMI medium (*see* **Note 41**).
2. Dispense 800 μL of PBMC per well in 24-well plate. Gently shake plates to allow all components to mix.
3. Dispense 200 μL of positive control samples into corresponding wells of 24-well plate containing 800 μL of PBMC for a total of 1 mL per well.

   (a) Immunophenotyping Panel #1: Positive Control = PMA/Ionomycin.

   (b) Immunophenotyping Panel #2: Positive Control = ODN2216/PHA-M.

4. Incubate for about 24 h in a humidified 37 °C, 5% CO_2 incubator.

3.6 Flow Cytometry Control Experiments with PBMC and FMO Controls: Day 2, FMO Controls (see Note 40)

3.6.1 Prepare Heat Shocked PBMC

1. Obtain needed quantity of treated PBMC from the 24-well plate and transfer to a 15 mL conical tube (*see* **Note 42**).
2. Centrifuge sample for 10 min at 400 × *g*. Aspirate the supernatant and resuspend the pellet in 1 mL 1X PBS.
3. Transfer sample to microcentrifuge tube and place sample in digital heat block set to 70 °C for at least 1.5 h.

3.6.2 Harvest and Prepare PBMC

1. Transfer 750 μL of cell sample (0.75×10^6 cells/well) to appropriately labeled microcentrifuge tubes (*see* **Note 43**).
2. Centrifuge samples at 400 × *g* for 7 min. Aspirate each supernatant and resuspend each pellet in 1 mL 1X PBS.
3. Add 62.5 μL heat shocked PBMC (0.25×10^6 dead PBMC) to each sample (*see* **Note 44**).

3.6.3 Stain Cells

1. Centrifuge each sample at 400 × *g* for 7 min. Aspirate each supernatant and resuspend each pellet in 49.2 μL of 1X PBS.
2. Make fivefold dilution of Zombie Aqua dye for staining samples with 1X PBS as the diluent (*see* **Note 45**).

Table 6
Volumes of fivefold antibody dilutions needed for FMO

Antibody dilution determined to be optimal in step 6.1	Final antibody dilution	Volume of fivefold dilution antibody needed (μL) per 100 μL sample
Dil 25	Dil ~250	8
Dil 50	Dil ~500	4
Dil 100	Dil ~1000	2
Dil 125	Dil ~1250	1.6

Prepare enough diluted antibodies for n + 2. Each antibody is only needed at one previously determined optimal dilution (see Table 5), i.e., for Immunophenotyping Panel #1, the antibody dilution that was determined to lead to the optimal concentration of CD8-FITC was Dil 25; therefore, 8 μL of the fivefold dilution of CD8-FITC is needed per FMO control

3. Add 0.8 μL of a fivefold diluted Zombie Aqua dye to all samples except unstained sample and Zombie Aqua FMO sample.
4. Incubate samples for 30 min at room temperature in the dark.
5. Wash each sample two times with 500 μL staining buffer.
6. Resuspend each sample in 40 μL staining buffer.
7. Prepare fivefold antibody dilutions in staining buffer in labeled microcentrifuge tubes (one dilution per antibody needed; Table 6).
8. Make a master mix (MM) for each FMO control in an appropriately labeled microcentrifuge tube (*see* **Note 46**).
9. Add appropriate FMO MM (60 μL) to appropriate cell sample (40 μL) from step 6 above.
10. Incubate samples for 30 min at room temperature in the dark.
11. Wash each sample two times with 500 μL staining buffer.
12. Fix cells by resuspending each cell sample in 100 μL 2% PFA and incubating for 15 min at room temperature in the dark.
13. Wash each sample two times with 500 μL staining buffer.
14. Resuspend each sample in 500 μL staining buffer and transfer samples to appropriately labeled flow tubes.
15. Briefly vortex samples.

3.6.4 Data Acquisition with NovoCyte 3005 for FMO Controls

1. Turn on computer → NovoExpress → username and password.
2. Turn on flow cytometer.
3. Save file.
4. Check the experimental parameters.

(a) Mode: Custom; 24 tube rack.
(b) Indicate samples to run horizontal.
(c) Click on proper squares to create sample. Rename each sample accordingly in work list.
(d) Mix: 1 cycle, every 2 wells (1500 rpm; 10 s); Rinse: 1 cycle, every 3 wells.
(e) Experimental Run. Stop conditions: 300 μL. Fast (Apply changes to all samples).
(f) Review Cytometer Settings: Parameters (Table 4) and make sure channels and laser intensity are the same as the single stain control experiment.

5. Insert samples in tube rack.
6. Click run plate. Highlight all boxes that are to be run (all) → run → Okay.
7. Save file.
8. Export FCS files.

3.6.5 Data Analysis of FMO Controls with NovoExpress (see **Note 47**)

1. Apply compensation matrix to FMO controls (*see* **Note 48**).
2. Make a copy of the single stain control file (.ncf) that correlates to the FMO control file.
3. Upload the FCS files from the appropriate FMO control experiment into the single stain control file (.ncf).
   (a) Right click on file name in Experimental Manager.
   (b) Click on Import FCS files.
   (c) Browse for the folder with appropriate FCS files.
4. Apply the compensation matrix from the single stain controls to the FMO samples.
   (a) Under "Compensation Specimen" click on the "+" for any sample.
   (b) Right click on Compensation and select Copy.
   (c) Right click on overarching FMO control experimental sample and click on Paste.
   (d) "Are you sure you want to paste Compensation to all samples in experimental sample name?" Yes.
   (e) Can then delete the single stain Compensation Specimen.
5. Rename the file with the compensated FMO samples and save.

3.6.6 Analysis Method/Compensation Adjustment of FMO Controls

1. Make plots of each FMO sample (Fig. 4).
   (a) Make a plot of FSC-H vs. SSC-H and gate main the cell population "P1."

(b) Make a plot of FSC-A vs. FSC-H from "P1" and gate the single cells "P2."

2. Adjust the compensation using the fully stained sample.
   (a) Make a plot of FSC-H vs. SSC-H and gate main the cell population "P1."
   (b) Make a plot of FSC-A vs. FSC-H from "P1" and gate the single cells "P2."
   (c) Make a plot of each fluorophore pair (example FITC vs. PE; FITC vs. PE-Cy-7; FITC vs. APC; etc.) from "P2" (*see* **Note 49**).
   (d) Right click on pairwise plots and click Edit overlay. Add FMO sample that corresponds to the fluorophore on the x-axis. Apply.
   (e) For plots that have large compensation, check the compensation. Adjust the compensation of the parameters using the quick compensation adjustments on the fully stained sample. Adjust plots to eliminate or minimize the slanting in the FMO overlay.
   (f) Once compensation of the fully stained sample is complete, apply the compensation to all the FMO samples.
3. Save compensated file and export compensated FCS files.

4 Notes

1. Technical specifications of individual flow cytometers vary. This protocol should be followed for the Novocyte 3005 flow cytometer. If another cytometer is used, the procedure described herein would require optimization according to the technical specifications of that instrument. It is, therefore, recommended to review the technical specification of a cytometer available in each research laboratory prior to planning the experiment.
2. Collection of primary human blood requires institutional approval. The blood can also be obtained from commercial sources.
3. All work with human blood must be performed in a biological safety cabinet and using personal protective equipment (lab coat, safety goggles, gloves).
4. For optimal results, whole blood must be processed within 2 h after collection.
5. Heat inactivation is intended to eliminate the effects of bovine complement proteins on human blood cells. Heat inactivation is performed by incubating FBS at 56 °C for 30 min. It is recommended to prepare single use aliquots and store them at −20 °C before use.

6. It is recommended to store the complete medium at 2–8 °C protected from light for no longer than 1 month and warm it in a water bath before each use.
7. It is recommended to prepare single use 20 μL aliquots and store them at −20 °C.
8. It is recommended to thaw a frozen aliquot of the stock at room temperature on the day of the experiment. Thawed PMA should be diluted in culture media to a final concentration of 50 ng/mL when used as a positive control.
9. It is recommended on the day of the experiment to thaw an aliquot at room temperature and dilute in culture media to a final concentration of 1 μg/mL.
10. It is recommended on the day of the experiment to thaw an aliquot at room temperature and dilute in culture media to a final concentration of 5 μg/mL.
11. It is recommended to prepare single use aliquots and store them at a nominal temperature of −20 °C. It is also recommended to avoid repeated freezing/thawing. On the day of the experiment, the stock PHA-M should be diluted in cell culture medium to a final concentration of 10 μg/mL.
12. Alternatively, the blood/PBS mixture may be slowly layered over the Ficoll-Paque solution. It is recommended to use 3 mL of Ficoll-Paque solution per 4 mL of blood/PBS mixture. For example, 15 mL Ficoll-Paque per 20 mL of diluted blood in a 50 mL tube.
13. To maintain Ficoll-blood interface it is helpful to hold the tube at a 45° angle.
14. It is important to turn centrifuge brakes off. Using brakes will result in vibrations which will disturb a buffy coat. Depending on the type of a centrifuge, one may also need to set acceleration speed to minimum.
15. Usually 4 mL of blood/PBS mixture results in ~2 mL of mononuclear layer and requires at least 6 mL of HBSS for the wash step. It is recommended to use 10 mL of HBSS per each 2 mL of cells.
16. The expected cell viability after this step is at least 90%. If the viability is lower, it is not recommended to use the cells for subsequent experimental procedure.
17. The cells isolated during this step can also be used to prepare Live: Dead cell controls.
18. Typical antibody concentrations in commercial stocks vary between 50 μg/mL and 400 μg/mL.

19. It is recommended to prepare a 6-point calibration curve by using the stock antibody and performing five fivefold serial dilutions. The calibration curve should include the antibody at its stock concentration and five fivefold serial dilutions (dils) of the stock labeled Dils 5; 25; 125; 625; and 3125. It is recommended to check with the manufacturer to determine if the concentration for a given antibody remains unchanged between different lots of the antibody.
20. Antibody titration needs to be performed to establish the antibody/dye dilutions for the panels that will be run. Consult the information in Tables 2 and 3 for the antibody-fluorophore combinations used for Immunophenotyping Panels #1 and #2, respectively. Once the panel is established, the antibody titrations do not need to be run again unless there are issues encountered with a new antibody lot or the stock concentration changed. Antibody dilutions have already been determined for panels described herein.
21. It is highly recommended to vortex beads.
22. Six (6) wells of beads are needed per antibody.
23. It is helpful to review Tables 2 and 3 for antibodies used in Immunophenotyping Panels #1 and #2 and Fig. 2 for an example of plate setup.
24. Six (6) wells are recommended per dye.
25. Ideally, this sample should contain 1:1 mixture of live and dead PBMC at a total concentration of approximately 5×10^5 cells/mL. An example Plate 1 setup may be helpful (Fig. 2).
26. Freshly isolated or freeze thawed PBMC work well. If necessary, cells can be heat shocked at 70 °C to obtain about 50% cell death.

	1	2	3	4	5	6	7	8	9	10	11	12
A	Dil 10 Beads	Dil 10 Beads	Dil 10 Beads	Dil 10 Beads	Dil 10 Beads	Dil 10 Beads	Dil 10 Beads	Dil 10 Beads	Dil 10 Beads	Dil 10 Beads	Dil 10 Beads	Live/Dead PBMC
B	Dil 10 Beads	Dil 10 Beads	Dil 10 Beads	Dil 10 Beads	Dil 10 Beads	Dil 10 Beads	Dil 10 Beads	Dil 10 Beads	Dil 10 Beads	Dil 10 Beads	Dil 10 Beads	Live/Dead PBMC
C	Dil 10 Beads	Dil 10 Beads	Dil 10 Beads	Dil 10 Beads	Dil 10 Beads	Dil 10 Beads	Dil 10 Beads	Dil 10 Beads	Dil 10 Beads	Dil 10 Beads	Dil 10 Beads	Live/Dead PBMC
D	Dil 10 Beads	Dil 10 Beads	Dil 10 Beads	Dil 10 Beads	Dil 10 Beads	Dil 10 Beads	Dil 10 Beads	Dil 10 Beads	Dil 10 Beads	Dil 10 Beads	Dil 10 Beads	Live/Dead PBMC
E	Dil 10 Beads	Dil 10 Beads	Dil 10 Beads	Dil 10 Beads	Dil 10 Beads	Dil 10 Beads	Dil 10 Beads	Dil 10 Beads	Dil 10 Beads	Dil 10 Beads	Dil 10 Beads	Live/Dead PBMC
F	Dil 10 Beads	Dil 10 Beads	Dil 10 Beads	Dil 10 Beads	Dil 10 Beads	Dil 10 Beads	Dil 10 Beads	Dil 10 Beads	Dil 10 Beads	Dil 10 Beads	Dil 10 Beads	Live/Dead PBMC
G												
H												

Fig. 2 Example plate map for bead and live/dead PBMC plate. Sample position on a plate is shown; other ways to organize samples on the plate are possible

	1	2	3	4	5	6	7	8	9	10	11	12
A												
B	CD8-FITC Ab Dil 5	CD4-PE Ab Dil 5	CD19-PE/Cy7 Ab Dil 5	CCR4-APC Ab Dil 5	CD45RA-AF700 Ab Dil 5	TCR-γδ-APC/Fire750 Ab Dil 5	CD45-PB Ab Dil 5	CD3-BV570 Ab Dil 5	CD25-BV650 Ab Dil 5	CD154-BV711 Ab Dil 5	CD127-BV785 Ab Dil 5	Zombie Aqua dye Dil 5
C	CD8-FITC Ab Dil 25	CD4-PE Ab Dil 25	CD19-PE/Cy7 Ab Dil 25	CCR4-APC Ab Dil 25	CD45RA-AF700 Ab Dil 25	TCR-γδ-APC/Fire750 Ab Dil 25	CD45-PB Ab Dil 25	CD3-BV570 Ab Dil 25	CD25-BV650 Ab Dil 25	CD154-BV711 Ab Dil 25	CD127-BV785 Ab Dil 25	Zombie Aqua dye Dil 25
D	CD8-FITC Ab Dil 125	CD4-PE Ab Dil 125	CD19-PE/Cy7 Ab Dil 125	CCR4-APC Ab Dil 125	CD45RA-AF700 Ab Dil 125	TCR-γδ-APC/Fire750 Ab Dil 125	CD45-PB Ab Dil 125	CD3-BV570 Ab Dil 125	CD25-BV650 Ab Dil 125	CD154-BV711 Ab Dil 125	CD127-BV785 Ab Dil 125	Zombie Aqua dye Dil 125
E	CD8-FITC Ab Dil 625	CD4-PE Ab Dil 625	CD19-PE/Cy7 Ab Dil 625	CCR4-APC Ab Dil 625	CD45RA-AF700 Ab Dil 625	TCR-γδ-APC/Fire750 Ab Dil 625	CD45-PB Ab Dil 625	CD3-BV570 Ab Dil 625	CD25-BV650 Ab Dil 625	CD154-BV711 Ab Dil 625	CD127-BV785 Ab Dil 625	Zombie Aqua dye Dil 625
F	CD8-FITC Ab Dil 3125	CD4-PE Ab Dil 3125	CD19-PE/Cy7 Ab Dil 3125	CCR4-APC Ab Dil 3125	CD45RA-AF700 Ab Dil 3125	TCR-γδ-APC/Fire750 Ab Dil 3125	CD45-PB Ab Dil 3125	CD3-BV570 Ab Dil 3125	CD25-BV650 Ab Dil 3125	CD154-BV711 Ab Dil 3125	CD127-BV785 Ab Dil 3125	Zombie Aqua dye Dil 3125
G												
H												

Fig. 3 Example of plate map for antibody/dye titration plate. In this example, antibodies used in the Immunophenotyping Panel 1 are shown. A similar layout can be used for Panel 2

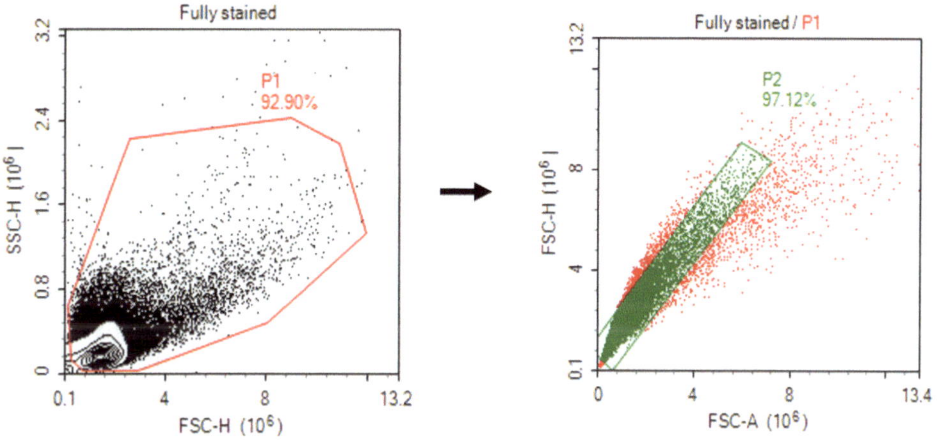

Fig. 4 Gating FMO samples. Left: Gate of the cell population; Right: Gate of the singlet population

27. It is recommended to use five wells in a column prepared for each antibody/dye (Wells B-F) (Fig. 3).
28. This step has to be repeated for each antibody/dye.
29. Another ~tenfold-dilution applied to each dilution from the titration (5 μL antibody into 50 μL diluent) that will be taken into account in subsequent calculations.
30. Plate is now ready to read.

31. After samples are created, it is important to make sure that the experimental settings for each specimen are the same. Whenever a change is made, change should be applied to all and saved.
32. More than two log differences are needed. The negative sample has to be as low as possible without a large shift in positive population.
33. This step must be completed each time a new antibody lot is obtained. One replicate per condition is normally used. Tables 2 and 3 can be used to view antibody-fluorophore combinations used in single stain controls for Immunophenotyping Panels #1 and #2, respectively.
34. It is recommended to use one well per antibody.
35. At the end of this step, one should have two wells total: One well prepared for Zombie Aqua dye and one well prepared for unstained PBMC.
36. Freshly isolated or freeze-thawed PBMC work well. If necessary, cells can be heat shocked at 70 °C for 2 h to obtain about 50% cell death.
37. Zombie Aqua dye dilution is to be prepared in 1X PBS.
38. Panel #1 de-select Area and Height Parameters for PE-Cy5 and PerCP-eFluor710; Panel #2 de-select PE-Cy5, PerCP-eFluor710, and QDot705.
39. It is recommended to use 2500 positive events for the software to compensate correctly.
40. This step takes 2 days per panel and must be completed each time a new antibody lot is obtained. It is recommended to review antibody-fluorophore combinations used in single stain controls for Immunophenotyping Panels #1 and #2 (Tables 2 and 3). An unstained sample and a fully stained labeling antibody sample with all antibodies/dyes are also required. One replicate per condition. FMOs for panels may be processed concurrently or separately.
41. This procedure is optimized for 24-well plates. If other plates are used, the volumes may need to be adjusted accordingly.
42. The number of heat shocked PBMC needed per sample is 0.25×10^6; remaining cells are used in **step 2**.
43. Samples needed in this step include unstained control, FMO control for each labeling antibody/dye, and fully stained control containing all labeling antibodies/dyes.

Table 7
Immunophenotyping panel #1 FMO control samples

		FMO controls												Fully stained labeling Ab sample	Unstained
		1 CD8-FITC FMO MM	2 CD4-PE FMO MM	3 CD19-PE/Cy7 FMO MM	4 CCR4-APC FMO MM	5 CD45RA-AF700 FMO MM	6 TCR-γδ APC-Fire750 FMO MM	7 CD45-PacBlue FMO MM	8 L/D Zombie Aqua FMO MM	9 CD3-BV570 FMO MM	10 CD25-BV650 FMO MM	11 CD154-BV711 FMO MM	12 CD127-BV785 FMO MM	13 MM	14 MM
Antibodies added to each sample	**1**—CD8-FITC Dil 5	X	8 μL	8 μL	8 μL	8 μL	8 μL	8 μL	8 μL	8 μL	8 μL	8 μL	8 μL	8 μL	X
	2—CD4-PE Dil 5	1.6 μL	X	1.6 μL	1.6 μL	1.6 μL	1.6 μL	1.6 μL	1.6 μL	1.6 μL	1.6 μL	1.6 μL	1.6 μL	1.6 μL	X
	3—CD19-PE/Cy7 Dil 5	1.6 μL	1.6 μL	X	1.6 μL	1.6 μL	1.6 μL	1.6 μL	1.6 μL	1.6 μL	1.6 μL	1.6 μL	1.6 μL	1.6 μL	X
	4—CCR4-APC Dil 5	1.6 μL	1.6 μL	1.6 μL	X	1.6 μL	1.6 μL	1.6 μL	1.6 μL	1.6 μL	1.6 μL	1.6 μL	1.6 μL	1.6 μL	X
	5—CD45RA-AF700 Dil 5	8 μL	8 μL	8 μL	8 μL	X	8 μL	8 μL	8 μL	8 μL	8 μL	8 μL	8 μL	8 μL	X
	6—TCR- γδ APC Fire750 Dil 5	4 μL	4 μL	4 μL	4 μL	4 μL	X	4 μL	4 μL	4 μL	4 μL	4 μL	4 μL	4 μL	X
	7—CD45-PacBlue Dil 5	8 μL	8 μL	8 μL	8 μL	8 μL	8 μL	X	8 μL	8 μL	8 μL	8 μL	8 μL	8 μL	X
	8—L/D Zombie Aqua Dil 5	X	X	X	X	X	X	X	X	X	X	X	X	X	X
	9—CD3-BV570 Dil 5	8 μL	8 μL	8 μL	8 μL	8 μL	8 μL	8 μL	8 μL	X	8 μL	8 μL	8 μL	8 μL	X

(continued)

Table 7
(continued)

	FMO controls												Fully stained labeling Ab sample	Unstained
	1 CD8-FITC FMO MM	2 CD4-PE FMO MM	3 CD19-PE/Cy7 FMO MM	4 CCR4-APC FMO MM	5 CD45RA-AF700 FMO MM	6 TCR-γδ APC-Fire750 FMO MM	7 CD45-PacBlue FMO MM	8 L/D Zombie Aqua FMO MM	9 CD3-BV570 FMO MM	10 CD25-BV650 FMO MM	11 CD154-BV711 FMO MM	12 CD127-BV785 FMO MM	13 MM	14 MM
10—CD25-BV650 Dil 5	2 μL	2 μL	2 μL	2 μL	2 μL	2 μL	2 μL	2 μL	2 μL	X	2 μL	2 μL	2 μL	X
11—CD154-BV711 Dil 5	4 μL	4 μL	4 μL	4 μL	4 μL	4 μL	4 μL	4 μL	4 μL	4 μL	X	4 μL	4 μL	X
12—CD127-BV785 Dil 5	8 μL	8 μL	8 μL	8 μL	8 μL	8 μL	8 μL	8 μL	8 μL	8 μL	8 μL	X	8 μL	X
Total volume (μL)	46.8	53.2	53.2	53.2	46.8	50.8	46.8	54.8	46.8	52.8	50.8	46.8	54.8	0
Staining buffer (μL)	13.2	6.8	6.8	6.8	13.2	9.2	13.2	5.2	13.2	7.2	9.2	13.2	5.2	60

Table 8
Immunophenotyping panel #2 FMO control samples

		FMO controls											Fully stained labeling Ab sample	Unstained
		1 CD56-FITC FMO MM	2 CD14-PE FMO MM	3 CD19-PE/Cy7 FMO MM	4 CD123-APC FMO MM	5 CD54-AF700 FMO MM	6 CD20 APC-Fire750 FMO MM	7 CD45-PacBlue FMO MM	8 L/D zombie aqua FMO MM	9 CD3-BV570 FMO MM	10 CD69-BV650 FMO MM	11 CD11c-BV785 FMO MM	12 MM	13 MM
Antibodies added to each sample	1—CD56-FITC Dil 5	X	1.6 μL	1.6 μL	1.6 μL	1.6 μL	1.6 μL	1.6 μL	1.6 μL	1.6 μL	1.6 μL	1.6 μL	1.6 μL	X
	2—CD14-PE Dil 5	1.6 μL	X	1.6 μL	1.6 μL	1.6 μL	1.6 μL	1.6 μL	1.6 μL	1.6 μL	1.6 μL	1.6 μL	1.6 μL	X
	3—CD19-PE/Cy7 Dil 5	1.6 μL	1.6 μL	X	1.6 μL	1.6 μL	1.6 μL	1.6 μL	1.6 μL	1.6 μL	1.6 μL	1.6 μL	1.6 μL	X
	4—CD123-APC Dil 5	8 μL	8 μL	8 μL	X	8 μL	8 μL	8 μL	8 μL	8 μL	8 μL	8 μL	8 μL	X
	5—CD54-AF700 Dil 5	1.6 μL	1.6 μL	1.6 μL	1.6 μL	X	1.6 μL	1.6 μL	1.6 μL	1.6 μL	1.6 μL	1.6 μL	1.6 μL	X
	6—CD20 APC-Fire750 Dil 5	8 μL	8 μL	8 μL	8 μL	8 μL	X	8 μL	8 μL	8 μL	8 μL	8 μL	8 μL	X
	7—CD45-PacBlue Dil 5	8 μL	8 μL	8 μL	8 μL	8 μL	8 μL	X	8 μL	8 μL	8 μL	8 μL	8 μL	X
	8—L/D Zombie Aqua Dil 5	X	X	X	X	X	X	X	X	X	X	X	X	X

(continued)

Table 8 (continued)

	FMO controls											Fully stained labeling Ab sample	Unstained
	1 CD56-FITC FMO MM	2 CD14-PE FMO MM	3 CD19-PE/Cy7 FMO MM	4 CD123-APC FMO MM	5 CD54-AF700 FMO MM	6 CD20 APC-Fire750 FMO MM	7 CD45-PacBlue FMO MM	8 L/D zombie aqua FMO MM	9 CD3-BV570 FMO MM	10 CD69-BV650 FMO MM	11 CD11c-BV785 FMO MM	12 MM	13 MM
9—CD3-BV570 Dil 5	8 μL	8 μL	8 μL	8 μL	8 μL	8 μL	8 μL	8 μL	X	8 μL	8 μL	8 μL	X
10—CD69-BV650 Dil 5	8 μL	8 μL	8 μL	8 μL	8 μL	8 μL	8 μL	8 μL	8 μL	X	8 μL	8 μL	X
11—CD11c-BV785 Dil 5	8 μL	8 μL	8 μL	8 μL	8 μL	8 μL	8 μL	8 μL	8 μL	8 μL	X	8 μL	X
Total volume (μL)	52.8	52.8	52.8	46.4	52.8	46.4	46.4	54.4	46.4	46.4	46.4	54.4	0
Staining buffer (μL)	7.2	7.2	7.2	13.6	7.2	13.6	13.6	5.6	13.6	13.6	13.6	5.6	60

44. Total of 1×10^6 PBMC in each sample ($\sim 0.75 \times 10^6$ live cells and $\sim 0.25 \times 10^6$ dead cells).
45. One will need 0.8 μL of fivefold diluted Zombie Aqua dye x sample number (n) = 0.8 (n + 2) μL of the fivefold dilution.
46. Add fivefold dilution antibodies to the appropriate microcentrifuge tube as indicated in Tables 7 and 8. Add necessary amount of staining buffer to make final volume 60 μL.
47. The necessary data from this analysis is the compensation/spillover matrices. The FMO control samples need to be compensated with the compensation matrix from the single stain control samples. The compensation matrix may then need to be adjusted to account for the differences between beads (single stain controls) and cells (FMO controls).
48. There are multiple ways to apply the compensation matrix of the single stain controls to the FMO controls. This protocol describes one of them.
49. Plots can auto-populate by clicking on the dot plot symbol until all pairwise fluorophore plots are plotted.

Acknowledgement

This project has been funded in whole or in part by federal funds from the National Cancer Institute, National Institutes of Health, under contract 75N91019D00024. The content of this publication does not necessarily reflect the views or policies of the Department of Health and Human Services, nor does mention of trade names, commercial products, or organizations imply endorsement by the US Government.

References

1. Oughton JA, Kerkvliet NI (2005) Immune cell phenotyping using flow cytometry. Curr Protoc Toxicol 23(1):18.18.11–18.18.24
2. McCoy J, Philip J (2019) Immunophenotyping. Springer. ISBN:1493996495
3. McKinnon KM (2018) Flow cytometry: an overview. Curr Protoc Immunol 120:5 1 1–5 1 11
4. Newton HS, Dobrovolskaia MA (2022) Immunophenotyping: analytical approaches and role in preclinical development of nanomedicines. Adv Drug Deliv Rev. 185:114281
5. Dobrovolskaia MA, McNeil SE (2007) Immunological properties of engineered nanomaterials. Nat Nanotechnol 2(8):469–478
6. Acea BioScience Inc. (2017) NovoCyte® Flow Cytometer Technical Specifications
7. Agilent Technologies (2022) NovoCyte 3 Lasers 3005 Model 3005-default configuration

Chapter 25

Immunophenotyping, Part II: Analysis of Nanoparticle Effects on the Composition and Activation Status of Human Peripheral Blood Mononuclear Cells

Hannah S. Newton, Jenny Zhang, and Marina A. Dobrovolskaia

Abstract

The use of nanoparticles as drug delivery carriers requires analysis of their safety, which among other tests, includes immunotoxicity. Nanoparticles are also increasingly used for applications intended to specifically activate, inhibit, or modify the immune system's responses to improve the treatment of inflammatory and autoimmune disorders, cancer immunotherapy, and vaccines targeting cancer cells and viral and bacterial pathogens. In addition to the safety, the analysis of nanoparticles intended for immune system targeting includes mechanistic immunology investigations. Immunophenotyping provides researchers with a tool to assess the immune cell viability and activation status. These results provide mechanistic insights into nanoparticle efficacy and toxicity and therefore are of interest to the biomedical nanotechnology field. However, no standardized approaches exist due to the breadth of methods and instruments available for this analysis. This chapter provides detailed instructions for applying this methodology to analyze nanoparticle effects on subsets of immune cells present in peripheral blood. While this experimental strategy is specific to the NovoCyte 3005 flow cytometer, it can be adapted to other instruments. Instructions for instrument setup, calibration, and antibody qualification are described in this book's Chapter 24, Immunophenotyping, part I.

Key words Nanoparticles, immunophenotyping, PBMC, instrument setup, calibration, reagent qualification

1 Introduction

Immunophenotyping utilizes antibodies specific to antigens present on the surface or inside the cells. Distinct immune cell subsets express a unique set of antigens not found in other cells. This property allows the use of these antigens to determine the identity and activation statuses of immune cells [1–4]. This technique is commonly used to characterize the cell makeup in many diseases and identify treatment-related effects, such as immunotoxicity or intended immunomodulatory properties of nanotechnology carriers and nanoformulations carrying therapeutic payloads

Jeffrey D. Clogston et al. (eds.), *Characterization of Nanoparticles Intended for Drug Delivery*, Methods in Molecular Biology, vol. 2789, https://doi.org/10.1007/978-1-0716-3786-9_25,

[5]. Since many nanoparticles are designed to modify immune responses, while others may cause unintended immune-mediated responses, it is essential to have a methodology capable of quickly identifying individual immune cell subsets and determining the activation status of these cells [4, 5]. Currently, multicolor flow cytometry is a common choice for doing such studies [2, 3]. Herein, we describe a protocol for the analysis of nanoparticle effects on the subsets of myeloid and lymphoid immune cells present in normal human peripheral blood.

This protocol covers two separate immunophenotyping panels utilizing 11–12 antibody-fluorophore conjugates:

Immunophenotyping Panel #1 (or Lymphocyte Panel): This panel includes antibody-fluorophore conjugates that allow for the analysis of different lymphocyte populations, including B cells and T cells ($CD8^+$ T cells, $CD4^+$ T cells, regulatory T (Treg) cells, naïve T cells, and γδ TCR T cells). This panel also determines cellular CD25 and CD154 expression, which are markers of proliferation and co-stimulation/presentation, respectively.

Immunophenotyping Panel #2 (or Monocyte, Dendritic Cell (DC), Natural Killer (NK) Cell Panel): This panel includes antibody-fluorophore conjugates that allow for analysis of different cell populations, including $CD14^+$ monocytes, DCs (plasmacytoid (p) and myeloid (m) DCs), and NK cells along with NK T cells. This panel also examines cellular CD69 and CD54 expression, which are markers of early activation and adhesion, respectively.

When used in conjunction with other immunoassays, this protocol aids in establishing efficacy and safety profiles of engineered nanoparticles used for vaccine or drug delivery. The protocol has two parts: Part I described in Chapter 24 of this book is intended for instrument calibration, and part II, described herein, is intended for the analysis of nanoparticle-treated cells.

Herein, we use peripheral blood mononuclear cells (PBMC) derived from healthy donor volunteers and culture these specimens in the presence of controls and nanoparticles to assess the effect of nanoparticles on various immune cell populations and their activation statuses. The changes in cell populations and activation status between the untreated negative control samples, the positive control samples, and the treatment samples will allow one to determine the effects of nanoparticles on immune cell status. Prior to conducting this study, a separate protocol—Part I, Chapter 24—must be used to properly calibrate the instrument and qualify the performance of monoclonal antibodies.

It is also important to recognize that the complexity of immune cell phenotypes assessed by immunophenotyping depends on the available flow cytometry instrumentation. This protocol is

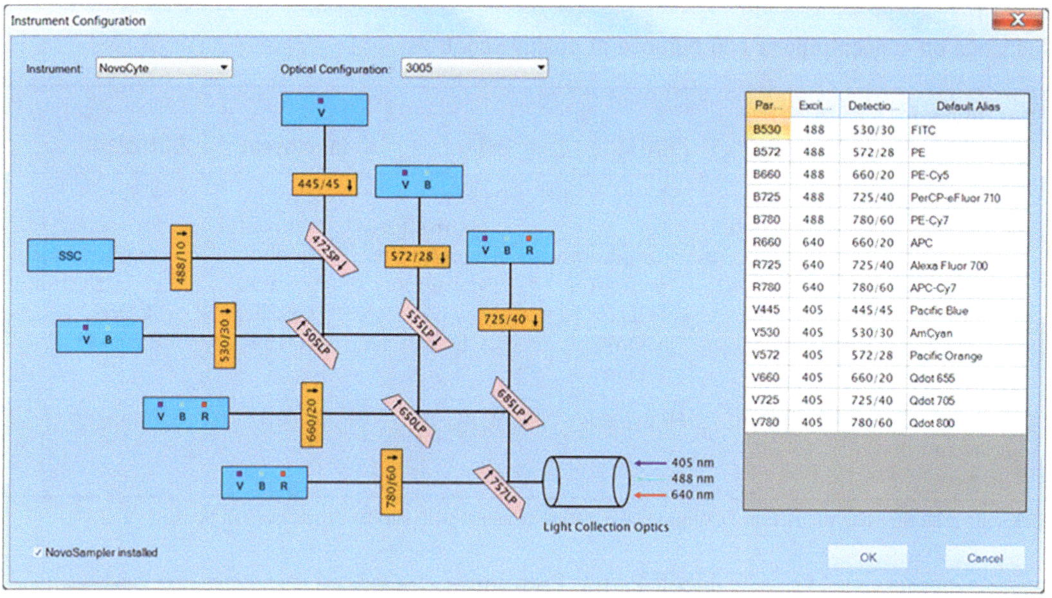

Fig. 1 Instrument configuration and settings. Parameters used on a NovoCyte 3005 instrument are shown from reference [9] (*see* **Note 1**)

optimized for the flow cytometer NovoCyte 3005 by Acea Biosciences Inc. (part of Agilent Technologies), which has the capacity to acquire data from 14 distinct fluorophores simultaneously. NovoCyte 3005 is equipped with 3 lasers (405 nm, 488 nm, and 640 nm) and 6 detectors (445/45 nm, 530/30 nm, 572/28 nm, 660/20 nm, 725/40 nm, and 780/60 nm) [6] (Fig. 1, *see* **Note 1**).

This protocol was successfully used to analyze both clinical grade and research nanomaterials [7]. However, nanomaterials with intrinsic fluorescent properties (e.g., quantum dots) or loaded with drugs possessing fluorescent properties (e.g., doxorubicin) may interfere with the assay. Therefore, the analysis of unstained, drug-treated cells prior to the immunophenotyping experiment is instrumental in identifying the interference [7].

2 Materials

2.1 Reagents for PBMC Cultures

1. Human blood anti-coagulated with Li-heparin and obtained from healthy donors (*see* **Notes 2–4**).
2. Phosphate buffered saline (PBS).
3. RPMI 1640.
4. Fetal bovine serum (FBS).
5. Penicillin streptomycin solution.
6. L-glutamine.

Table 1
Guidance on concentrations and purpose of positive controls

Description/ control	LPS	PHA-M	PMA	Ionomycin	ODN2216
Primary purpose	Positive control for lymphocyte activation (CD25)	Positive control for lymphocyte activation (CD25; CD69; CD54)	Positive control for lymphocyte activation (CD25; CD154)	Positive control for lymphocyte activation (CD25; CD154)	Positive control for monocyte and dendritic cell activation (CD69; CD54)
Final concentration in assay	20 ng/mL	10 μg/mL	50 ng/mL	1 μg/mL	5 μg/mL

Reagents used for assay validation are summarized. Other reagents with similar properties can be used

7. Ficoll Paque Premium.
8. Hank's balanced salt solution (HBSS).
9. Any reagent for live and dead cell counting.
10. Ultrapure LPS from K12 E.col strain (Table 1).
11. Phytohemagglutinin M (PHA-M).
12. Phorbol 12-myristate 13-acetate (PMA).
13. Ionomycin.
14. Oligodeoxyribonucleotide, Human TLR9 ligand (ODN2216).

2.2 Reagents for Flow Cytometry

1. eBioscience Flow Cytometry Staining Buffer (Invitrogen, 00-4222-26).
2. Paraformaldehyde (PFA) 20% Solution.
3. NovoRinse (Agilent Technologies, 872B603).
4. NovoClean (Agilent Technologies, 872B602).
5. NovoFlow (Agilent Technologies, 871B607).
6. Antibodies/Dyes (Tables 2, 3, 4, and 5).

2.3 General Supplies and Equipment

1. Pipettes covering a range of 0.05–10 mL.
2. 24-well flat bottom plates.
3. Polypropylene tubes, 15 and 50 mL.
4. Microcentrifuge tubes.
5. 5 mL Polystyrene round bottom tubes; 12 × 75 mm.
6. Centrifuge.
7. Cell culture incubator with 5% CO_2 and 95% humidity.

Table 2
Anti-human labeling antibodies used in the immunophenotyping panel 1

(Lymphocyte panel)						
Fluorochrome	**Marker**	**Cell type**	**Purpose**	**Company**	**Concentration (μg/mL)**	**Catalog #**
FITC	CD8a	T cells	Cytotoxic T cells	BioLegend	200	300905 or 300906
PE	CD4	T cells	T-helper	BioLegend	100	317409 or 317409
PE-Cy7	CD19	B cells	LPS co-receptor	BioLegend	100	302215 or 302216
APC	CCR4	T-reg	CD194	BioLegend	50	359407 or 359408
AF700	CD45RA	Naïve cells	Memory status	Invitrogen	50	56-0458-42
APC-Fire 750	TCR-γδ	TCR- γδ T cells	TCR- γδ receptor	BioLegend	400	331227 or 331228
PacBlue	CD45	All leukocytes	Pan-leukocyte	BioLegend	100	368539 or 368540
Zombie Aqua	Live/dead (L/D)	Dead cells	Excludes dead cells	BioLegend	Not Available	423101 or 423102
BV570	CD3	T cells	T cells	BioLegend	80	300435 or 300436
BV650	CD25	T-reg and activation status of cells	IL-2 receptor α	BioLegend	100	302633 or 302634
BV711	CD154	Lymphocytes	CD40L	BioLegend	100	310837 or 310838
BV785	CD127	T-reg	IL-7 receptor α	BioLegend	50	351329 or 351330

8. Biohazard safety cabinet approved for level II handling of biological material.
9. Water bath.
10. Vortex.
11. Instrument for cell counting.
12. Flow Cytometer (Acea Novocyte 3005).

Table 3
Isotype antibodies used in the immunophenotyping Panel 1

(Lymphocyte panel)					
Fluorochrome	**Isotype**	**Clone**	**Company**	**Concentration (μg/mL)**	**Catalog #**
FITC	Mouse IgG1, K	MOPC-21	BioLegend	200	400109 or 400110
PE	Mouse IgG2b, K	MPC-11	BioLegend	200	400313 or 400314
PE-Cy7	Mouse IgG1, K	MOPC-21	BioLegend	200	400125 or 400126
APC	Mouse IgG1, K	MOPC-21	BioLegend	200	400121 or 400122
AF700	Mouse IgG2b kappa	eBMG2b	Invitrogen	200	56-4732-80
APC-Fire 750	Mouse IgG1, K	MOPC-21	BioLegend	200	400195 or 400196
PacBlue	Mouse IgG1, K	MOPC-21	BioLegend	500	400131
BV570	Mouse IgG1, K	MOPC-21	BioLegend	100	400159 or 400160
BV650	Mouse IgG1, K	MOPC-21	BioLegend	100	400163 or 400164
BV711	Mouse IgG1, K	MOPC-21	BioLegend	100	400167 or 400168
BV785	Mouse IgG1, K	MOPC-21	BioLegend	100	400169 or 400170

3 Methods

3.1 Preparation of Reagents and Controls

1. Prepare Complete RPMI 1640 medium by supplementing stock RPMI medium with the following reagents: 10% heat inactivated FBS (*see* **Note 5**), 2 mM L-glutamine, 100 U/mL penicillin,100 μg/mL streptomycin (*see* **Note 6**).
2. Prepare 1 mg/mL LPS stock by adding 1 mL of sterile water to 1 mg of LPS in the provided vial and vortex to mix (*see* **Note 7**).
3. Prepare PMA stock by reconstituting the dry powder in DMSO to a stock concentration of 1 mg/mL (*see* **Notes 8** and **9**).
4. Ionomycin stock is supplied as 1% free acid and 99% ethyl alcohol, and a concentration of 10 mg/mL (*see* **Note 10**).
5. Prepare 1 mg/mL ODN 2216 stock by reconstituting the supplied powder in pyrogen-free, nuclease-free water to a concentration of 1 mg/mL (*see* **Note 11**).
6. Prepare 1 mg/mL PHA-M stock by adding 1 mL of sterile PBS or cell culture medium per 1 mg of PHA-M to the vial and gently rotating the vial to mix (*see* **Note 12**).
7. Prepare nanoparticles and corresponding vehicle controls (*see* **Note 13**).

Table 4
Anti-human labeling antibodies for immunophenotyping panel 2

(Monocyte, DC, NK cell panel)						
Fluorochrome	**Marker**	**Cell type**	**Purpose**	**Company**	**Concentration (μg/mL)**	**Catalog #**
FITC	CD56	NK cells	Neural cell adhesion molecule	BioLegend	200	362545 or 362546
PE	CD14	Monocytes	LPS co-receptor	BioLegend	200	301805 or 301806
PE-Cy7	CD19	B cells	LPS co-receptor	BioLegend	100	302215 or 302216
APC	CD123	pDCs	Interleukin-3 receptor	BioLegend	100	306011 or 306012
AF700	CD54	Lymphocytes and monocytes	Adhesion	BioLegend	400	353125 or 353126
APC-Fire 750	CD20	B cells	B cells	BioLegend	200	302357 or 302358
PacBlue	CD45	All leukocytes	Pan-leukocyte	BioLegend	100	368539 or 368540
Zombie Aqua	Live/dead	Dead cells	Excludes dead cells	BioLegend	Not available	423101 or 423102
BV570	CD3	T cells	T cells	BioLegend	80	300435 or 300436
BV650	CD69	Neutrophils/monocytes	Early activation	BioLegend	50	310933 or 310934
BV785	CD11c	mDCs	Integrin α x	BioLegend	160	301643 or 301644

3.2 PBMC Isolation and Treatment with Nanoparticles and Control (Day 1)

1. Place freshly drawn blood into 15 or 50 mL conical centrifuge tubes, add an equal volume of room-temperature 1X PBS, and mix well (*see* **Notes 2–4**).
2. Slowly layer the Ficoll-Paque solution underneath the blood/PBS mixture by placing the tip of the pipet containing Ficoll-Paque at the bottom of the blood sample tube (*see* **Notes 14** and **15**).
3. Centrifuge 30 min at 900 × *g* and 18–20 °C (*see* **Note 16**).
4. Using a sterile pipet, remove the upper layer containing plasma and platelets and discard it.
5. Using a fresh sterile pipet, transfer the mononuclear cell layer into another centrifuge tube.

Table 5
Isotype antibodies for immunophenotyping panel 2

(Monocyte, DC, NK cell panel)					
Fluorochrome	**Isotype**	**Clone**	**Company**	**Concentration (μg/mL)**	**Catalog #**
FITC	Mouse IgG1, K	MOPC-21	BioLegend	200	400109 or 400110
PE	Mouse IgG2a, K	MOPC-173	BioLegend	200	400211
PE-Cy7	Mouse IgG1, K	MOPC-21	BioLegend	200	400125 or 400126
APC	Mouse IgG1, K	MOPC-21	BioLegend	200	400121 or 400122
AF700	Mouse IgG1, K	MOPC-21	BioLegend	500	400143
APC-Fire 750	Mouse IgG2b, K	MPC-11	BioLegend	200	400371 or 400372
PacBlue	Mouse IgG1, K	MOPC-21	BioLegend	500	400131
BV570	Mouse IgG1, K	MOPC-21	BioLegend	100	400159 or 400160
BV650	Mouse IgG1, K	MOPC-21	BioLegend	100	400163 or 400164
BV785	Mouse IgG1, K	MOPC-21	BioLegend	100	400169 or 400170

6. Wash cells by adding an excess of HBSS and centrifuging for 10 min at 400 × *g*, 18–20 °C. The HBSS volume should be ~3 times the volume of mononuclear layer (*see* **Note 17**).
7. Discard supernatant and repeat wash step once more.
8. Resuspend cells in complete RPMI 1640 medium and determine cell viability using any available live-dead stain reagent (*see* **Note 18**).
9. Adjust the concentration of PBMC to 1.25×10^6 viable cells/mL using complete RPMI medium (*see* **Note 19**, Tables 2, 3, 4, 5, 6, and 7).
10. Dispense 800 μL of PBMC from **step 9** per well in a 24-well plate. Gently shake plates to allow all components to mix.
11. Dispense 200 μL of blank complete media (baseline negative control) or positive control into corresponding wells of a 24-well plate containing 800 μL of PBMCs for a total of 1 mL per well.
12. Dispense 100 μL of 1X PBS (negative control), vehicle control or test sample into corresponding wells of a 24-well plate containing 800 μL of PBMCs. Add 100 μL of fresh complete media for a total of 1 mL per well (*see* **Note 20**).
13. Incubate for about 24 h in a humidified 37 °C, 5% CO_2 incubator.

Table 6
Conditions for immunophenotyping panel #1

Unstained samples	Isotype control samples	Fully stained labeling antibody samples
Negative control: Untreated	Negative control: Untreated (×2)	Negative control: Untreated (×2)
Positive control: LPS/PHA-M	Positive control: LPS/PHA-M (×2)	Positive control: LPS/PHA-M (×2)
Positive control: PMA/Ionomycin	Positive control: PMA/Ionomycin (×2)	Positive control: PMA/Ionomycin (×2)
Vehicle controls	Vehicle controls (×2)	Vehicle controls (×2)
Test samples (MRD 10)	Test samples (MRD 10) (×2)	Test samples (MRD 10) (×2)

Table 7
Conditions for immunophenotyping panel #2

Unstained samples	Isotype control samples	Fully stained labeling antibody samples
Negative control: Untreated	Negative control: Untreated (×2)	Negative control: Untreated (×2)
Positive control: ODN2216/PHA-M	Positive control: ODN2216/PHA-M (×2)	Positive control: ODN2216/PHA-M (×2)
Vehicle controls	Vehicle controls (×2)	Vehicle controls (×2)
Test samples (MRD 10)	Test samples (MRD 10) (×2)	Test samples (MRD 10) (×2)

3.3 Master Mix Preparation, PBMC Harvesting, and PBMC Staining (Day 2)

1. Prepare master mixes (MM) for the isotype controls and the antibody-stained samples in labeled microcentrifuge tubes based on the previously determined antibody concentrations (see Chapter 24) (*see* **Notes 21–23**).
2. Collect cultured PBMC in properly labeled 1.5 mL microcentrifuge tubes and spin in a centrifuge at 400 × *g* for 5 min.
3. Transfer supernatants into fresh tubes and store at −80 °C (*see* **Note 24**).
4. Wash each cell sample twice with 1X PBS by adding 1 mL of 1X PBS to each tube, centrifuging the tube for 7 min at 400 × *g*, and aspirating and discarding supernatants in each wash cycle.
5. Resuspended each cell sample in 49.2 μL 1X PBS.
6. Dilute Zombie Aqua Dye five-fold using 1X PBS as the diluent (*see* **Note 25**).
7. Add 0.8 μL of the Dil 5 Zombie Aqua dye to all fully stained labeling Ab samples in both Immunophenotyping Panels #1 and Panel #2 (Tables 6 and 7).

8. Incubate samples for 30 min at room temperature in the dark.
9. Wash each sample twice with 500 μL staining buffer.
10. Resuspend samples in staining buffer.
    (a) Panel #1: Unstained samples: 100 μL staining buffer; isotype control samples: 92.4 μL staining buffer; fully stained labeling antibody samples: 89.04 μL staining buffer.
    (b) Panel #2: Unstained samples: 100 μL staining buffer; isotype control samples: 91.58 μL staining buffer; fully stained labeling antibody samples: 89.12 μL staining buffer.
11. Add appropriate MM volume and staining buffer to appropriate cell sample keeping the final volume of 100 μL.
    (a) Panel #1: Isotype control samples: 7.6 μL MM; fully stained labeling antibody samples: 10.96 μL MM.
    (b) Panel #2: Isotype control samples: 8.42 μL MM; fully stained labeling antibody samples: 10.88 μL MM.
12. Incubate samples for 30 min at room temperature in the dark.
13. Wash each sample twice with 500 μL staining buffer.
14. Fix cells by resuspending each cell sample in 100 μL 2% PFA and incubating for 15 min at room temperature in dark.
15. Wash each sample twice with 500 μL staining buffer.
16. Resuspend each sample in 500 μL staining buffer (*see* **Note 26**).

3.4 Sample Preparation Before Analysis on a Flow Cytometer (Day 3)

1. Remove the Panel #1 or Panel #2 samples from the 4 °C fridge and let samples come to room temperature (*see* **Note 27**).
2. Obtain flow tubes and label them appropriately.
3. Transfer 500 μL of each sample to the appropriately labeled flow tube. Briefly vortex samples (*see* **Note 28**).

3.5 Data Acquisition on Novocyte 3005 (Day 3)

1. Turn on computer → NovoExpress → username and password.
2. Turn on flow cytometer.
3. Make sure all parameters (channels and laser intensity) are the corresponding FMO experiment (*see* **Note 29**).
4. Save file.
5. Check the experimental parameters:
    (a) Plate manager: Mode—custom, 24-tube rack, indicate samples to run horizontal, select proper squares to create samples, rename each sample according to the work list. Mix—1 cycle every 2 wells (1500 rpm; 10 s). Rinse: 1 cycle every 3 wells.

Table 8
Cytometer parameters

Parameter	Alias	Gain	A	H	Parameter	Alias	Gain	A	H
FSC	FSC	–	☒	☒	FSC	FSC	–	☒	☒
SSC	SSC	–	☒	☒	SSC	SSC	–	☒	☒
B530	CD8-FITC	460	☒	☒	B530	CD56-FITC	460	☒	☒
B572	CD4-PE	397	☒	☒	B572	CD14-PE	397	☒	☒
B660	PE-Cy5	537	☐	☐	B660	PE-Cy5	537	☐	☐
B725	PerCP-eFluor 710	648	☐	☐	B725	PerCP-eFluor 710	648	☐	☐
B780	CD19-PE-Cy7	478	☒	☒	B780	CD19-PE-Cy7	478	☒	☒
R660	CCR4-APC	537	☒	☒	R660	CD123-APC	537	☒	☒
R725	CD45RA-Alexa Fluor 700	648	☒	☒	R725	CD54-Alexa Fluor 700	648	☒	☒
R780	TCR-g/d-APC-Fire750	478	☒	☒	R780	CD20-APC-Fire750	478	☐	☒
V445	CD45-Pacific Blue	425	☒	☒	V445	CD45-Pacific Blue	425	☒	☒
V530	L/D-Zombie Aqua	460	☒	☒	V530	L/D-Zombie Aqua	460	☒	☒
V572	CD3-BV570	397	☒	☒	V572	CD3-BV570	397	☒	☒
V660	CD25-BV650	537	☒	☒	V660	CD69-BV650	537	☒	☒
V725	CD154-BV711	648	☐	☐	V725	Qdot 705	648	☒	☒
V780	CD127-BV785	478	☒	☒	V780	CD11c-BV785	478	☒	☒

Left: Panel #1; Right: Panel #2. Parameters used on the Novocyte 3005 flow cytometer (*see* **Note 1**)

(b) Experimental run stop conditions: 300 μL. Fast (*see* **Note 30**).

(c) Cytometer settings (Table 8).

6. Insert samples in a tube rack.
7. Click run plate. Highlight all boxes that are to be run (all).
8. Save file.
9. Export FCS files.

3.6 Applying Compensation Matrix to Immunophenotyping Panels for Data Analysis (See Note 31)

1. Make a copy of the compensated FMO control file (.ncf) that correlates to the immunophenotyping file.
2. Upload the FCS files from the appropriate immunophenotyping experiment into the FMO control file (.ncf):

(a) Right click on file name in Experimental Manager.

(b) Click on Import FCS files.

(c) Browse for the folder with appropriate FCS files.

3. Apply the compensation matrix from the FMO control sample to the immunophenotyping samples:
   (a) Under "Compensation Specimen" click on the "+" for any sample.
   (b) Right click on Compensation and select Copy.
   (c) Right click on overarching immunophenotyping sample and click on Paste.
   (d) "Are you sure you want to paste Compensation to all samples in experimental sample name" Yes.
   (e) Delete the FMO control sample.
4. Rename the file with the compensated immunophenotyping samples and save.

3.7 Data Analysis of Immunophenotyping Panel 1

1. Make sure that FCS Express is set up to use the compensation of the FCS files: File → Options → Choose default compensation options → FCS file.
2. Import the compensated FCS files to appropriate FCS Express file (*see* **Note 32**).
3. Set up histograms and dot plots as per cell definition for fully stained samples (Figs. 2, 3, 4, and 5, *see* **Note 33**).
4. Save all files.

3.8 Data Analysis of Immunophenotyping Panel 2

1. Make sure that FCS Express is set up to use the compensation of the FCS files: File → Options → Choose default compensation options → FCS file.

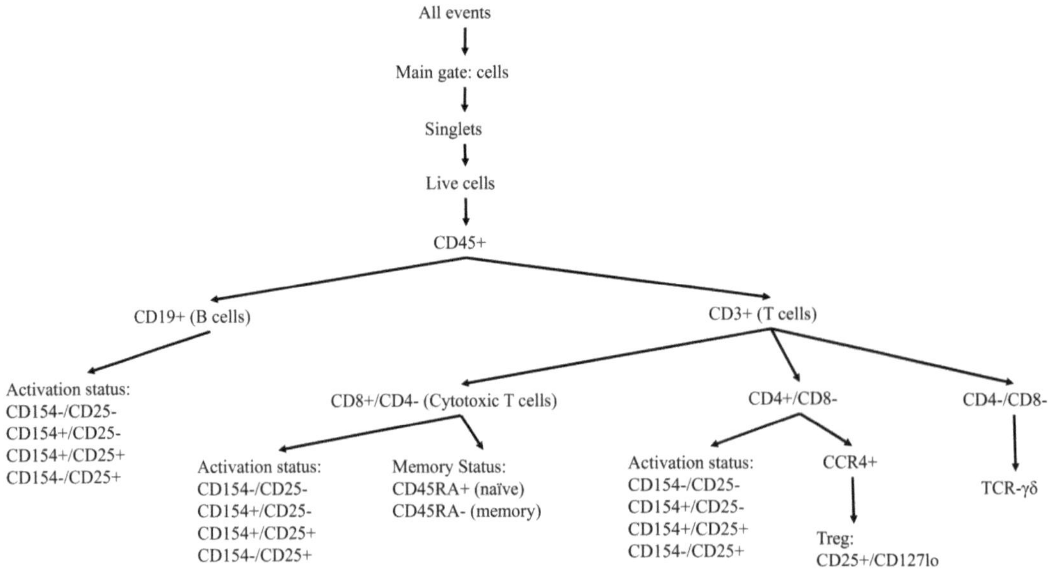

Fig. 2 Gating strategy for immunophenotyping panel #1

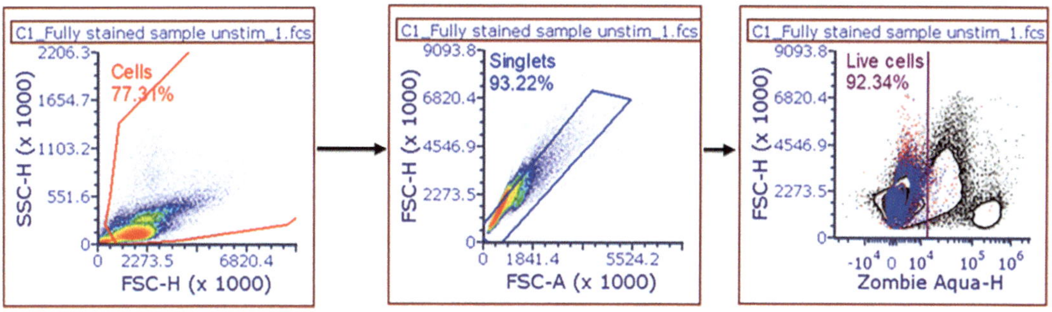

Fig. 3 Live cell gating. Sequential gates are used to determine live cell population. Cells: SSC-H v FSC-H → Singlets: FSC-H v FSC-A → Live cells: FSC-H v L/D-Zombie Aqua: Gate dead cells and exclude gate

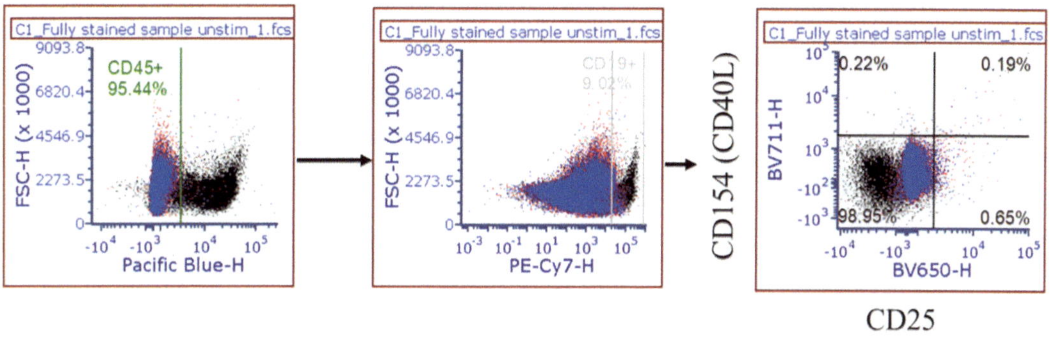

Fig. 4 B Cell gating. Sequential gates are used to determine CD19$^+$ B cell population and activation status. Cells → Singlets → Live cells → CD45+: FSC-H v CD45-PacBlue-H → CD19$^+$ (B cells): FSC-H v CD19-PE-Cy7-H. Activation (CD154 v CD25): CD154-BV711-H v CD25-BV650-H: CD25$^-$/CD154$^-$, CD25$^-$/CD154$^+$, CD25$^+$/CD154$^+$, CD25$^+$/CD154$^-$

2. Import the compensated FCS files to appropriate FCS Express file (*see* **Note 32**).
3. Set up histograms and dot plots as per cell definition for fully stained samples (Figs. 6, 7, 8, 9, 10, and 11, *see* **Note 33**).
4. Save all files.

4 Notes

1. Technical specifications of individual flow cytometers vary. This protocol should be followed for the Novocyte 3005 flow cytometer. If another cytometer is used, the procedure described herein would require optimization according to the technical specifications of that instrument. It is, therefore, recommended to review the technical specification of a cytometer available in each research laboratory prior to planning the experiment.

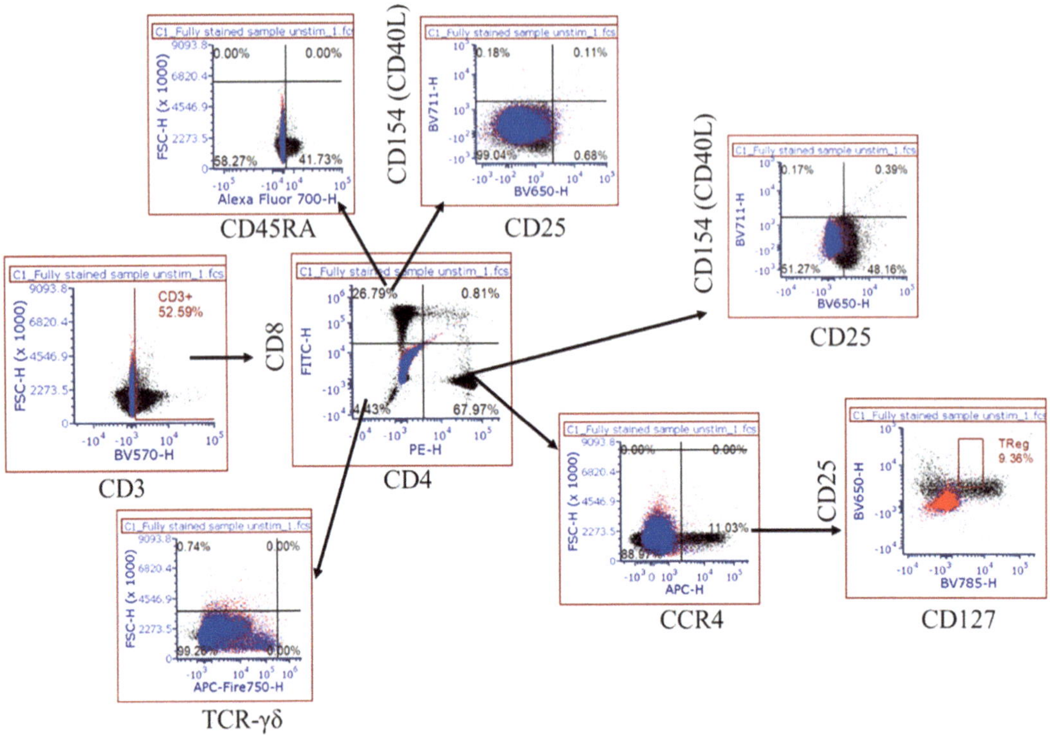

Fig. 5 T Cell gating to determine the different T cell populations and activation statuses. Cells → Singlets → Live cells → CD45+ → CD3+: FSC-H v CD3-BV570-H. *$CD8^+$ vs $CD4^+$ T cells*: Cells → Singlets → Live cells → $CD45^+$ → $CD3^+$ → CD8 v CD4: CD8-FITC-H v CD4-PE-H: $CD4^-/CD8^-$, $CD4^-/CD8^+$, $CD4^+/CD8^+$, $CD4^+/CD8^-$, $CD4^-/CD8^+$ → Activation (CD154 v CD25): CD154-BV711-H v CD25-BV650-H: $CD25^-/CD154^-$, $CD25^-/CD154^+$, $CD25^+/CD154^+$, $CD25^+/CD154^-$, $CD4^+/CD8^-$ → Activation (CD154 v CD25): CD154-BV711-H v CD25-BV650-H: $CD25^-/CD154^-$, $CD25^-/CD154^+$, $CD25^+/CD154^+$, $CD25^+/CD154^-$. *CD8+ Naïve T cells*: Cells → Singlets → Live cells → $CD45^+$ → $CD3^+$ → $CD8^+/CD4^-$ → $CD45RA^+$: FSC-H v CD45RA-AF700-H. *Regulatory T cells (Treg)*: Cells → Singlets → Live cells → $CD45^+$ → $CD3^+$ → $CD4^+/CD8^-$ → $CCR4^+$: FSC-H v CCR4-APC-H → Treg ($CD25^+$, CD127low): CD25-BV650-H v CD127-BV785-H. ***TCR g/d cells***: Cells → Singlets → Live cells → $CD45^+$ → $CD3^+$ → $CD4^-/CD8^-$ → TCR g/d: FSC-H v TCRg/d-APC-Fire750-H

2. Collection of primary human blood requires institutional approval. The blood can also be obtained from commercial sources.
3. All work with human blood must be performed in a biological safety cabinet and using personal protective equipment (lab coat, safety goggles, gloves).
4. For optimal results, whole blood must be processed within 2 h after collection.
5. Heat inactivation is intended to eliminate the effects of bovine complement proteins on human blood cells. Heat inactivation is performed by incubating FBS at 56 °C for 30 min. It is recommended to prepare single use aliquots and store them at −20 °C before use.

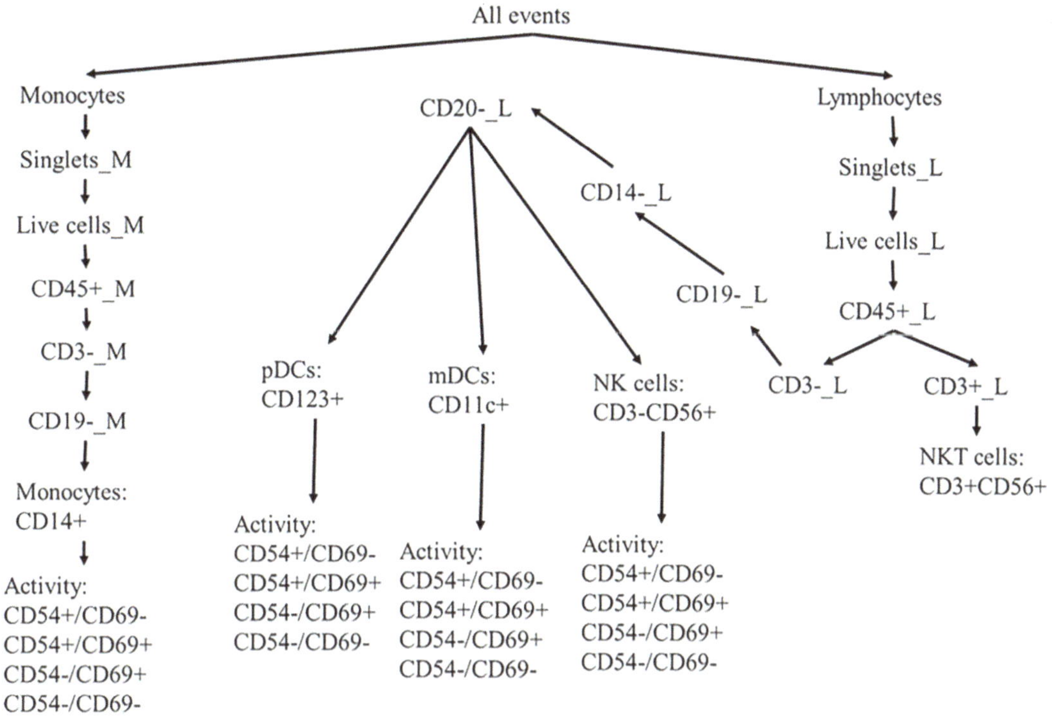

Fig. 6 Gating strategy for immunophenotyping panel #2

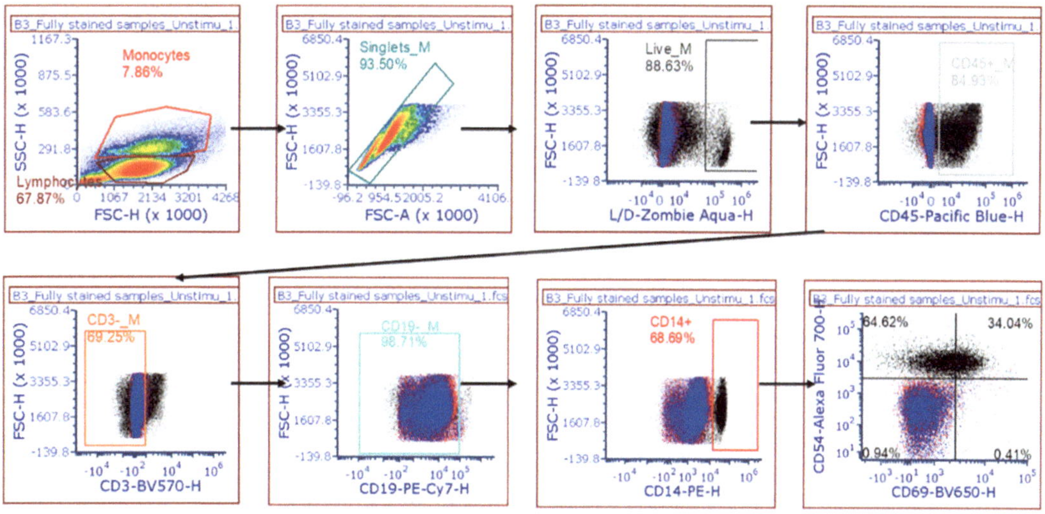

Fig. 7 Monocyte gating. Sequential gating is used to determine monocyte population and activation status. Monocyte population ($CD14^+$) via scatter: SSC-H v FSC-H → Singlets_M: FSC-H v FSC-A → Live_M: FSC-H v L/D-Zombie Aqua: Gate dead cells and exclude gate → $CD45^+$_M: FSC-H v CD45-PacBlue-H → $CD3^-$_M: FSC-H v CD3-BV570-H → $CD19^-$_M: FSC-H v CD19-PE-Cy7 → Monocytes ($CD14^+$): FSC-H v CD14-PE-H. Activation: (CD54 v CD69): CD54-AF700-H v CD69-BV650-H: $CD54^-/CD69^-$, $CD54^-/CD69^+$, $CD54^+/CD69^+$, $CD54^+/CD69^-$

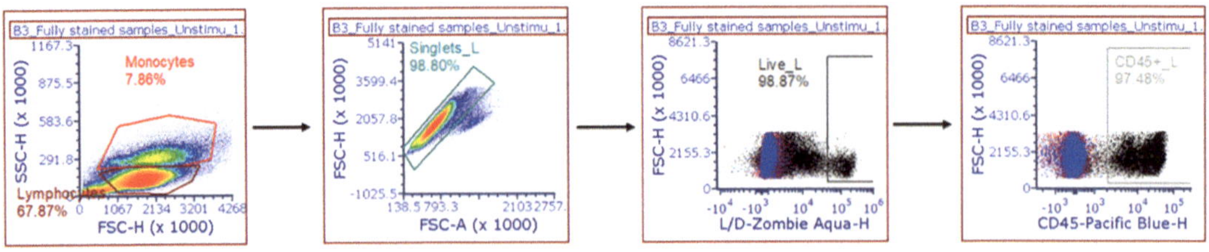

Fig. 8 CD45$^+$ Lymphocyte gating. Sequential gating is used to identify CD45$^+$ Lymphocyte population. Lymphocyte population via scatter: SSC-H v FSC-H → Singlets_L: FSC-H v FSC-A → Live_L: FSC-H v L/D-Zombie Aqua: Gate Dead cells and exclude gate → CD45$^+$_L: FSC-H v CD45-PacBlue-H

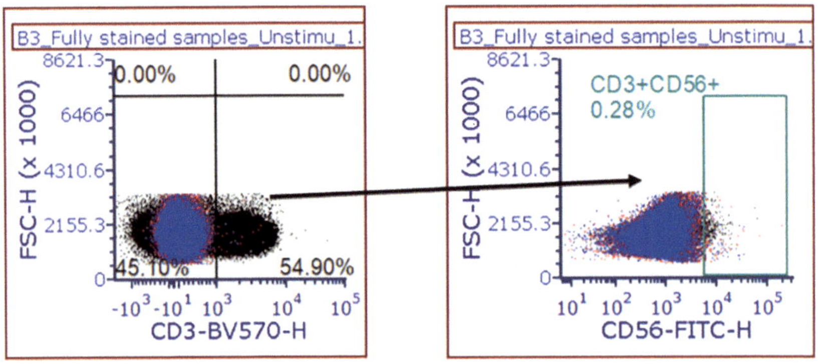

Fig. 9 NK T cell gating. Gating to determine the CD3$^+$ CD56$^+$ T cell population is shown. Lymphocyte population via scatter → Singlets_L → Live_L → CD45$^+$_L → CD3$^+$_L: FSC-H v CD3-BV570-H → CD56$^+$: FSC-H v. CD56-FITC-H

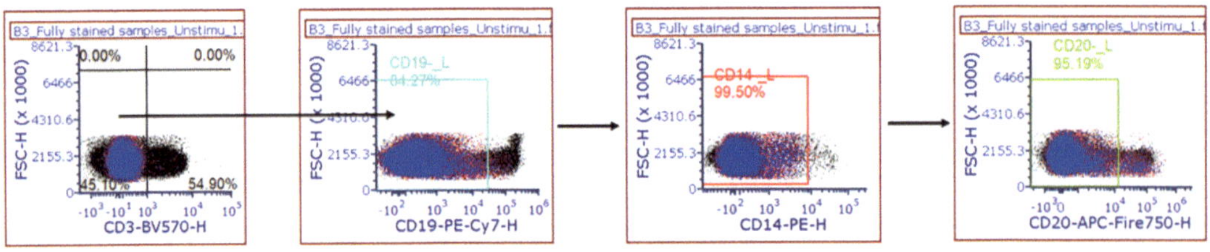

Fig. 10 CD45$^+$ CD3$^-$ CD14$^-$ CD20$^-$ Cell population gating. Lymphocyte population via scatter → Singlets_L → Live_L → CD45$^+$_L → CD3$^-$_L: FSC-H v CD3-BV570-H → CD19$^-$_L: FSC-H v CD19-PE-Cy7 → CD14$^-$_L: FSC-H v CD14-PE-H → CD20$^-$_L: FSC-H v CD20-APC-Fire750-H

6. It is recommended to store the complete medium at 2–8 °C protected from light for no longer than 1 month and warm it in a water bath before each use.

7. Stocks with higher concentration (5–10 mg/mL) can also be prepared. It is recommended to prepare 20 μL aliquots and store them at a nominal temperature of −20 °C. It is also recommended to avoid repeated freeze-thaw cycles. On the day of the experiment thaw one aliquot and use it at a final of 20 ng/mL concentration in PBMC or WB culture.

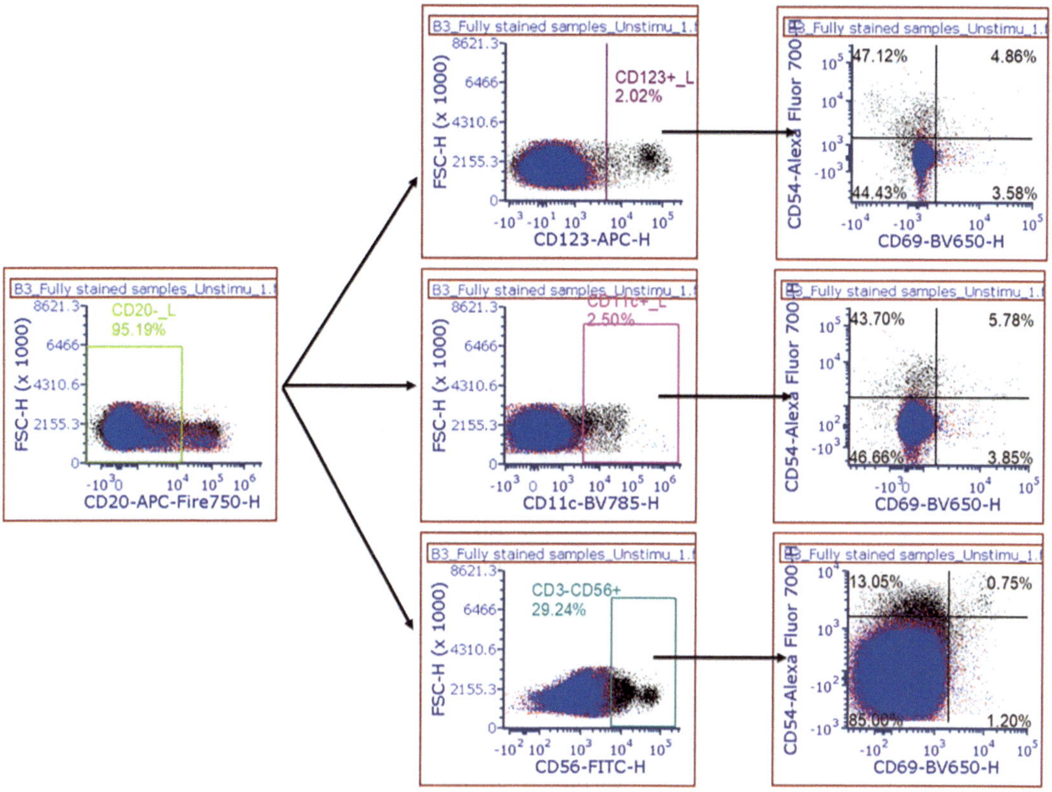

Fig. 11 Gating for pDCs, mDCS, NK cells. Gating strategy for different lymphocyte populations is shown. *pDCs*: Lymphocyte population via scatter → Singlets_L → Live_L → $CD45^+$_L → $CD3^-$_L→ CD19-_L → $CD14^-$_L → $CD20^-$_L: → $CD123^+$_L: FSC-H v $CD123^-$APC-H. Activation: (CD54 v CD69): CD54-AF700-H v CD69-BV650-H: $CD54^-/CD69^-$, $CD54^-/CD69^+$, $CD54^+/CD69^+$, $CD54^+/CD69^-$. *mDCs*: Lymphocyte population via scatter → Singlets_L → Live_L → $CD45^+$_L → $CD3^-$_L→ $CD19^-$_L → $CD14^-$_L→ $CD20^-$_L→ $CD11c^+$_L: FSC-H v CD11c-BV785-H. Activation: (CD54 v CD69): CD54-AF700-H v CD69-BV650-H: $CD54^-/CD69^-$, $CD54^-/CD69^+$, $CD54^+/CD69^+$, $CD54^+/CD69^-$. *NK cells*: Lymphocyte population via scatter → Singlets_L → Live_L → $CD45^+$_L → $CD3^-$_L→ $CD19^-$_L → $CD14^-$_L→ $CD20^-$_L→ $CD56^+$: FSC-H v CD56-FITC-H. Activation: (CD54 v CD69): CD54-AF700-H v CD69-BV650-H: $CD54^-/CD69^-$, $CD54^-/CD69^+$, $CD54^+/CD69^+$, $CD54^+/CD69^-$

8. It is recommended to prepare single use 20 μL aliquots and store them at −20 °C.
9. It is recommended to thaw a frozen aliquot of the stock at room temperature on the day of the experiment. Thawed PMA should be diluted in culture media to a final concentration of 50 ng/mL when used as a positive control.
10. It is recommended on the day of the experiment to thaw an aliquot at room temperature and dilute in culture media to a final concentration of 1 μg/mL.
11. It is recommended on the day of the experiment to thaw an aliquot at room temperature and dilute in culture media to a final concentration of 5 μg/mL.

12. It is recommended to prepare single use aliquots and store them at a nominal temperature of –20 °C. It is also recommended to avoid repeated freezing/thawing. On the day of the experiment, the stock PHA-M should be diluted in cell culture medium to a final concentration of 10 μg/mL.
13. The total amount of nanoparticle sample needed for testing these two immunophenotyping panels is at least 1.0 mL per donor at testing concentration (when minimal required dilution (MRD) is 10). The nanoparticle concentration is selected based on the plasma concentration of the nanoparticle at the intended therapeutic dose. For the purpose of this protocol this concentration is called "theoretical plasma concentration." Considerations for estimating theoretical plasma concentration were reviewed elsewhere [8]. Buffer used to store nanoparticles is used as a vehicle control.
14. Alternatively, the blood/PBS mixture may be slowly layered over the Ficoll-Paque solution. It is recommended to use 3 mL of Ficoll-Paque solution per 4 mL of blood/PBS mixture. For example, 15 mL Ficoll-Paque per 20 mL of diluted blood in a 50 mL tube.
15. To maintain Ficoll-blood interface, it is helpful to hold the tube at a 45° angle.
16. It is important to turn centrifuge brakes off. Using brakes will result in vibrations which will disturb a buffy coat. Depending on the type of a centrifuge, one may also need to set acceleration speed to minimum.
17. Usually 4 mL of blood/PBS mixture results in ~2 mL of mononuclear layer and requires at least 6 mL of HBSS for the wash step. It is recommended to use 10 mL of HBSS per each 2 mL of cells.
18. The expected cell viability after this step is at least 90%. If the viability is lower, it is not recommended to use the cells for subsequent experimental procedure.
19. Experimental procedure takes at least 2 days but realistically takes 3 days per donor as each donor needs its own unstained and isotype control samples. Day 1 consists of isolation and treatment of PBMC. Day 2 involves staining of PBMC for flow cytometry. Flow cytometry samples can be run on Day 2 or 3. It is advised to review Tables 6 and 7 for the conditions/samples needed for Immunophenotyping Panels #1 and #2 (two replicates per condition). If there are more than 65 samples in total (between the two immunophenotyping panels), it is suggested to perform the two immunophenotyping panels separately. It is also advised to review Tables 2, 3, 4, and 5 for antibodies/dyes included in isotype control samples and fully stained labeling antibody samples.

Table 9
Immunophenotyping human panel #1 labeling antibody MM

Optimal antibody (Ab) dilution (ITA-37.1 Step 6.1)	Final antibody dilution	Volume of stock Ab needed (μL) per 100 μL samples	Number of samples ($n + 2$)	Total volume of stock Ab needed (μL)	Number of Abs in panel	Total MM volume (μL)
Dil 25	Dil ~250	1.6	$n + 2$	$(n + 2)(1.6)$	5	$5(n + 2)(1.6)$
Dil 50	Dil ~500	0.8	$n + 2$	$(n + 2)(0.8)$	2	$2(n + 2)(0.8)$
Dil 100	Dil ~1000	0.4	$n + 2$	$(n + 2)(0.4)$	1	$1(n + 2)(0.4)$
Dil 125	Dil ~1250	0.32	$n + 2$	$(n + 2)(0.32)$	3	$3(n + 2)(0.32)$
Total MM volume (μL)						**Sum of volumes in above column**
Total MM volume added to each sample = (total MM volume)/($n + 2$)						**10.96 μL**

20. If these volumes are used, the MRD of vehicle control or test sample is 10. If MRD of 5 is to be used, 200 μL of vehicle control or test sample should be added to the 800 μL of PBMC. Also, if one wants to understand whether nanoparticle may potentiate or inhibit cellular response to the assay positive control (LPS, PHA-M, PMA, ionomycin, and/or ODN2216), the positive control should be co-cultured with nanoparticles in the presence of cells.
21. Prepare amount of MM for sample number (n) + 2. Concentrations to be used for the isotype control antibodies are the same as the concentration of corresponding fluorescent labeling antibody. It is recommended to review Tables 9 and 10 for Immunophenotyping Panel #1 and Tables 11 and 12 for Immunophenotyping Panel #2.
22. Human Immunophenotyping Panel #1 Antibody Titration Concentration Determined: Dil 25: CD8-FITC, CD45RA-AF700, CD45-PacBlue, CD3-BV570, CD127-BV785; Dil 50: TCR-gamma/delta APC-Fire750, CD154-BV711; Dil 100: CD25-BV650; Dil 125: CD4-PE, CD19-PE-Cy7, CCR4-APC.

Table 10
Immunophenotyping human panel #1 isotype control MM

Isotype control Ab	Conc. of isotype control Ab (μg/mL)	Conc. of corresponding labeling Ab (μg/mL)	Ratio of isotype Ab conc: Labeling Ab conc	Quantity of stock labeling Ab added to labeling Ab MM (μL)	Quantity of stock isotype Ab to add to isotype control MM (μL)
FITC Mouse IgG1, K	200	200	1:1	$(n + 2)(1.6)$	$(n + 2)(1.6)/1$
PE Mouse IgG2b, K	200	100	2:1	$(n + 2)(0.32)$	$(n + 2)(0.32)/2$
PE-Cy7 Mouse IgG1, K	200	100	2:1	$(n + 2)(0.32)$	$(n + 2)(0.32)/2$
APC Mouse IgG1, K	200	50	4:1	$(n + 2)(0.32)$	$(n + 2)(0.32)/4$
AF700 Mouse IgG2b, K	200	50	4:1	$(n + 2)(1.6)$	$(n + 2)(1.6)/4$
APC-Fire750 Mouse IgG1, K	200	400	1:2	$(n + 2)(0.8)$	$(n + 2)(0.8)/(1/2)$
PacBlue Mouse IgG1, K	500	100	5:1	$(n + 2)(1.6)$	$(n + 2)(1.6)/5$
BV570 Mouse IgG1, K	100	80	1:0.8	$(n + 2)(1.6)$	$(n + 2)(1.6)/(1/0.8)$
BV650 Mouse IgG1, K	100	100	1:1	$(n + 2)(0.4)$	$(n + 2)(0.4)/1$
BV711 Mouse IgG1, K	100	100	1:1	$(n + 2)(0.8)$	$(n + 2)(0.8)/1$
BV785 Mouse IgG1, K	100	50	2:1	$(n + 2)(1.6)$	$(n + 2)(1.6)/2$
Total MM volume (μL)					**Sum of volumes in above column**
Total MM volume added to each sample = (total MM volume)/$(n + 2)$					**7.6 μL**

Table 11
Immunophenotyping human panel #2 labeling antibody MM

Optimal antibody (Ab) dilution (ITA-37.1 Step 5.1)	Final antibody dilution	Volume of stock Ab needed (μL) per 100 μL	Number of samples (*n* + 2)	Total volume of stock Ab needed (μL)	Number of Abs in panel	Total master mix volume (μL)
Dil 25	Dil ~250	1.6	*n* + 2	(*n* + 2)(1.6)	6	6(*n* + 2)(1.6)
Dil 125	Dil ~1250	0.32	*n* + 2	(*n* + 2)(0.32)	4	4(*n* + 2)(0.32)
Total MM volume (μL)						**Sum of volumes in above column**
Total MM volume added to each sample = (Total MM volume)/(*n* + 2)						**10.88 μL**

23. Human Immunophenotyping Panel #2 Antibody Titration Concentration Determined: Dil 25: CD123-APC, CD20-APC-Fire750, CD45-PacBlue, CD3-BV570, CD69-BV650, CD11c-BV785; Dil 125: CD56-FITC, CD14-PE, CD19-PE-Cy7, CD54-AF700.
24. This step is optional. Supernatants can be stored for ELISAs to be performed at later time point, if desired.
25. To complete the experiment, one needs 0.8 μL of dil 5 Zombie Aqua dilution x *n* conditions = 0.8(*n* + 2) μL of dil 5 dilution. *N* = the number of fully stained labeling antibody samples in both Panel #1 and Panel #2.
26. Samples can be stored at 2–8 °C and analyzed on the following day.
27. It is recommended to process panel 1 and panel 2 samples separately.
28. Samples are ready for flow cytometry.
29. The FMO control experiment can be saved as a template from which the new experiment can be made.
30. Apply changes to all samples.
31. There are multiple ways to apply the compensation matrix of the FMO controls to the immunophenotyping panels. This protocol describes one such method. Compensation should be performed for both Immunophenotyping Panel #1 and #2 using the appropriate FMO controls.

Table 12
Immunophenotyping human panel #2 isotype control MM

Isotype control Ab	Conc. of isotype control Ab (μg/mL)	Conc. of corresponding labeling Ab (μg/mL)	Ratio of isotype Ab conc: Labeling Ab conc	Quantity of stock labeling Ab added to labeling Ab MM (μL)	Quantity of stock isotype Ab to add to isotype control MM (μL)
FITC Mouse IgG1, K	200	200	1:1	$(n + 2)(0.32)$	$(n + 2)(0.32)/1$
PE Mouse IgG2a, K	200	200	1:1	$(n + 2)(0.32)$	$(n + 2)(0.32)/1$
PE-Cy7 Mouse IgG1, K	200	100	2:1	$(n + 2)(0.32)$	$(n + 2)(0.32)/2$
APC Mouse IgG1, K	200	100	2:1	$(n + 2)(1.6)$	$(n + 2)(1.6)/2$
AF700 Mouse IgG1, K	500	400	1.25:1	$(n + 2)(0.32)$	$(n + 2)(0.32)/1.25$
APC-Fire750 Mouse IgG2b, K	200	200	1:1	$(n + 2)(1.6)$	$(n + 2)(1.6)/1$
PacBlue Mouse IgG1, K	500	100	5:1	$(n + 2)(1.6)$	$(n + 2)(1.6)/5$
BV570 Mouse IgG1, K	100	80	1.25:1	$(n + 2)(1.6)$	$(n + 2)(1.6)/1.25$
BV650 Mouse IgG1, K	100	50	2:1	$(n + 2)(1.6)$	$(n + 2)(1.6)/2$
BV785 Mouse IgG1, K	100	160	1:1.6	$(n + 2)(1.6)$	$(n + 2)(1.6)/(1/1.6)$
Total MM volume (μL)					**Sum of volumes in above column**
Total MM volume added to each sample = (Total MM volume)/$(n + 2)$					**8.42 μL**

32. Negative control, positive controls, vehicle controls, and test sample files should be imported into separate FCS Express files and analyzed separately.
33. Corresponding isotype controls should be overlaid onto each plot and used to set the gates which are then applied to the fully stained labeling antibody samples.

Acknowledgment

This project has been funded in whole or in part by federal funds from the National Cancer Institute, National Institutes of Health, under contract 75N91019D00024. The content of this publication does not necessarily reflect the views or policies of the Department of Health and Human Services, nor does mention of trade names, commercial products, or organizations imply endorsement by the US Government.

References

1. Oughton JA, Kerkvliet NI (2005) Immune cell phenotyping using flow cytometry. Curr Protoc Toxicol 23(1):18.18.11–18.18.24
2. McCoy J, Philip J (2019) Immunophenotyping. Springer. ISBN:1493996495
3. McKinnon KM (2018) Flow cytometry: an overview. Curr Protoc Immunol 120:5.1.1–5.1.11
4. Newton HS, Dobrovolskaia MA (2022) Immunophenotyping: analytical approaches and role in preclinical development of nanomedicines. Adv Drug Deliv Rev 185:114281
5. Dobrovolskaia MA, McNeil SE (2007) Immunological properties of engineered nanomaterials. Nat Nanotechnol 2(8):469–478
6. Acea BioScience Inc. (2017) NovoCyte® Flow Cytometer Technical Specifications
7. Newton HS, Zhang J, Donohue D, Unnithan R, Cedrone E, Xu J, Vermilya A, Malys T, Clogston JD, Dobrovolskaia MA (2023) Multicolor flow cytometry-based immunophenotyping for preclinical characterization of nanotechnology-based formulations: an insight into structure activity relationship and nanoparticle biocompatibility profiles. Front Allergy 4:1126012
8. Dobrovolskaia MA, McNeil SE (2013) Understanding the correlation between in vitro and in vivo immunotoxicity tests for nanomedicines. J Control Release 172(2):456–466
9. Agilent Technologies (2022) NovoCyte 3 Lasers 3005 Model 3005-default configuration

Chapter 26

Understanding the Role of Scavenger Receptor A1 in Nanoparticle Uptake by Murine Macrophages

Edward Cedrone, Martin Schuster, Rosemarie Preyer, and Marina A. Dobrovolskaia

Abstract

Nanoparticles can be cleared from the circulation and taken up by tissue-resident macrophages. This property can be beneficial when drug or antigen delivery to macrophages is desired; however, rapid clearance of nanoparticles not intended for delivery to immune cells may reduce nanoparticle circulation time and affect the efficacy of nanoparticle-formulated drug products. Therefore, understanding nanoparticles' uptake by macrophages is an essential step in the preclinical development of nanotechnology-based drug products. Understanding the route of nanoparticle uptake by macrophages may also provide mechanistic insights into the immunotoxicity of nanomaterials. The protocol described herein can be used to assess the nanoparticles' uptake by macrophages and understand the involvement of scavenger receptor A1 to inform mechanistic studies.

Key words Nanoparticles, Phagocytosis, Scavenger receptors, SR-A1, Macrophages

1 Introduction

Phagocytic cells take up nanoparticles via multiple routes, including phagocytosis and pinocytosis [1, 2]. Nanoparticles with an anionic surface or made of anionic polymers are often phagocytosed via scavenger receptor A1 (SR-A1); multiple uptake routes may also be involved in the uptake of the same particle [3]. Detecting nanoparticle uptake by macrophages and understanding the route of nanoparticle uptake help provide mechanistic insight into biological responses to and biodistribution of nanoparticles. The protocol described herein utilizes the wild-type murine macrophage cell line RAW 264.7, expressing the SR-A1 receptor, and its engineered clone ½, deficient in SR-A1 expression. The protocol can be used for two purposes: (1) to identify whether the given nanoparticle is taken up by macrophages and (2) to understand if the uptake occurs via the scavenger receptor SR-A1. If the only goal of the

Jeffrey D. Clogston et al. (eds.), *Characterization of Nanoparticles Intended for Drug Delivery*, Methods in Molecular Biology, vol. 2789, https://doi.org/10.1007/978-1-0716-3786-9_26,

experiment is to monitor nanoparticle uptake, the parent RAW cells and Alexa488-, FITC-, Cy3-, or Cy5-labeled particles can be used. If, however, the involvement of SR-A1 in uptake is of interest, then both the wild type and SR-A1 deficient cells (clone ½) are needed along with the Cy5-labeled particles. Clone ½ cells have inherent low-level green fluorescence due to the GFP-tagged lentiviral construct incorporated into the genome of these cells [3]; therefore, this green autofluorescence interferes with the accurate assessment of the uptake of nanoparticles tagged with a green fluorophore. The AID reader described in this protocol is equipped with filters for monitoring FITC, Cy3, and Cy5 fluorophores. Other methods, such as flow cytometry, can be used to monitor particle uptake. In this case, the fluorophores attached to the particles should match the specifications of the flow cytometer used in the experiment. The uptake of unlabeled metallic nanoparticles by parent RAW cells and clone ½ can also be tracked using elemental analysis such as inductively coupled plasma-mass spectrometry (ICP-MS).

When an AID multi spot reader is used to monitor uptake, the analysis results in two readouts—number of spots and activity. The number of spots reflects the number of cells that took up nanoparticles; this measure is equivalent to the percent positive cells readout in flow cytometry. The activity readout reflects the brightness of the fluorescent signal and, therefore, indicates the efficiency of uptake by a single cell; this measure is equivalent to the mean fluorescent intensity readout in flow cytometry. Experimental steps involving instrument settings, plate reading, and analysis are based on a series of technical documents by Autoimmun Diagnostika Gmb [4–7].

2 Materials

1. Roswell park memorial institute (RPMI) 1640.
2. RPMI 1640, phenol red free.
3. Penicillin streptomycin solution.
4. L-glutamine.
5. β-mercaptoethanol.
6. Fetal bovine serum (FBS).
7. Phosphate buffered saline (PBS).
8. Trypan blue solution.
9. Distilled or sterile water.
10. RAW 264.7 (ATCC, TIB-71).
11. RAW 264.7 SAR1 KO (Clone ½) [3] (*see* **Note 1**).
12. AID multispot reader.

3 Methods

3.1 Preparation of Reagents and Controls

1. Prepare Complete RPMI 1640 medium by adding 10% heat-inactivated FBS, 2 mM L-glutamine, 100 U/mL penicillin, 100 μg/mL streptomycin sulfate, 50 μM β-mercaptoethanol and store at 2–8 °C protected from light for no longer than 1 month. Before use, warm the complete medium in a water bath (*see* **Note 2**).
2. Prepare heat-inactivated fetal bovine serum. Thaw a bottle of FBS at room temperature, or overnight at 2–8 °C and allow to equilibrate to room temperature. Incubate 30 min at 56 °C in a water bath, mixing every 5 min (*see* **Note 3**).
3. Use PBS as a negative control.
4. Prepare vehicle control (*see* **Note 4**).

3.2 Preparation of Study Samples

1. Prepare nanoparticles according to the manufacturer's instructions (*see* **Notes 5** and **6**).

3.3 Preparation of Cells

1. Grow cells in complete culture media, using conditions specific for the individual cell lines. On the day of the experiment, count the cells; if cell viability is ≥80%, concentrate the cells in complete culture media to a concentration of 5.0×10^4 cells/mL (*see* **Note 7**).

3.4 Experimental Procedure

1. Add 100 μL of cells (either RAW 264.7 or the SRA1 KO clone) at a concentration of 5.0×10^4 cells/mL to the appropriate wells in a 96-well plate (5000 cells/well) (Fig. 1).
2. Allow the cells to attach at room temperature for 20–30 min, then incubate overnight at 37 °C and 95% CO_2.
3. Prepare the test samples at 2× the final concentrations to be analyzed.
4. Seed 100 μL of prepared samples to the appropriate wells (*see* **Note 8**).
5. Incubate for 24 h at 37 °C and 95% CO_2 (*see* **Note 9**).
6. After incubation, wash the cells twice with phenol red-free medium (*see* **Note 10**).
7. Add 50 μL of phenol red-free medium and read the plate using the EliSpot reader (*see* **Notes 11** and **12**).

3.5 Calculations and Data Interpretation

1. A percent coefficient of variation should be calculated for each control or test according to the following formula:

$$\frac{\text{Standard deviation}}{\text{Mean}} \times 100\%$$

%CV for each control and test sample should be less than 30%.

	1	2	3	4	5	6	7	8	9	10	11	12
A	PC											PC
B												
C		Untreated	Untreated	T1	T1	T2	T2	T3	T3	T4	T4	
D		Untreated	Untreated	T1	T1	T2	T2	T3	T3	T4	T4	
E		Untreated	Untreated	T1	T1	T2	T2	T3	T3	T4	T4	
F		Untreated	Untreated	T1	T1	T2	T2	T3	T3	T4	T4	
G												
H												PC

Fig. 1 Example platemap. This figure outlines an example of sample position on a test plate. Wells A1, A12, and H12 include RAW 264.7 cells and PC for stage calibration; Rows C and D are for RAW 264.7 cells treated with Cy5-labeled nanoparticles. Rows E and F are for RAW 264.7 SRA1 KO cells (Clone ½) treated with Cy5-labeled nanoparticles. If the nanoparticle uptake by wildtype RAW 264.7 cells is the only desirable outcome of this experiment, FITC- or Alexa488-labeled nanoparticles can be used

2. If positive control or negative control fails to meet the acceptance criterion in **step 1** above, the assay should be repeated.
3. Within the acceptable assay, if two of three replicates of the unknown sample fail to meet the acceptance criterion described in **step 1**, this unknown sample should be re-analyzed.
4. Statistical analysis can be used for comparing the number of spots and activity in control and test samples.
5. For comparison between the wild-type and SR-A1 deficient cells, the number of spots and activity are compared for the same sample in two cell lines (Fig. 2).

4 Notes

1. Raw 264.7 Clone ½ cells are available for non-commercial, research purposes and can be requested by email at ncl@mail.nih.gov.

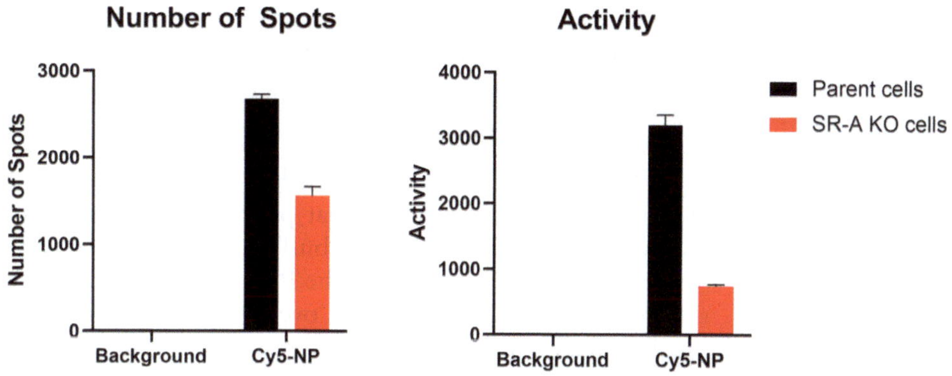

Fig. 2 Example of data generated using this protocol. The number of spots and activity measurements are above the baseline in both wild type RAW264.7 and SR-A1 deficient (SRA-KO) cells, indicative of nanoparticle uptake by both cell lines. The uptake in SR-A1 deficient cells is lower than that in the control cell line for the same type of nanomaterial, suggesting that SR-A1 is partially involved in the uptake of these nanoparticles by macrophages

2. Working with cells and nanomaterials requires personal protective equipment (safety goggles, gloves, lab coat) and a biological safety cabinet.
3. Fifty (50) mL single-use aliquots may be stored at 2–8 °C for up to 1 month or at a nominal temperature of −20 °C indefinitely.
4. Vehicle control is the buffer or media used to formulate test nanomaterials. Common excipients used in nanoformulations are trehalose, sucrose, and albumin. However, other reagents and materials are also used alone or in combination. Vehicle control should match the formulation buffer of the test nanomaterial by both composition and concentration. This control can be skipped if nanoparticles are stored in PBS.
5. This assay requires 1 mL of fluorescently-labeled nanoparticles. If uptake by the parent cell line is of interest, FITC, Alexa488, Cy3, and Cy5 can be used. However, if analysis of the role of SR-A1 is of interest, Cy5 is a more optimal label.
6. Each nanoparticle sample is plated four times. Concentration or a series of concentrations of nanoparticles is selected for each nanoparticle experimentally. The highest non-cytotoxic concentration with fluorescence sufficient for the AID reader to detect is recommended for this protocol. Parent RAW cells and control nanoparticles, e.g., FITC or Cy-5 labeled, drug-free liposomes with a total lipid concentration of ~0.5–1 mg/mL, can be used for such preliminary studies.
7. Cells concentration may need to be adjusted accordingly for different treatment times.

8. It is advisable to use positive samples in wells A1, A12, and H12 to facilitate stage calibration (Fig. 1).
9. Incubation time may differ depending on nanoparticles and the research needs of individual projects.
10. Add and aspirate the medium gently to avoid loss of cells.
11. Alternately, after washes, treat plates with 4% paraformaldehyde (PFA) (or 10% neutral buffered formalin) for 15 min at room temperature, aspirate the PFA, quickly rinse with water, and blot on a paper towel. Cover with foil and keep at 4 °C until ready to read.
12. It is recommended to consult the operator's manual of the ELISpot reader available in a given laboratory. The instructions for operating the instrument used in the authors' lab are available online at https://www.cancer.gov/nano/research/ncl/protocols-capabilities/ncl-method-ita-40.pdf and in references [4–7].

Acknowledgement

This project has been funded in whole or in part by federal funds from the National Cancer Institute, National Institutes of Health, under contract 75N91019D00024. The content of this publication does not necessarily reflect the views or policies of the Department of Health and Human Services, nor does mention of trade names, commercial products, or organizations imply endorsement by the US Government.

References

1. Liu Y, Hardie J, Zhang X, Rotello VM (2017) Effects of engineered nanoparticles on the innate immune system. Semin Immunol 34:25–32
2. Dobrovolskaia MA, McNeil SE (2007) Immunological properties of engineered nanomaterials. Nat Nanotechnol 2(8):469–478
3. França A, Aggarwal P, Barsov EV, Kozlov SV, Dobrovolskaia MA, González-Fernández Á (2011) Macrophage scavenger receptor A mediates the uptake of gold colloids by macrophages in vitro. Nanomedicine (Lond) 6(7): 1175–1188
4. AID (2021) User guide AID EliSpot 8.0
5. AID (2021) Quick guide for the AID EliSpot Software 7.0
6. AID (2021) Introduction to the AID multiSpot
7. AID (2023) Autoimmun diagnostika GmbH, introduction to the AID cytoSpot Reader

Part IV

Pharmacology

Chapter 27

Assessment of Temperature-Dependent Drug Release of Solubilizing Nanoformulations Using the SITUA

Stephan T. Stern, Sarah L. Skoczen, and Young Hwan Park

Abstract

The evaluation of temperature-dependent drug release for solubilizing nanoformulations uses a modification of the stable isotope ultrafiltration assay (SITUA). This method is specific to parenterally administered solubilizing nanomedicines and can be used to assess drug release from the total dosage form for regulatory purposes of lot release. The principle upon which this method is based is the relationship between drug solubility and temperature in a plasma simulating media, 4.5% human serum albumin, that allows for discrimination of passing and failing lots based upon the release characteristics.

Key words Stable Isotope Tracer Ultrafiltration Assay (SITUA), Nanomedicine, Drug release

1 Introduction

Requirements for regulatory dissolution/in vitro drug release assays intended for nanomedicine lot release include: (1) preparation of the entire dosage form as intended for administration; (2) at least 85% of the active ingredient should be released; and, (3) the assay should discriminate between passing and failing lots, i.e., batches with formulation and manufacturing differences that may alter the clinical performance [1]. In addition, for a nanomedicine, the dissolution/in vitro drug release assay should be conducted in a physiologically relevant media, as the media can substantially affect release, and requires separation of encapsulated and unencapsulated drug fractions [2]. Development of a discriminating assay for solubilizing nanomedicine formulations is challenging because dissolution/release is often instantaneous based on the solubility of the drug in the assay media, so temporal differences in release between nanomedicine lots cannot be evaluated. This current method utilizes temperature-dependent solubility to differentiate lot-to-lot drug release characteristics.

Jeffrey D. Clogston et al. (eds.), *Characterization of Nanoparticles Intended for Drug Delivery*, Methods in Molecular Biology, vol. 2789, https://doi.org/10.1007/978-1-0716-3786-9_27,

We have previously developed a drug release method for fractionation of nanomedicine encapsulated and unencapsulated drug in plasma, the stable isotope ultrafiltration assay (SITUA) [3]. The SITUA method utilizes a stable isotope tracer that is spiked into plasma containing the nanomedicine and allows for accurate determination of drug protein binding, which then allows for determination of unencapsulated drug concentration based upon measurement of the unbound, filterable drug concentration. The principle upon which the current drug release method is based is the relationship between drug solubility and temperature. For this purpose, the previous SITUA method has been modified to evaluate temperature-dependent drug release from a solubilizing parenteral nanomedicine dosage form in physiologically relevant 4.5% w/v human albumin. Temperature-dependent drug release allows for potential assay discrimination of passing and failing lots based upon release characteristics. The method explained below details how to conduct an in vitro temperature-dependent drug release study of a solubilizing nanomedicine in human albumin, comparing two failing lots and one passing lot of a micellar formulation containing SN-38 as a relevant example.

2 Materials

1. 10 kDa molecular weight cutoff (MWCO) cellulose membrane ultrafiltration device, 0.5 mL capacity.
2. 10 kDa MWCO centrifugal filter units, 4 mL capacity.
3. Human serum albumin (HSA).
4. Methanol.
5. Acetonitrile (ACN).
6. Formic acid.
7. SN-38.
8. ^{13}C-SN-38.
9. $^{2}H_3$-SN-38.
10. SN-38 micelle test nanoparticle, passing and failing lots (Development Code: SNB-101).
11. Phosphate buffered saline (PBS).
12. Magnetic stirrer hotplate with temperature probe and control.
13. Liquid chromatography-mass spectrometry (LC-MS) instrumentation.
14. C18 high performance liquid chromatography column, 2.1 mm × 100 mm and matching C18 guard column, 2.1 mm × 10 mm.

15. Amber glass screw top HPLC vial with fixed Teflon insert and cap.
16. Evaporator and concentrator workstation.

3 Methods

This method gives an example of an in vitro temperature-dependent drug release experiment in human serum albumin, using an SN-38 micelle as a model nanomedicine formulation to demonstrate that this assay can distinguish between passing and failing lots. According to the nanomedicine guidance, passing lots of the drug product are prepared under the intended "target conditions" and failing lots prepared with variation in physicochemical parameters, such as particle size, drug loading, and types and/or amounts of excipients [1]. In the current method, the passing lot is labeled "sample 1," and the two failing lots are labeled "sample 2" and "sample 3." The failing lots, samples 2 and 3, were prepared with different amounts of excipients than the target specifications. The temperature range chosen was 25 °C to 37.5 °C. A temperature at slightly above ambient room temperature was chosen as the starting temperature so that the temperature could be maintained for the 10-min incubation period. Release was tested at 2.5 °C increments up to physiological temperature, 37.5 °C. For specifics of the SITUA technique and assay validation, please refer to Skoczen et al. [3]. Assay validation should be performed prior to running the method to ensure compatibility with the drug and formulation.

3.1 Temperature-Dependent Drug Release in 4.5% HSA

1. Prepare 1 L of 4.5% HSA for each test sample in triplicate and additional 4.5% HSA and filtered 4.5% HSA for preparing standard curves (*see* **Note 1**). To prepare filtered 4.5% HSA, transfer 4.5% HSA into a 4 mL centrifugal filer unit with a 10 kDa MWCO, centrifuge at 5000 × *g* for 30 min, and collect filtrate.
2. Equilibrate each beaker of 1 L 4.5% HSA to 25 °C for 1 h on a stir plate.
3. Spike ^{13}C-SN-38 tracer into the equilibrated 4.5% HSA at a final concentration of 50 ng/mL (*see* **Note 2**). Equilibrate for 10 min at 25 °C on a stir plate with constant stirring.
4. Add entire dosage vial of the nanoformulation, prepared as intended for clinical use, to 4.5% HSA beakers containing ^{13}C-SN-38 tracer, with final drug concentrations of 10 μg/mL SN-38 for samples 1, 2, and 3. Allow to equilibrate at 25 °C for 10 min with constant stirring.

5. After 10 min equilibration, take 50 μL of sample for analysis by LC-MS as described below, and transfer 400 μL to 10 kDa MWCO centrifuge device and centrifuge at 6000 × *g* for 10 min at 25 °C. Analyze 50 μL of ultrafiltrate by LC-MS as described below.
6. The hotplate temperature is increased incrementally by 2.5 °C at a time. Once the target temperature is reached, **step 5** is repeated at each temperature increment, up to 37.5 °C.

3.2 LC-MS Set Up

1. Set HPLC conditions at: 10 μL injection volume, water-ACN/0.1% formic acid gradient (hold at 20% ACN/0.1% formic acid from 0–2 min, then linear increase to 90% ACN/0.1% formic acid from 2–3 min, hold at 90% ACN/0.1% formic acid from 3–5 min, linear decrease to 20% ACN/0.1% formic acid from 5–5.5 min), flow rate of 0.35 mL/min, and column temperature of 32 °C. The column regeneration time between injections is 4.5 min.
2. Use an MS instrument with an electrospray ionization source in positive ion mode. Set detector voltage at 3.5 kV, capillary temperature to 280 °C, and auxiliary gas heater temperature to 320 °C. SN-38, ^{13}C-SN-38, and $^{2}H_3$-SN-38 elution times were all 3.6 min, and m/z ion transitions monitored by parallel reaction monitoring (PRM) were 393.14 → 293.13, 349.15; 399.16 → 299.15, 355.17; and 396.16 → 267.14, 296.15, 352.17, respectively (Fig. 1).

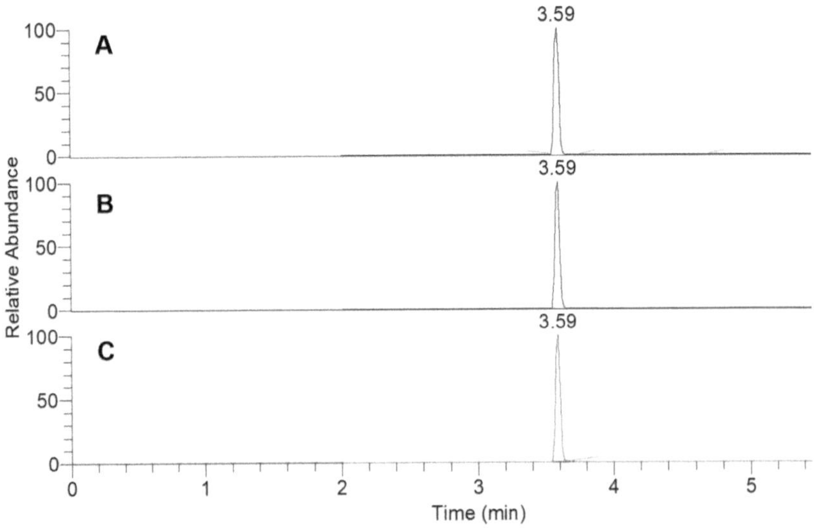

Fig. 1 LC-Q Orbitrap parallel reaction monitoring chromatograms of analytes and internal standards. Displayed are chromatograms of (**a**) SN-38 (PRM 393.14 → 293.13, 349.15), (**b**) ^{13}C-SN-38 (PRM 399.16 → 299.15, 355.17), and (**c**) $^{2}H_3$-SN-38 (PRM 396.16 → 267.14, 296.15, 352.17)

Table 1
Example of SN-38 and ^{13}C-SN-38 calibration curve data in filtered 4.5% HSA

Filtered 4.5% HSA SN-38 standard curve			Filtered 4.5% HSA ^{13}C-SN-38 standard curve		
	Conc. (ng/mL)	Accuracy (%)		Conc. (ng/mL)	Accuracy (%)
50.0 ng/mL	48.7	97	5.0 ng/mL	5.16	103
50.0 ng/mL	48.5	97	5.0 ng/mL	5.28	106
100 ng/mL	105.6	106	10.0 ng/mL	8.68	87
100 ng/mL	101.9	102	10.0 ng/mL	9.74	97
500 ng/mL	544	109	50.0 ng/mL	47.1	94
500 ng/mL	526	105	50.0 ng/mL	48.7	97
1000 ng/mL	1053	105	100 ng/mL	97.2	97
1000 ng/mL	1036	104	100 ng/mL	101.4	101
5000 ng/mL	4616	92	500 ng/mL	527	105
5000 ng/mL	5058	101	500 ng/mL	514	103
10,000 ng/mL	10,111	101	1000 ng/mL	1011	101
10,000 ng/mL	9163	92	1000 ng/mL	1004	100
20,000 ng/mL	17,894	89	2000 ng/mL	2124	106
20,000 ng/mL	19,846	99	2000 ng/mL	2020	101
QC			QC		
QCL	53.8	108	QCL	4.48	90
QCL	54.3	109	QCL	4.73	95
QCM	984	98	QCM	92.9	93
QCM	965	96	QCM	99.8	100
QCH	19,805	99	QCH	2167	108
QCH	18,986	95	QCH	2187	109

Percent accuracy between the theoretical and actual values are shown along with quality control (QC) samples

3. Measure the peak area ratio of analyte to internal standard, $^{2}H_3$-SN-38, and use it to interpolate SN-38 and ^{13}C-SN-38 concentrations of unknowns from a linear fit of calibration curves (Tables 1 and 2). Calibration curves are made for SN-38 and ^{13}C-SN-38 in 4.5% HSA and filtered 4.5% HSA. The calibration curve ranges can vary depending on the assay analyte concentrations and percent binding of the drug. For each calibration run, include quality control samples from the low, mid, and high points of the calibration curve, prepared in

Table 2
Example of SN-38 and ^{13}C-SN-38 calibration curve data in 4.5% HSA

4.5% HSA SN-38 standard curve			4.5% HSA ^{13}C-SN-38 standard curve		
	Conc. (ng/mL)	Accuracy (%)		Conc. (ng/mL)	Accuracy (%)
50.0 ng/mL	50.7	101	1.00 ng/mL	5.35	107
50.0 ng/mL	47.9	96	1.00 ng/mL	5.25	105
100.0 ng/mL	93.8	94	10.0 ng/mL	8.80	88
100.0 ng/mL	108.1	108	10.0 ng/mL	8.99	90
500 ng/mL	543	109	50.0 ng/mL	46.9	94
500 ng/mL	546	109	50.0 ng/mL	49.0	98
1000 ng/mL	952	95	100.0 ng/mL	95.7	96
1000 ng/mL	1070	107	100.0 ng/mL	98.5	98
5000 ng/mL	4936	99	500 ng/mL	509	102
5000 ng/mL	4960	99	500 ng/mL	512	102
10,000 ng/mL	9170	92	1000 ng/mL	1009	101
10,000 ng/mL	9661	97	1000 ng/mL	1001	100
20,000 ng/mL	17,578	88	2000 ng/mL	2179	109
20,000 ng/mL	21,350	107	2000 ng/mL	2199	110
QC			QC		
QCL	53.5	107	QCL	4.68	94
QCL	49.5	99	QCL	5.11	102
QCM	1044	104	QCM	102.3	102
QCM	1078	108	QCM	101.1	101
QCH	22,364	112	QCH	2202	110
QCH	18,565	93	QCH	2089	104

Percent accuracy between the theoretical and actual values are shown along with quality control (QC) samples

appropriate assay matrix. For specifics as to how the calibration and quality control samples are to be prepared, refer to Subheading 3.3 below.

3.3 Calibration and Quality Control Standards Preparation

1. Initial stock solutions of SN-38, ^{13}C-SN-38, and $^{2}H_3$-SN-38 are prepared at 1 mg/mL in DMSO, and further working stocks are diluted in methanol. These stocks are used to prepare calibration and quality control standards.

2. Prepare SN-38 and ^{13}C-SN-38 calibration standards in 4.5% HSA and filtered 4.5% HSA at concentrations ranging from 5 to 20,000 ng/mL. $^{2}H_3$-SN-38 is used as an internal standard at a concentration of 25 ng/mL (*see* **Note 3**).
3. Prepare SN-38 in 4.5% HSA and filtered 4.5% HSA in duplicate with concentrations of 50, 1000, and 20,000 ng/mL as low, medium, and high quality control standards, respectively. Prepare ^{13}C-SN-38 in 4.5% HSA and filtered 4.5% HSA in duplicate with concentrations of 5, 100, and 2000 ng/mL as low, medium, and high-quality control standards, respectively. The average of the measured duplicate quality control concentrations should be within 15% of theoretical.

3.4 Sample Preparation with Controls

1. Add 50 μL of sample, calibration standard, or quality control standard to Eppendorf tubes containing 200 μL of ice-cold methanol with 0.1% formic acid and 25 ng/mL $^{2}H_3$-SN-38 internal standard and vortex.
2. Place the sample at −80 °C for 10 min, then thaw at room temperature to precipitate proteins.
3. Centrifuge the thawed sample at 18,000 × *g* for 20 min at 4 °C to pellet precipitated protein.
4. Transfer the supernatant to a clean Eppendorf tube and dry in an evaporator and concentrator workstation at 50 °C.
5. Resuspend the dried residue in 150 μL 20% ACN with 0.1% formic acid.
6. Transfer the supernatant to a 1.5 mL amber glass screw top HPLC vial with fixed insert and cap and place in an HPLC autosampler vial rack.
7. Run quality control samples with each calibration curve.
8. Follow LC-MS method in Subheading 3.2.

3.5 Calculations

1. For calculations of percent released drug at each temperature increment using the SITUA method, refer to the SITUA protocol in Skoczen et al. [3]. The calculations for the current method can be found in Table 3.
2. The percent SN-38 drug release for each temperature and test sample are displayed in Fig. 2. The failing lots, samples 2 and 3, had SN-38 temperature release profiles that were dissimilar to the passing sample 1, with <60% release at 25 °C for the failing lots and 80% release for the passing lot. At 37.5 °C, all three lots showed ~100% release. Based on these data, the SITUA total dosage form release assay was able to discriminate between passing and failing lots and could be used as a lot release assay upon determination of appropriate acceptance criteria.

Table 3
SN-38 analytical data for samples 1, 2, and 3

	Filtrate SN-38 (ng/mL)	Reservoir SN-38 (ng/mL)	%Unbound SN-38 =(Filtrate SN-38 / Reservoir SN-38 *100)	% Bound =(100-% unbound SN-38)	Filtrate ^{13}C-SN-38 (ng/mL)	Reservoir ^{13}C-SN-38 (ng/mL)	%Unbound ^{13}C-SN-38 =(Filtrate ^{13}C-SN-38/ Reservoir^{13}C-SN-38 * 100)	% Bound ^{13}C-SN-38 =(100 -% unbound ^{13}C-SN-38)	Unencapsulated (ng/mL) =(Filtrate SN-38/ (1-(%bound ^{13}C-SN-38/100)))	Encapsulated (ng/mL) =(Reservoir SN-38-unencapsulated)	% Release = (Unencapsulated/ Reservoir SN-38) *100	Avg % release	SD
A. Sample 1													
25 °C	1270	8119	15.6	84	9.2	50.1	18	82	6919	1201	85	80	10.4
	1325	9300	14.3	86	9.8	59.6	16	84	8095	1205	87		
	1072	9657	11.1	89	9.3	57.3	16	84	6581	3076	68		
27.5 °C	1373	9291	14.8	85	9.6	55.5	17	83	7971	1320	86	86	0.3
	1663	11,950	13.9	86	11.4	70.8	16	84	10,317	1633	86		
	1496	10,451	14.3	86	9.9	59.0	17	83	8936	1516	85		
30 °C	1644	9735	16.9	83	9.2	49.0	19	81	8809	927	90	94	4.5
	1682	9177	18.3	82	10.2	51.2	20	80	8448	730	92		
	1632	8181	19.9	80	9.8	49.6	20	80	8235	−54	101		
32.5 °C	1804	8924	20.2	80	9.7	47.4	20	80	8807	117	99	97	7.0
	2218	10,435	21.3	79	10.9	53.0	20	80	10,833	−397	104		
	1961	10,801	18.2	82	10.4	49.9	21	79	9407	1394	87		
35 °C	2162	7767	27.8	72	11.6	45.1	26	74	8426	−659	108	107	4.4
	2343	12,070	19.4	81	11.2	58.8	19	81	12,253	−183	102		
	1968	9253	21.3	79	10.3	54.5	19	81	10,362	−1109	112		
37.5 °C	2618	9452	27.7	72	12.3	45.5	27	73	9656	−205	102	104	4.2
	2377	10,564	22.5	77	12.8	57.1	22	78	10,581	−17	100		
	2225	9954	22.4	78	11.4	56.0	20	80	10,937	−983	110		
B. Sample 2													
25 °C	594	8904	6.7	93	7.7	79.7	10	90	6141	2763	69	57	11.2
	621	9605	6.5	94	6.7	57.2	12	88	5331	4274	56		
	539	10,297	5.2	95	7.6	68.2	11	89	4811	5486	47		
27.5 °C	777	9215	8.4	92	8.8	58.7	15	85	5178	4037	56	66	10.3
	913	11,711	7.8	92	8.2	64.8	13	87	7192	4520	61		
	894	10,793	8.3	92	7.6	73.0	10	90	8645	2148	80		

30 °C	1051	8931	11.8	88	10.6	55.2	19	81	5492	3438	61	64	4.2
	***1256**	**8305**	**15.1**	**85**	-	**65.4**	-	-	-	-	-		
	937	10,385	9.0	91	8.4	62.6	13	87	7001	3384	67		
32.5 °C	1088	8992	12.1	88	9.1	64.8	14	86	7716	1276	86	72	10.7
	1088	10,817	10.1	90	10.2	73.2	14	86	7781	3037	72		
	1188	14,028	8.5	92	9.7	68.0	14	86	8346	5682	59		
35 °C	1368	9601	14.2	86	10.3	60.1	17	83	8001	1600	83	82	8.0
	1219	10,468	11.6	88	10.0	61.7	16	84	7514	2954	72		
	1365	10,313	13.2	87	10.1	69.8	14	86	9422	891	91		
37.5 °C	1579	9285	17.0	83	9.1	62.8	14	86	10,948	−1663	118	100	16.1
	1488	10,908	13.6	86	11.1	63.9	17	83	8589	2319	79		
	1585	9471	16.7	83	11.1	67.9	16	84	9730	−259	103		
C. Sample 3													
25 °C	621	9911	6.3	94	7.3	55.1	13	87	4704	5207	47	50	8.2
	746	9577	7.8	92	15.8	88.9	18	82	4200	5377	44		
	726	10,880	6.7	93	8.9	79.5	11	89	6478	4402	60		
27.5 °C	888	8833	10.1	90	8.5	53.3	16	84	5576	3257	63	60	6.4
	975	10,878	9.0	91	11.4	64.5	18	82	5515	5363	51		
	976	8550	11.4	89	10.0	57.0	17	83	5578	2972	65		
30 °C	1152	9262	12.4	88	10.0	65.3	15	85	7540	1722	81	67	15.8
	1061	9092	11.7	88	13.6	52.8	26	74	4132	4959	45		
	1155	9342	12.4	88	10.0	61.1	16	84	7058	2284	76		
32.5 °C	1312	9456	13.9	86	9.8	54.5	18	82	7259	2198	77	68	8.0
	1517	9852	15.4	85	16.6	62.0	27	73	5670	4182	58		
	1418	10,403	13.6	86	10.4	53.8	19	81	7316	3087	70		
35 °C	1812	9157	19.8	80	11.7	58.3	20	80	9001	156	98	93	4.8
	1647	9549	17.2	83	10.7	58.7	18	82	9024	525	94		
	1620	9796	16.5	83	11.0	57.6	19	81	8497	1299	87		
37.5 °C	2260	9373	24.1	76	12.4	53.9	23	77	9791	−418	104	103	1.0
	2218	8109	27.3	73	11.9	44.6	27	73	8284	−175	102		
	2065	9522	21.7	78	12.9	61.5	21	79	9875	−353	104		

Presented are the dosage form temperature-dependent SITUA SN-38 drug release analytical data for Sample 1 (**A**), Sample 2 (**B**), and Sample 3 (**C**). The concentration of test samples used was 10 μg SN-38/mL, with samples taken at temperatures 25, 27.5, 30, 32.5, 35, and 37.5 °C. *Sample was excluded from analysis.

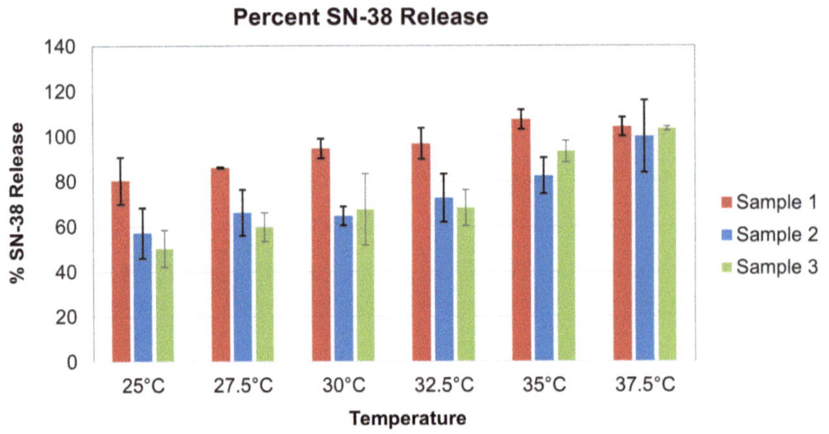

Fig. 2 Dosage form temperature-dependent drug release of SN-38 in 4.5% HSA. Displayed are the calculated percent release of SN-38 from samples 1, 2, and 3 (Mean ± SD, $N = 3$)

4 Notes

1. The volume of the beaker of 4.5% HSA should allow for sink conditions, such that the active drug is completely soluble at physiological temperature, 37.5 °C, when the total dosage form is added.
2. The stable isotope tracers should be three amu separated from the normoisotopic drug and bioanalytical internal standard. Tracer addition to the test beakers is performed before addition of nanoformulation to prevent tracer precipitation. The concentration of the tracer should be selected based on the bioanalytical sensitivity limits of the filtered 4.5% HSA standard curve, i.e., lower limit of quantitation (LLOQ), and the % binding of the drug to the 4.5% HSA. For example, in the case of this study, the 50 ng/mL concentration of both tracers was selected because it allowed for quantitation of the filtered tracer, as the unbound fractions of ^{13}C-SN-38 in 4.5% HSA were 20 and 60%, respectively, and LLOQ for filtered 4.5% HSA for both isotopes was 5 ng/mL.
3. An additional requirement of the method is a second stable isotope that can be used as an internal standard to allow for accurate quantitation of the normoisotopic drug and stable isotope tracer by mass spectrometry. This stable isotope should be at least three amu different from both the normoisotopic drug and the stable isotope tracer in order to allow for accurate mass separation and quantitation.

Acknowledgement

This project has been funded in whole or in part with Federal funds from the National Cancer Institute, National Institutes of Health, under Contract No. 75N91019D00024. The content of this publication does not necessarily reflect the views or policies of the Department of Health and Human Services, nor does mention of trade names, commercial products, or organizations imply endorsement by the US Government.

References

1. Food and Drug Administration (FDA) Center for Drug Evaluation and Research (CDER) and Center for Biologics Evaluation and Research (CBER) (2022) Drug products, including biological products, that contain nanomaterials guidance for industry. https://www.fda.gov/media/157812/download. Accessed 11 Sept 2023
2. Ambardekar VV, Stern ST (2015) NBCD pharmacokinetics and drug release methods. In: Crommelin DJA, de Vlieger JSB (eds) Non-biological complex drugs; the science and the regulatory landscape, AAPS advances in the pharmaceutical sciences series, vol 20, 1st edn. Springer International Publishing, pp 261–287. https://doi.org/10.1007/978-3-319-16241-6
3. Skoczen SL, Stern ST (2018) Improved ultrafiltration method to measure drug release from nanomedicines utilizing a stable isotope tracer. Methods Mol Biol 1682:223–239

Chapter 28

Evaluation of Nanomedicine Tissue Distribution and Stability by Alexa Fluor 488 and PEG Immunohistochemistry

Teagan D. Ware, Donna Butcher, Elijah Edmondson, and Stephan T. Stern

Abstract

A primary issue with nanomedicine biological evaluation is determination of nanoparticle carrier tissue distribution and stability. Here we present a method to evaluate nanomedicine distribution in tissues that is applicable to most nanomedicine constructs. This method utilizes immunohistochemical (IHC) analysis of an Alexa Fluor 488-tag and/or polyethylene glycol (PEG), a very common nanomedicine component, for tissue localization. Using specific Alexa Fluor 488- and/or PEG antibody-based IHC staining procedures allows evaluation of high-resolution nanoparticle tissue distribution, nanoparticle tissue stability, and also allows correlation of distribution with morphological changes. This protocol outlines the methods to follow to ensure proper tissue collection and optimized immunohistochemical staining of Alexa Fluor 488-tag and PEG in tissues.

Key words Nanomedicine, Immunohistochemistry, Biodistribution, Alexa488, PEG

1 Introduction

A primary issue with nanomedicine biological evaluation is determination of nanoparticle carrier tissue distribution and stability. While many methods are available to measure drug encapsulated in or released from nanomedicines within the systemic circulation and tissues, measurement of the nanoparticle carrier is limited [1]. Here we present a method to evaluate nanomedicine distribution in tissues that is applicable to most nanomedicine constructs. This method utilizes immunohistochemical (IHC) analysis of Alexa Fluor 488-tag and/or polyethylene glycol (PEG) for tissue localization [2]. PEG is a very common component of nanomedicines, as it prevents aggregation in solution and optimization and clearance by the immune system [3].

Jeffrey D. Clogston et al. (eds.), *Characterization of Nanoparticles Intended for Drug Delivery*, Methods in Molecular Biology, vol. 2789, https://doi.org/10.1007/978-1-0716-3786-9_28,

Using specific Alexa Fluor 488 and/or PEG antibody-based IHC staining procedures allows evaluation of nanoparticle tissue distribution at cellular resolution. This can allow for determination of target tissue specificity of the nanomedicine, as well as off-target tissues of potential concern. Combined with standard histopathological evaluation, this method allows correlation of nanoparticle tissue distribution within microanatomic structures and evaluation of the histologic features of cells, including any potential histopathologic changes. Further, by tagging Alexa Fluor 488 to a non-PEG component of a PEG containing nanomedicine and conducting both Alexa Fluor 488 and PEG IHC, stability of the nanomedicine in vivo and within microanatomic compartments can be inferred. This protocol outlines the methods to ensure proper tissue collection and optimized IHC staining of Alexa Fluor 488-tag and PEG in tissues.

2 Materials

1. 4% paraformaldehyde aqueous solution (PFA) (stable for up to 24 h at 4 °C): 32% PFA concentrate, 1× phosphate buffered saline (PBS).
2. 80%, 95%, and 100% ethanol (EtOH).
3. Positively charged frosted glass slides.
4. Hydrophobic barrier pen (PAP pen).
5. Glass coverslips.
6. Mounting medium.
7. Xylene.
8. Distilled water.
9. 3% Hydrogen peroxide (*see* **Note 1**).
10. T-TBS wash solution: 0.1% tween 20 detergent.
11. 10% N-hydroxy succinimide (NHS)
    (a) Anti-Alexa Fluor 488 antibody (Invitrogen Life Technologies, #A-11094).
    (b) Anti-PEG antibody (Abcam, #ab51257).
12. Non-specific rabbit IgG antibody (isotype control).
13. Bond Polymer Refine Kit (Leica Biosystems #DS9800), with omission of the Post Primary Reagent.
14. Commercial IHC diluent.
15. Chromogen substrate: 3,3′-Diaminobenzidine (DAB).
16. Gill's hematoxylin counterstain solution.
17. Saturated lithium carbonate solution.
18. Citrate buffer, pH 6.

3 Methods

3.1 Tissue Sample Preparation

1. Collect tissues at necropsy immediately following euthanasia and place in fixative. To ensure adequate fixation, section larger tissues so that thickness of tissues is not greater than 1 cm (*see* **Note 2**).
2. Place fresh tissue into appropriate volume of freshly prepared 4% neutral-buffered paraformaldehyde (PFA) or 10% neutral buffered formalin (NBF) and fix for 24 h at room temperature (*see* **Notes 3** and **4**).
3. Decant fixative from tissues and add the equivalent volume of 80% ethanol to the samples. Store at 4 °C (*see* **Note 5**).
4. Trim tissues for histologic examination. Process tissue and embed in paraffin blocks as appropriate to show lesions or regions of interest within the tissue.
5. At microtomy, create 5 μm thick sections onto positively charged frosted glass slides. Allow slides to dry adequately prior to completing immunohistochemistry staining procedures.

3.2 Alexa Fluor 488 Immunohistochemistry (Bench Staining Procedure)

1. Along with test sections, including the following controls allows for optimal interpretation: Test tissue Alexa Fluor 488 antibody IHC; test tissue negative control (replacing primary antibody with nonspecific antibody from the same species and same antibody isotype); positive tissue control (known positive tissue for IHC target) (not shown in provided figures, since not available); positive tissue, negative control (replacing primary antibody with nonspecific antibody from the same species and same antibody isotype) (Not shown in provided figures, since not available.).
2. If applicable, bake the slides in the oven at 60 °C for 1 h.
3. Deparaffinize tissue section slides with xylene (*see* **Note 6**).
4. Rehydrate slides through a series of graded alcohols (100→95% EtOH) down to distilled water (dH_2O).
5. Wash in running dH_2O for 5 min.
6. Submerge slides in citrate buffer solution and process in a pressure cooker at 122–125 °C for 30 s to unmask antigens.
7. Remove slides from pressure cooker and allow to cool for 20 min in the buffer solution at room temperature.
8. Rinse in dH_2O.
9. Perform peroxidase block by adding 3% hydrogen peroxide and incubate slides for 10 min (*see* **Note 1**).

10. Rinse slides in T-TBS; add wash buffer to slides, agitate, and decant. Repeat three times within 10 min.
11. Use a hydrophobic barrier PAP pen to outline tissue sections on the slides (*see* **Note 7**).
12. Perform normal serum block by submerging slides in a 10% NHS in T-TBS solution and incubate for 60 min.
13. Tap/wipe excess serum from slides. Do not rinse the slides.
14. Dilute anti-Alexa Fluor 488 primary antibody at a ratio of 1: 500 in 0.1% BSA in PBS or 10% NHS in T-TBS (*see* **Note 8**).
15. Dilute IgG polyclonal rabbit isotype control to correlating concentration of the primary antibody, using the same diluent (*see* **Note 9**).
16. Add diluted antibody or isotype control and incubate overnight (approximately 20 h total incubation time) at 4 °C.
17. Rinse slides in T-TBS; add wash buffer to slides, agitate, and decant. Repeat three times within 10 min.
18. Apply peroxidase substrate solution (DAB) and incubate 5 min (*see* **Note 10**).
19. Rinse in running tap water.
20. Perform hematoxylin counterstain by briefly submerging slides in Gill's Hematoxylin solution.
21. Rinse in running tap H_2O.
22. Perform bluing of counterstain by submerging slides in a saturated lithium carbonate solution until desired coloring is achieved (~1 min) (*see* **Note 11**).
23. Rinse in running tap H_2O for a minimum of 5 min.
24. Dehydrate slides through a series of graded alcohols (95% → 100%), clear in xylene, and coverslip.
25. Scan slides at high magnification (200×) to visualize whole-tissue sections. See Figs. 1 and 2 for examples of anti-Alexa Fluor 488 IHC tissue staining results for nanomedicine treated tissue and isotope control. *NB: positive control tissue was not available for the Alexa Fluor 488 IHC.*

3.3 PEG (Polyethylene Glycol) Immunohistochemistry (Autostainer Procedure)

1. Along with test sections, include the following controls for optimal interpretation: Test tissue Alexa Fluor 488 antibody IHC; test tissue negative control (replacing primary antibody with nonspecific antibody from the same species and same antibody isotype) (not shown in provided figures.); positive tissue control (known positive tissue for IHC target); positive tissue, negative control (replacing primary antibody with nonspecific antibody from the same species and same antibody isotype).

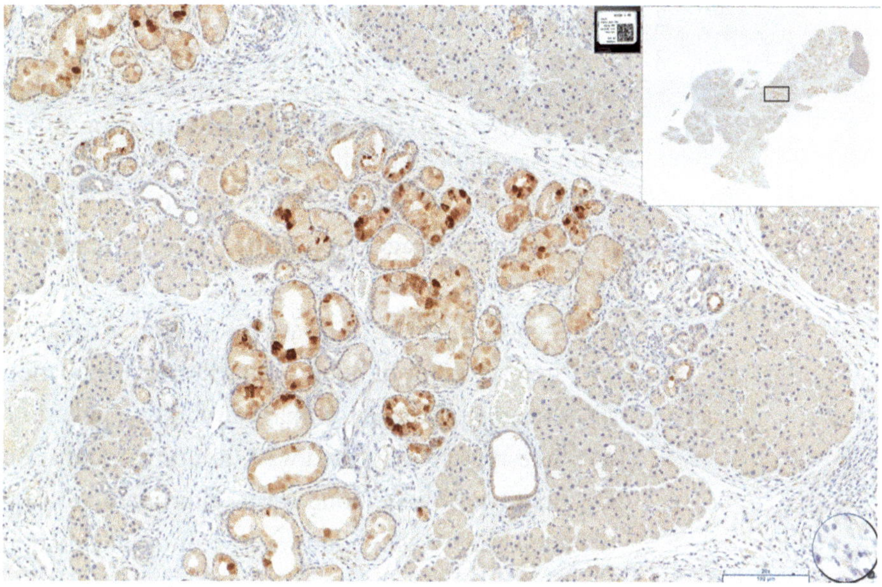

Fig. 1 Anti-Alexa Fluor 488 IHC tissue staining results for nanomedicine treated tissue. This photomicrograph shows anti-Alexa Fluor 488 positive staining in pancreatic intraductal neoplasia (PanIN) lesions from a LSL-KrasG12D/+, P48-Cre (KC) mouse model treated with an Alexa Fluor 488-labeled, CCK-BR receptor-targeted polyplex nanoparticle [2]

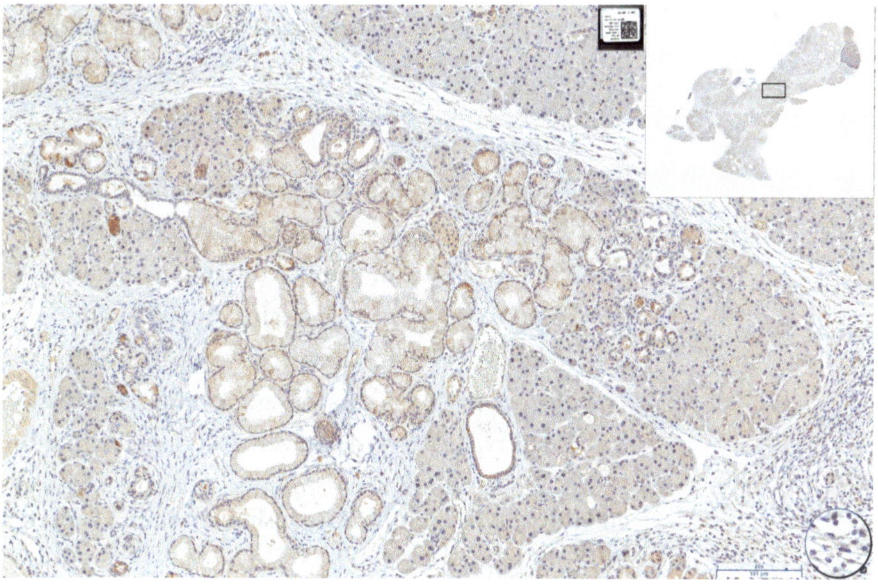

Fig. 2 Isotype control. This photomicrograph shows lack of isotope control staining in pancreas from a LSL-KrasG12D/+, P48-Cre (KC) mouse model treated with an Alexa Fluor 488-labeled, CCK-BR receptor-targeted polyplex nanoparticle [2]. The same tissue section as in the anti-Alexa Fluor 488 IHC shown in Fig. 1 was used

2. If applicable, bake the slides at 60 °C for 1 h.
3. Program the PEG staining protocol into the autostainer software using the steps as outlined below. Select heat-induced epitope retrieval (HIER) using citrate buffer for 20 min. Select option for online deparaffinization (*see* **Note 12**).
   (a) Step type: Wash, bond wash solution, incubation 0 min.
   (b) Step type: Reagent, peroxide block, incubation 5 min.
   (c) Step type: Wash, bond wash solution, incubation 0 min.
   (d) Step type: Wash, bond wash solution, incubation 0 min.
   (e) Step type: Wash, bond wash solution, incubation 0 min.
   (f) Step type: Reagent, primary antibody, incubation 30 min.
   (g) Step type: Wash, bond wash solution, incubation 0 min.
   (h) Step type: Wash, bond wash solution, incubation 0 min.
   (i) Step type: Wash, bond wash solution, incubation 0 min.
   (j) Step type: Wash, bond wash solution, incubation 2 min.
   (k) Step type: Wash, bond wash solution, incubation 2 min.
   (l) Step type: Reagent, polymer reagent, incubation 8 min.
   (m) Step type: Wash, bond wash solution, incubation 2 min.
   (n) Step type: Wash, bond wash solution, incubation 2 min.
   (o) Step type: Wash, deionized water, incubation 0 min.
   (p) Step type: Reagent, mixed DAB refine, incubation 0 min.
   (q) Step type: Reagent, mixed DAB refine, incubation 10 min.
   (r) Step type: Wash, deionized water, incubation 0 min.
   (s) Step type: Wash, deionized water, incubation 0 min.
   (t) Step type: Wash, deionized water, incubation 0 min.
   (u) Step type: Reagent, hematoxylin, incubation 5 min.
   (v) Step type: Wash, deionized water, incubation 0 min.
   (w) Step type: Wash, bond wash solution, incubation 0 min.
   (x) Step type: Wash, deionized water, incubation 0 min.
4. Select and program a titer number for the primary antibody and the isotype control. Assign slides to titers as appropriate (*see* **Note 13**).
5. Dilute a sufficient volume of primary anti-PEG antibody 1: 1000 using commercial IHC diluent (*see* **Note 14**).
6. Dilute a sufficient volume of rabbit IgG monoclonal isotype control to correlating concentration of the primary antibody, using the same diluent (*see* **Notes 9** and **14**).

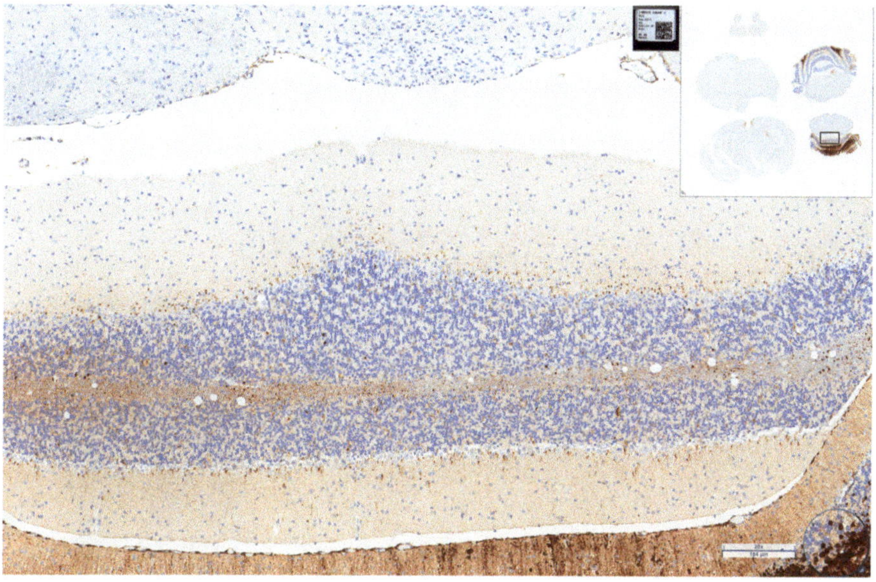

Fig. 3 Anti-PEG positive control. This photomicrograph shows anti-PEG staining in a brain section from a mouse treated intracranially with a PEG-containing nanomedicine

7. Add diluted antibodies to programmed titer vials and place on the bond.
8. Label slides with the barcode-generated labels from the instrument and load onto staining racks.
9. Place plastic slide covers over all slides and return slide racks to the instrument. Lower the racks into place.
10. Start the staining procedure, taking note of the estimated completion time.
11. When stain is complete, remove all slides from the autostainer and place into tap water. Do not allow the slides to dry out.
12. Dehydrate slides through a series of graded alcohols (95% → 100%), clear in xylene, and coverslip.
13. View slides. Scan slides at high magnification (200×) to visualize whole-tissue sections. See Figs. 3, 4, and 5 for examples of anti-PEG IHC tissue staining results for positive control tissue, positive isotype control, and nanomedicine treated tissue, respectively. *NB: Nanomedicine treated tissue isotype control was not available for the anti-PEG IHC.*

4 Notes

1. Prepare a 3% hydrogen peroxide solution from a 30% stock H_2O_2 solution by diluting 1:10 from stock in dH_2O.

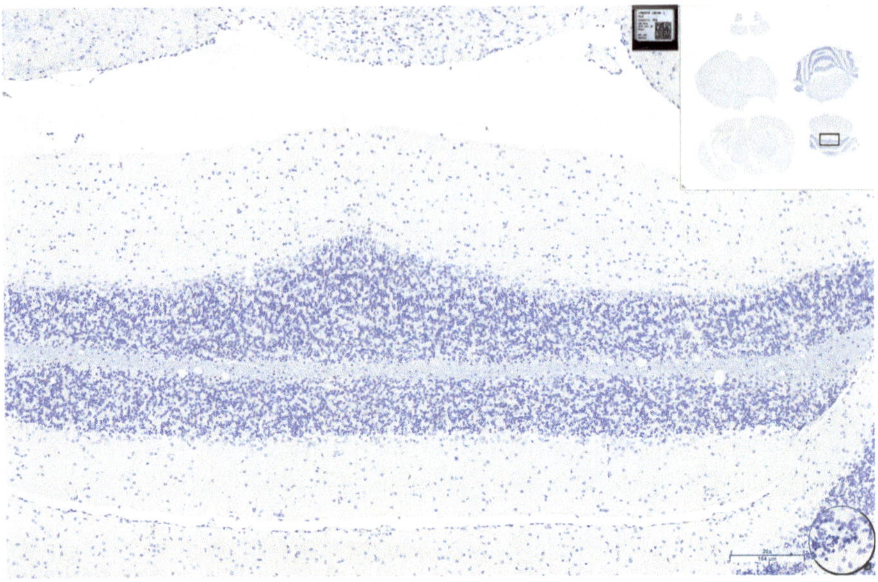

Fig. 4 Anti-PEG isotype control. This photomicrograph shows lack of isotype control staining in a brain section from a mouse treated intracranially with a PEG-containing nanomedicine. The same brain section as in the anti-PEG positive control shown in Fig. 3 was used

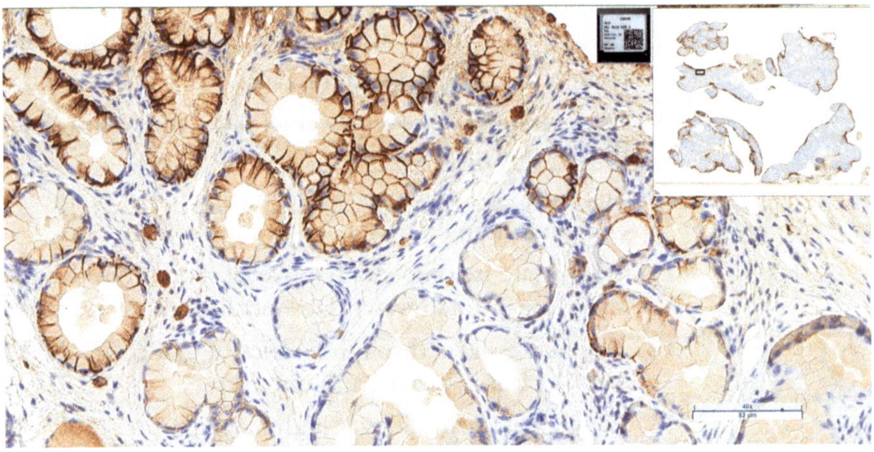

Fig. 5 Anti-PEG IHC tissue staining results for nanomedicine treated tissue. This photomicrograph shows anti-PEG staining in pancreatic intraductal neoplasia (PanIN) lesions from a LSL-KrasG12D/+, P48-Cre (KC) mouse model treated with a PEG containing, CCK-BR receptor-targeted polyplex nanoparticle [2]

2. Desired tissue size is 1–2 cm in diameter and 1–2 cm in depth. Multiple pieces may be collected but should not exceed this ideal size, to allow for proper fixation. If target tissues are GI tissue with contents inside (i.e., stomach, intestines), be sure to remove this material prior to embedding to avoid knife marks on collection sections.

3. Fixative should be added at a ratio of $\geq$20:1 PFA to tissue. Do not fix tissues in the 4% PFA for longer than 24 h.
4. Fixative should be prepared fresh at each collection. Make fixative fresh on morning of use, or evening before, and store overnight at 4 °C. If prepped and stored before use, bring the solution back to room temperature prior to adding tissues. To make 4% PFA, dilute 32% paraformaldehyde aqueous solution at a ratio of 1:8 in 1× PBS. For example, add 100 mL of 32% PFA to 700 mL of 1× PBS to make 800 mL of fixative. Check for proper pH of 7.4 prior to use.
5. Tissues will become brittle from long-term storage in EtOH. To avoid this, it is recommended to complete processing and paraffin embedding of the tissues within 3 days.
6. To adequately deparaffinize using xylene as a clearing agent, it is recommended to pass the slides through four separate containers of xylene. Slides should be fully submerged in each for a minimum of 8 min.
7. Use of a hydrophobic barrier pen is recommended to allow for uniform dispersity and proper containment of the antibody solution on top of the tissues and to avoid the solution from inadvertently dispersing across the entire slide. Be sure to carefully outline tissues and avoid drawing over any viable tissue with the PAP pen, as this will prevent adequate staining of the underlying tissue. PAP pen will clear from slides when processing through xylene at the end of the staining procedure.
8. Either listed diluent is acceptable to use and has been validated in the staining procedures for this antibody. Assess the total number of slides to stain and the size of the tissue to approximate the total amount of diluted antibody needed to efficiently cover all tissue on all the slides. As an example, to prepare 2 mL of diluted antibody at 1:500, add 4 μL of the antibody stock to 1996 μL of diluent.
9. Determining the dilution factor of the negative isotype control is dependent upon the specific concentration of the lot being used and may equate to a different ratio than was used for the primary antibody. Calculate the dilution ratio of isotype control based on the concentration of the material, which can typically be found on the vial or provided by the vendor.
10. As an example, for preparation of the chromogen substrate, consider the Vector labs ImmPACT DAB kit. To make the solution, add 1 drop of chromogen concentrate to 1 mL diluent and mix well. Be sure to follow 5-minute timeframe as closely as possible to avoid over-development of the chromogen.

11. Follow standard recommended preparation of a saturated lithium carbonate solution, which is typically considered to be 1% lithium carbonate in distilled H_2O [4]. Prepare a sufficient amount of solution to adequately submerge all slides.
12. Dispense volume for all step reagents is 150 μL, and all steps are performed at ambient temperature. Final deionized water wash step will repeat to keep slides from drying out until slides are removed from the stainer.
13. The antibody and corresponding isotype control reagents are placed into separate Bond Titration "Titer" containers and will be used during the marker incubation step of the Bond RX staining procedure.
14. A suggested IHC diluent would be Novocastra IHC diluent from Leica. When calculating total volume of diluted antibody needed to add to the programmed titer on the bond, be sure to account for sufficient volume based on 150 μL dispensed per slide plus dead space within the titer vials and transfer loss, if applicable.

Acknowledgement

This project has been funded in whole or in part with Federal funds from the National Cancer Institute, National Institutes of Health, under Contract No. 75N91019D00024. The content of this publication does not necessarily reflect the views or policies of the Department of Health and Human Services, nor does mention of trade names, commercial products, or organizations imply endorsement by the US Government.

References

1. Ambardekar VV, Stern ST (2015) NBCD pharmacokinetics and drug release methods. In: Crommelin DJA, de Vlieger JSB (eds) Non-biological complex drugs; the science and the regulatory landscape, AAPS advances in the pharmaceutical sciences series, vol 20, 1st edn. Springer International Publishing, pp 261–287. https://doi.org/10.1007/978-3-319-16241-6
2. Smith JP, Cao H, Edmondson EF, Dasa SSK, Stern ST (2021) Cholecystokinin-B receptor-targeted nanoparticle for imaging and detection of precancerous lesions in the pancreas. Biomolecules 11(12):1766
3. Shi D, Beasock D, Fessler A, Szebeni J, Ljubimova JY, Afonin KA, Dobrovolskaia MA (2022) To PEGylate or not to PEGylate: immunological properties of nanomedicine's most popular component, polyethylene glycol and its alternatives. Adv Drug Deliv Rev 180:114079
4. Berry R, Church CD, Gericke MT, Jeffery E, Colman L, Rodeheffer MS (2014) Imaging of adipose tissue. Methods Enzymol 537:47–73

Index

Jeffrey D. Clogston et al. (eds.), *Characterization of Nanoparticles Intended for Drug Delivery*, Methods in Molecular Biology, vol. 2789, https://doi.org/10.1007/978-1-0716-3786-9, © Leidos Biomedical Research Inc. 2024

MIX
Papier aus verantwortungsvollen Quellen
Paper from responsible sources
FSC® C105338

If you have any concerns about our products,
you can contact us on
ProductSafety@springernature.com

In case Publisher is established outside the EU,
the EU authorized representative is:
Springer Nature Customer Service Center GmbH
Europaplatz 3, 69115 Heidelberg, Germany

Printed by Libri Plureos GmbH
in Hamburg, Germany